Micro-Nanorobotic Manipulation Systems and Their Applications

Toshio Fukuda, Fumihito Arai,
and Masahiro Nakajima

Micro-Nanorobotic Manipulation Systems and Their Applications

Springer

Prof. Toshio Fukuda
Department of Micro-Nano Systems
Engineering
Nagoya University
Nagoya
Japan

Dr. Masahiro Nakajima
Center for Micro-Nano Mechatronics
Nagoya University
Nagoya
Japan

Prof. Fumihito Arai
Department of Micro-Nano Systems
Engineering
Nagoya University
Nagoya
Japan

ISBN 978-3-642-36390-0 ISBN 978-3-642-36391-7 (eBook)
DOI 10.1007/978-3-642-36391-7
Springer Heidelberg New York Dordrecht London

Library of Congress Control Number: 2013930324

Printed on acid-free paper

Springer is part of Springer Science+Business Media (www.springer.com)

Preface

Micro-Nano robotics and automation technologies have rapidly been grown by associated with growth of Micro and Nanotechnologies. This book represents a summary of principals in micro-nano scale engineering and the current state of art of related technologies.

Currently, in our daily life, "Microtechnology" is commonly used and embedded for various devices and systems, such as automobiles, computer peripheries, printers, cameras, amusements, robotics, automation, environmental monitoring, energy resource, and biological-medical treatments to increase their performance, accuracy and to improve energy consumption, cost performance and so on. "Micromechatronics" came up as the one of the important technology by miniaturizing the elemental devices on sensors, actuators, and computers in micro-scale. Recently, "Nanotechnology" comes up to have an important role in the industrial applications as an advanced field of mechatronics named as "Nanomechatronics".

On the other hand, a manipulation technology enables us to control the small objects locally or individually. For industrial applications, automation technology is important to improve efficiency for the creation of large-scale structural devices from small scale based on the micro-nano manipulation techniques.

In this book, we summarize some major techniques which are related with micro-nano robotics and micro-nano manipulation technologies, such especially micro-nano fabrication, micro-nano assembly, micro-nano control techniques, micro-nano materials, micro-nano evaluation techniques and so on. It is Various devices are investigated based on the micro-nano mechatronics technologies to realize high-efficiency, high-integration, high-functionality, low-energy consumption, low-cost, miniature, saving material usage and so on.

One of the application field is the "system cell engineering" which is a new a new academic field aiming at synthetic understanding of cell function using engineering (Fig. 1). The manipulation techniques are one of the key factors to analyze/control to reveal the functions of biological cell system. Recently, the research direction is moving to synthesis from analysis of biological cells based on the system cell engineering.

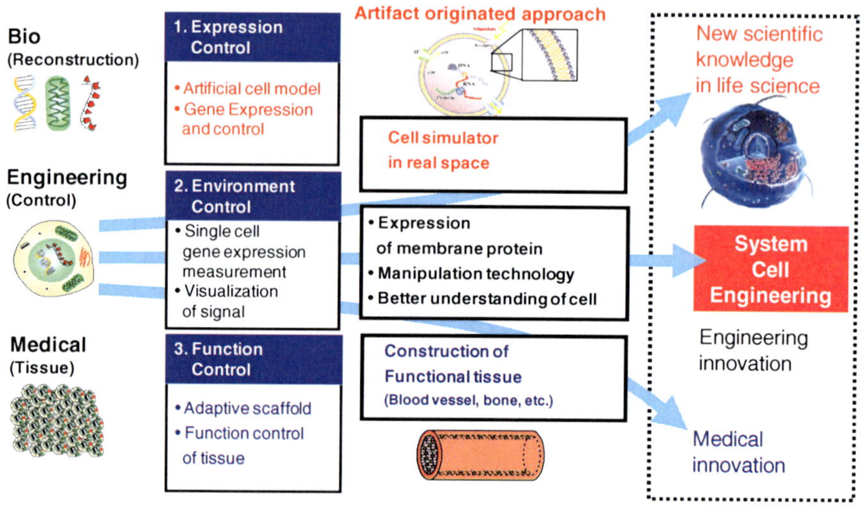

Fig. 1 System Cell Engineering by Micro-Nano Robotics and Automation

Through the micro-nano robotics and automation technologies, various research fields are investigated to achieve the next-generation science and engineering applications as shown in Fig. 2. In robotics field, the soft and wet robotics is key technology to mimic the biological system or combine the biological and artificial systems for example. The importance of micro-nano manipulation is that this technology is placed on the borderline between bottom-up and top-down fabrication and assembly technologies as shown in Fig. 3.

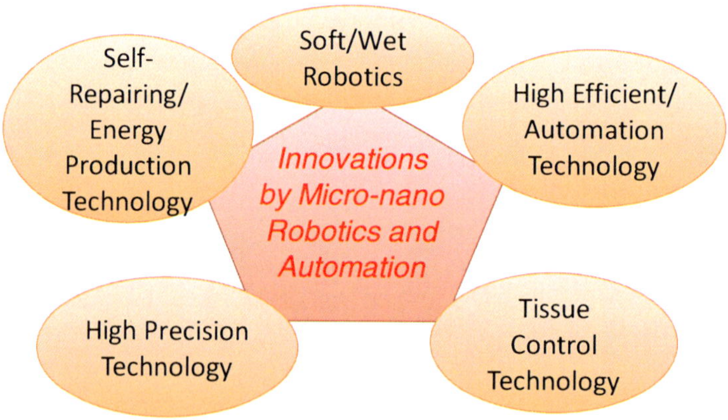

Fig. 2 Innovations by micro-nano robotics and automation

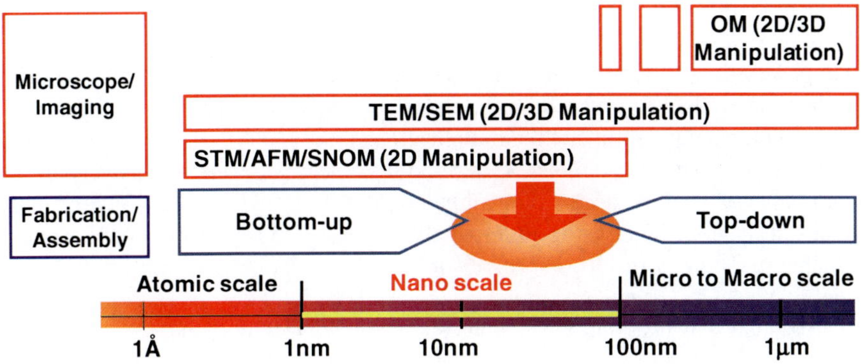

Fig. 3 Micro-Nanomanipulation based on imaging technologies for micro-nano fabrication and micro-nano assembly

In this book, we introduce the advanced technology from the basics and applications aspects of micro-nanorobotic manipulation system with their applications and related technologies. This book is organized as follows:

In Chapter 1, we introduce the overview and introduction of micro-nanorobotics and manipulation technology with some historical background and current achievements with related works.

In Chapter 2, the physics in micro-nano fields is explained in the scaling effects, mechanics, electronics, fluidics, surface interaction, and optical physics.

In Chapter 3, the related technologies in micro-nano scale are explained about the material and science, microscopes, fabrication technology, sensing and actuation, control techniques, and assembly techniques.

In Chapter 4, bio-manipulation system under optical microscope is explained under optical microscope by on-chip micro-fluidics devices, micro-tools, and laser micro-nano manipulations.

In Chapter 5, a rotational speed control of single bacterial flagellar motor is introduces as one of the applications of micromanipulation system under optical microscope.

In Chapter 6, nanomanipulation system is introduced under electron microscopes, with configurations of nano-robotics and manipulation system, hybrid nanorobotic manipulation system inside FE-SEM/TEM, nanorobotic manipulation system inside E-SEM, nano-tool exchanger system under hybrid microscope, automation of nanorobotic manipulation system inside E-SEM.

In Chapter 7, measurement, manipulation, and assembly of carbon nanotubes are explained under FE-SEM and TEM based on the nanomanipulation system. The techniques of picking up, cutting, peeling, bending, fixing of carbon nanotubes are presented as one of the applications of nanomanipulation system under optical microscope.

In Chapter 8, measurement and analysis of biological cells are explained under E-SEM for single cell nano-surgery system, especially mechanical property characterization, adhesion force measurement, electrical measurement, and cutting of single cell using various nano-tools such as nanoprobe/ nanofork/ nanoputter/ nanopicker/ dual nanoprobe with automation system.

In Chapter 9, we give a description of summary and applications of micro-nanorobotic manipulation and their applications finally.

We would like to acknowledge that the leading researchers extensively in the world to evolve the micro-nanomanipulation and related fields. We especially thanks to the excellent contributions of all students graduated from our laboratory to achieve fabulous results, especially for Dr. Lixin Dong, Dr. Ahmad Ridzuan, Dr. Liu Pou, Dr. Hisataka Maruyama, Dr. Akihiko Ichikawa, Dr. Akiko Kawaji, Dr. Kousuke Nogawa, and Dr. Yajing Shen. We express our sincere appreciation for the publication of this book supported by Nagoya University, the 21st COE program "Micro- and Nano- Mechatronics for Information-Based Society", and the global COE program "COE for Education and Research of Micro-Nano Mechatronics". This book would not have been possible without these generous supports.

Nagoya, January 2012

Toshio Fukuda
Fumihito Arai
Masahiro Nakajima

Contents

Chapter 1
Introduction of Micro-Nanorobotic Manipulation Systems

Technology has been moving toward greater control of the structure of matter for millennia. Progress in science and technology over the past decades suggest the feasibility of achieving thorough control of the molecular structure of matter. The possibility to control the structure of matter atom by atom was first discussed by Richard Feynman in 1959 seriously, which is now labeled "nanotechnology". Feynman wrote in a prophetic article on miniaturization [1]: "I am not afraid to consider the final question as to whether, ultimately—in the great future—we can arrange the atoms the way we want: the very atoms, all the way down!" He asserted that "At the atomic level, we have new kinds of forces and new kinds of possibilities, new kinds of effects. The problems of manufacture and reproduction of materials will be quite different. The principles of physics, as far as I can see, do not speak against the possibility of maneuvering things atom by atom."

The "great future" of Feynman began to be realized in the 1980s. Some of the capabilities he dreamed of have been demonstrated, while others are being developed. Although we are still far from having a general and reliable nanotechnology, progress in the last two decades or so has been tremendous. As the twenty-first century unfolds, the impact of nanotechnology on the health, wealth, and security of the world's people is expected to be at least as significant as the combined influences in the 20th century of antibiotics, the integrated circuit, and human-made polymers. Neal Lane stated in 1998, "If I were asked for an area of science and engineering that will most likely produce the breakthroughs of tomorrow, I would point to nanoscale science and engineering."[2] The great scientific and technological opportunities provided by or potentially would be provided by nanotechnology have stimulated extensive exploring of nano world ever since and initiated exciting worldwide competition especially after the publication of "National Nanotechnology Initiative" by U.S.A. government in 2000 [3].

Nanomanipulation is one of the most significant enabling technologies for nanotechnology, and might finally be the core-most part of nanotechnology if Drexler's machine-phase nanosystems based on self- replicative molecular assembler via mechanosynthesis would be realized [4].

Since the discovery, carbon nanotubes (CNTs) [5] have been extensively explored both theoretically and experimentally. The exceptional properties and broad potential applications of nanotubes make them become the most important and promising materials for nanotechnology discovered by so far.

This dissertation focuses on nanorobotic manipulations of carbon nanotubes. The main purpose is to provide an effective tool for the experimental exploration of this typical mesoscopic material, and for the construction of nanosystems with this exceptional nano building block. The long-term oriented aim is to develop a universal nanosystem builder with the abilities of instrumentation, fabrication, and assembly.

1.1 Background of Micro-Nanorobotic Manipulation Systems

Technological advancement on the top-down fabrication process, or micro machining, provides nanometer structures. On the other hand, the bottom-up fabrication process, or chemical synthesis such as self-assembly or super-molecule techniques, also provides nanometer structures. In fact, both approaches reach nanometer scale with the limitations of physical/chemical aspects at present. The "Nanotechnology" has an important role on the combinations of the top-down and bottom-up approaches. It is considered that the wide scale controlled devices from atomic scale to meter scale will be realized in the near future [6] (Figure 1. 1).

The essence of nanotechnology is the ability to work at these levels to generate larger structures with fundamentally new molecular organization. Such materials and systems can be rationally designed to exhibit novel and significantly improved physical, chemical, and biological properties, phenomena, and processes because of their size. When characteristic structural features are intermediate in extent between isolated atoms and bulk materials, the objects often display physical attributes substantially different from those displayed by either atoms or bulk materials.

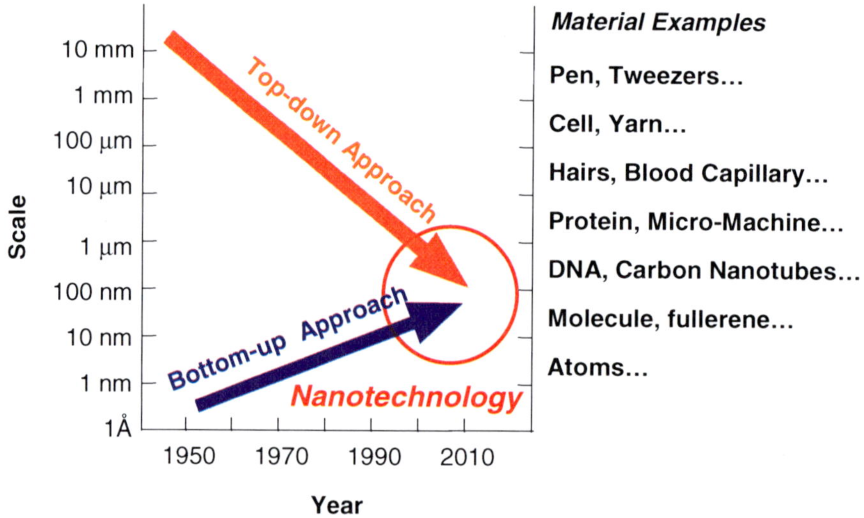

Fig. 1.1 Schematic diagram of "Nanotechnology" ("Top-Down" and "Bottom-Up")

1.1.1 *Why Micro to Nanotechnology?*

Control of matter on the nanometer scale has already played an important role in scientific disciplines as diverse as physics, chemistry, materials science, biology, medicine, engineering, and computer simulation. The potential benefits of nanotechnology are pervasive. Nanotechnology will be a strategic branch of science and engineering for the new century, one that will fundamentally restructure the technologies currently used for materials and manufacturing, computation and information, medicine and health, aeronautics and space exploration, environment and energy, biotechnology and agriculture, national defense and security, transportation and communication, and education.

For example, the ability to synthesize nanoscale building blocks with precisely controlled size and composition and then to assemble them into larger structures with unique properties and functions will revolutionize the way materials and devices being produced in the future. Some of the benefits that nanostructuring can bring include lighter, stronger, and programmable materials; innovative devices based on new principles and architectures; and use of molecular/cluster manufacturing, which takes advantage of assembly at the nanoscale level for a given purpose.

In the past several decades, there has been a nearly constant exponential growth in the capabilities of silicon-based microelectronics. However, it is unlikely that these advances will continue much because fundamental physical limitations, which prevent current designs from functioning reliably at the nanometer scale, will be reached while at the same time exponentially rising fabrication costs will make it prohibitive to raise integration levels. Now is the critical time for the development of workable nanostructures for the hardware necessary to satisfy continuing demands in information technology.

All natural materials and systems establish their foundation at the nanoscale; control of matter at molecular levels means tailoring the fundamental properties, phenomena, and processes exactly at the scale where the basic properties are determined. Therefore, by determining the novel properties of materials and systems at this scale, nanotechnology could impact the production of virtually every human-made object—everything from automobiles, tires, and computer circuits to advanced medicines and tissue replacements—and lead to the invention of objects yet to be imagined.

Properties of matter at the nanoscale are not necessarily predictable from those observed at larger scales. Important changes in behavior are caused not only by continuous modification of characteristics with diminishing size, but also by the emergence of totally new phenomena such as quantum size confinement, wave-like transport, and predominance of interfacial phenomena. Once it is possible to control feature size and shape, it is also possible to enhance material properties and device functions beyond what are already established. Currently known nanostructures include such remarkable entities as carbon nanotubes, proteins, deoxyribonucleic acid (DNA), and single-electron transistors that operate at room temperature. Rational fabrication and integration of nanoscale materials and devices herald a revolutionary age for science and technology, provided we can discover and fully utilize their underlying principles.

1.1.2 What Is Micro to Nanotechnology?

Like many emerging fields, there is no standard definition of nanotechnology yet, and the definition will be under change along with the advancement of the technology itself. The term nanotechnology has sometimes been used to refer to any technology giving some control at a substantially sub-micron scale. This might be misleading because it only emphasizes nanometer sizes, with which it is not enough to generate a revolutionary breakthrough.

Drexler used the term more narrowly to describe the ability to build structures to complex, atomic specifications by means of mechanosynthesis, and defined it as molecular nanotechnology, or briefly, nanotechnology. Mechanosynthesis is in turn defined as chemical synthesis controlled by mechanical systems operating with atomic-scale precision, enabling direct positional selection of reaction sites; synthetic applications of mechanochemistry [4]. Drexler's definition is a relative far future vision of nanotechnology.

At present, it is better to take the definition given in the "National Nanotechnology Initiative" [3]: nanotechnology is the creation and utilization of materials, devices, and systems through the control of matter on the nanometer scale (1 to 100 nm), i.e., at the level of atoms, molecules, and supramolecular structures. The main NNI activities on nanodevices, nanomanipulation, nanocharacterization, and nanomagnetics, are listed as Table 1.1.

1.1.3 What Is Micro-Nanorobotic Manipulation Systems?

It appears that until assemblers capable of replication can be built, the parallelism of chemical synthesis and self-assembly will be needed; groups of molecules can self-assemble quickly due to their thermal motions, which enable them to "explore" their environments and find (and bind to) complementary molecules. Given their key role in natural molecular machines, proteins are obvious candidates for early work in self-assembling artificial molecular systems. Recent work by Degrado [7] shows the feasibility of designing protein chains that fold predictably into solid molecular objects. Progress is also being made in artificial enzymes and other relatively small molecules that perform functions like those of natural proteins; the 1987 Nobel Prize in Chemistry went to Cram and Lehn for such work on supramolecular chemistry [8]. Several bottom-up strategies using self-assembly seem feasible [9]. M. Fujita's pioneering work has shown that self-assembly can be directed by adroitly exploiting the chemical and electrical bonds that hold natural molecules together, and hence get molecules to form desired nanometer-scale structures [10].

By the early 1980s, scanning tunneling microscopes (STMs) [11] radically changed the ways we interacted with and even regarded single atoms and molecules. The very nature of proximal probe methods encourages exploration of

Table 1.1 Main activities on National Nanotechnology Initiative (NNI)

Nanodevices	• Develop two new standard reference materials for semi-conductor, lap-on-a-chip, and other nanotechnologies. • Develop new quantitative measurement methods for analysis of physical and chemical properties of industrial nanoscale devices such as semiconductors. • Publish reference documentation on the development of measurement methods, standard reference materials, and calibration systems.
Nano manipulation	• Develop standards for autonomous atom assembly. • Develop modeling and simulation programs for condensed phases, including solutions, surfaces, solids, and interfaces. • Publish reference documentation on the development of manipulation and modeling data, algorithms, and research.
Nano characterization	• Develop three new standard reference materials for calibration and quality assurance of commercial analysis laboratories and instruments. • Develop several 3-D measurement methods for the analysis of physical and chemical at or near atomic spatial resolution. • Publish documentation on the development of measurement methods, standard reference materials.
Nanomagnetics	• Develop two new standard reference materials for semi-conductor, lap-on-a-chip, and other nanotechnologies. • Develop new quantitative measurement methods for analysis of physical and chemical properties of industrial nanoscale devices such as semiconductors. • Publish documentation on two university partnership papers per year on the measurements and standard reference materials for magnetic data storage.

the nanoworld beyond conventional microscopic imaging. Scanned probes now allow us to perform "engineering" operations on single molecules, atoms, and bonds, thereby providing a tool that operates at the ultimate limits of fabrication. They have also enabled exploration of molecular properties on an individual non-statistical basis.

STMs and other nanomanipulators are non-molecular machines but take bottom-up strategies. Although performing only one molecular reaction at a time will be impractical for making large amounts of a product, it is a promising way to provide the next generation nanomanipulators. Most important, it is possible to realize the directed assembly of molecules or supermolecules to build larger nanostructures through nanomanipulation. Hence, to certain extent, nanomanipulation leads to nanotechnology by starting from the middle or mesoscopic scale. The products produced by nanomanipulation will go bottom-up, and these products will be used to construct nanomachines including nanomanipulators. Nanomanipulators will take a top-down strategy to shrink their sizes, and finally, they will be used to make copies of themselves. This vision promises nanomanipulators, as non-molecular machines at present, to generate the next generations of nanomanipulators, step by step to molecular versions.

In industrial field, industrial robot is mainly as manipulator type controlled by automation technologies. As shown in Figure 1.2, the manipulator is used for mainly "painting", "soldering", "pick and place", "measurement", "assembly". To extend the ability of human worker, it has advantages for heavy duty task, long time operation, and high precision task to realize various manipulation or assembly tasks. Basic idea of micro-nanorobotic manipulation system is to adapt the manipulator for micro-nano meter scale material handling effectively as same as industrial manipulator robot.

Main application field of micro-nano manipulation is summarized in Figure 1. 3. Depending on the target of manipulation, the environmental fields will be decided on dry environment for inorganic targets to wet environment for organic targets.

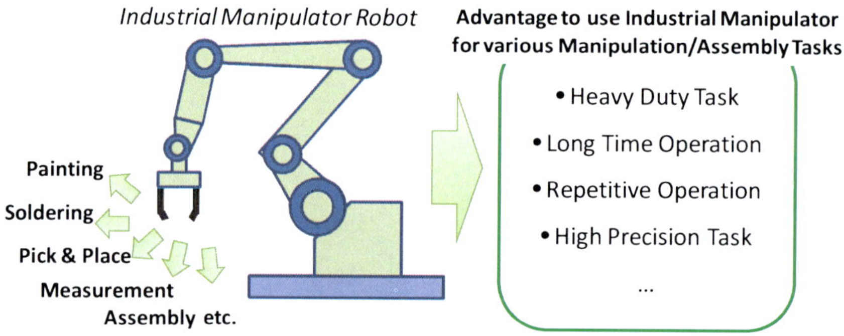

Fig. 1.2 Various industrial applications for manipulation and assembly tasks in macro-scale

Inorganic (Dry) ↕ Organic (Wet)

Application Field	Basics	Components	Applications
Mechanical	Nanocarbon, Battery,...	Fuel Cells, Micro-Nano Actuators, Inkjet System...	MEMS, Photonic Devices, Storage,...
Electrical	Photonic Crystals,...	Semi-conductor Devices,...	IC, Memory, Molecular Computing, ...
Biology/ Medical	Biomaterials, Artificial Cells,...	Bio-sensors, Bio-catalyst,	Bio-chips, DNA Computing,...

Fig. 1.3 Application fields of Micro-nano manipulation

From these aspects, the organization of this book is summarized depending on the application field environment for wet/dry/semi-wet environments as shown in Figure 1. 4. Based on the application fields and principles, we will be able to choose the micro-nano manipulation system adapting the environments under various microscopes. In the following chapters, why/what/how the environment or manipulation system is restrict and changed will be explained in detail.

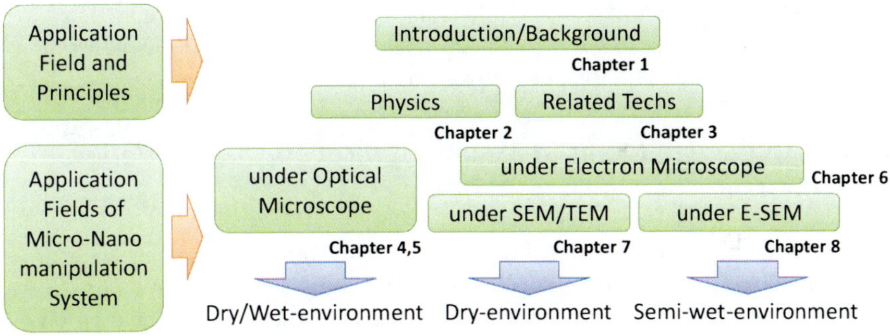

Fig. 1.4 Organization of this book

1.2 Strategies and Related Works of Micro-Nanorobotic Manipulation Systems

Two main strategies lead to use to realize the nanotechnology: top-down approach versus bottom-up approaches. The following sections mentioned in detail of these approaches.

1.2.1 Top-Down Approach

Top-down is an approach to atomic-scale control as outlined by Feynman [1]. He suggested that large machines would be used to build smaller machines, which would build yet smaller machines, working toward molecular dimensions. The sequence has been termed a "top-down" strategy. Modern microtechnology uses a different top-down approach—that of photolithography, thin films, and selective etching—to make microelectronic systems. These techniques are now being used to make mechanical structures such as gears and motors. Currently, such devices are tens of microns in diameter, many orders of magnitude larger than nanometer-scale.

1.2.2 Bottom-Up Approach

Chemists, in contrast, have started at the atomic level, building molecules precisely. Progress in molecular biology and synthetic chemistry has been steady and impressive: the operation of many naturally-occurring molecular machines has been elucidated, and some have been synthesized. Drexler [4] proposed a "bottom-up" approach to artificial molecular machines, ultimately including devices such as molecular assemblers, assembler-based replicators, and mechanical nanocomputers. The "bottom-up" style builds toward larger and more complex systems by starting at the molecular level and maintaining precise control of molecular structure. One approach would involve designing new protein-based devices which self-assemble to form molecular machines like those found in nature. Natural molecular devices show that such machines could include parts acting as struts, cables, fasteners, motors, bearings, and numerical control systems.

1.2.3 Applicable Fields of Micro-Nanorobotic Manipulation Systems

Micro-nano manipulations, or positional control at the micro-nanometer scale, are a key enabling technology for nanotechnology by filling the gap between top-down and bottom up strategies, and may lead to the appearance of replication-based assemblers, which has been proposed as general purpose manufacturing device that is able to build a wide range of useful products as well as copies of itself (self-replication). Micromanipulations maneuver objects in micrometer scale, or microstructures, with micrometer scale resolution, for example biological cells. "Nanostructures" have dimensions of about 1 to 100 nanometers, a size that is small by engineering standard, common by biological standards, and large to chemists. One of the approaches to fill the gap between top-down and bottom-up approaches is "Micro-nano manipulation", which realizes controlling the position at the micro-nanometer scale, is considered to be one of the promising ways (Figure 1. 5).

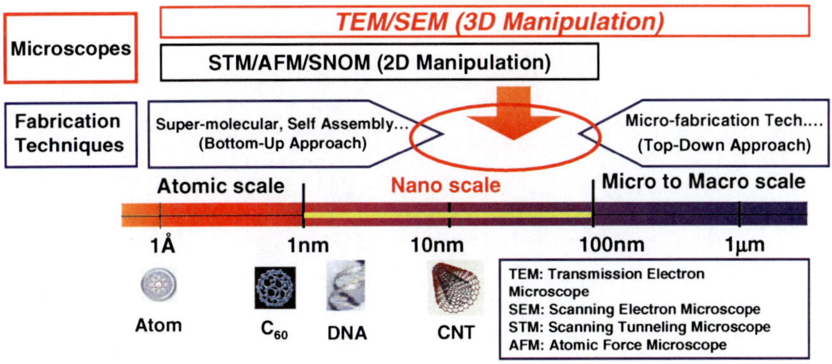

Fig. 1.5 Nanorobotic manipulation system for nanodevices constructions

The purpose of micro-nano manipulations is mainly 1) analysis of micro-nano object, 2) assembly of micro-nano systems, 3) . The Micro-nano manipulations technique involves the handling, the modification, and the connection of micro-nanostructures. Presently, nanomanipulation can be applied to the scientific exploration of mesoscopic phenomena and the construction of prototype nanodevices. It is a fundamental technology for property characterization of nano materials, structures and mechanisms, for the fabrication of nano building blocks, and for the assembly of nano devices.

Micro-nano manipulation system generally includes micro-nanomanipulator as the positioning device, kinds of microcopies as its "eyes", various end-effectors including cantilevers and tweezers, among others, as its "fingers", and types of sensors (force, displacement, tactile, strain, etc.) to facilitate the manipulations and/or to probe the properties of the objects. Key technologies for micro-nano manipulation include observation, actuation, measurement, system design and fabrication, calibration and control, communication, and human-machine interface, among others.

The Figure 1.6 depicts the microscopes, environments and strategies of micro-nano manipulations. In principle, optical microscopes (OMs) can be used for micro-meter scale, but not for nanometer scale (smaller than the wavelength of light) observation because of the limitation of diffraction. Thus, for example, scanning near-field OMs (SNOMs) breaks this limitation and shows promising as a real-time observation device for nanomanipulations, especially for ambient environment. SNOMs can be combined with atomic force microscopes (AFMs), and potentially with nanorobotic manipulator (NRMs) for nanoscale biomanipulations. Contact manipulation can be used in almost any environment but hardly for atomic manipulations, whereas noncontact mode is mainly applied for manipulations of atoms and molecules in UHV or bio-objects in liquid.

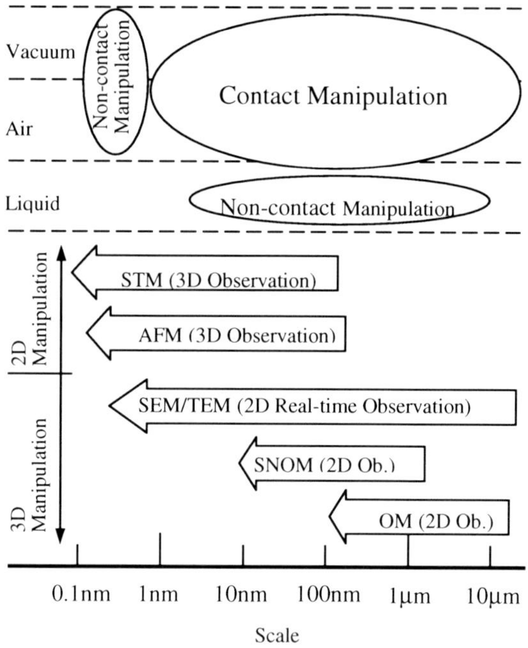

Fig. 1.6 Microscopes, Environments and Strategies for Micro-nano manipulations

In nano-scale, for the observations of manipulated objects, STMs can provide sub-angstrom imaging resolution, whereas AFMs can provide atomic one. Both of them can obtain 3D surface topology. Because AFMs can be used in ambient environment, it provides a powerful tool for biomanipulation that potentially demands a liquid environment. Resolution of scanning electron microscope (SEM) can hardly be better than 1nm, whereas field-emission SEM (FESEM) can. SEM/FESEM can be used for 2D real-time observation for both the objects and end-effectors of manipulators, their large ultra-high vacuum (UHV) sample chambers provide enough space to contain a NRM with many degrees-of-freedom inside for 3D nanomanipulations. However, the natural 2D observation makes the positioning along electron-beam direction cannot be detected. High-resolution transmission electron microscopes (HRTEM) can provide atomic resolution. However, the narrow UHV specimen chamber can hardly contain large manipulators inside it, but it might be the best one when complex nanomanipulators shrink enough.

Strategies for micro-nano manipulation can be broadly classified into three types: (1) lateral non-contact, (2) lateral contact, and (3) vertical manipulations. Generally, STMs, optical tweezers, and magnetic tweezers take lateral non-contact way, i.e., to use different kinds of fields to perform manipulations, AFMs usually adopt lateral contact manipulation mode, and NRMs apply the vertical process.

Figure 1.7 shows the processes of the three basic strategies. In the figure, A, B, C, … represent the positions of end-effector (e.g., a tip), A', B', C',… the positions of objects, 1, 2, 3, …the motions of end-effector , and 1', 2', 3',… the motions of objects. Tweezers can be used in pick-and-place to facilitate the picking-up, but are generally not necessarily helpful for placing.

Lateral Non-contact Manipulations Strategies

Motions of the lateral non-contact manipulations are shown in Figure 1.7 (a). Applicable effects [12] that can cause the motions of objects include long-range van der Waals (vdW) forces (attractive) generated by the proximity of the tip to the sample [55, 13], electric-field-induced fields by the voltage bias between the tip and the sample [14, 15], and tunneling current induced local heating or inelastic tunneling caused vibration [16,17]. With these methods, some nano devices and molecules have been assembled [18, 19] (Table 1.2).

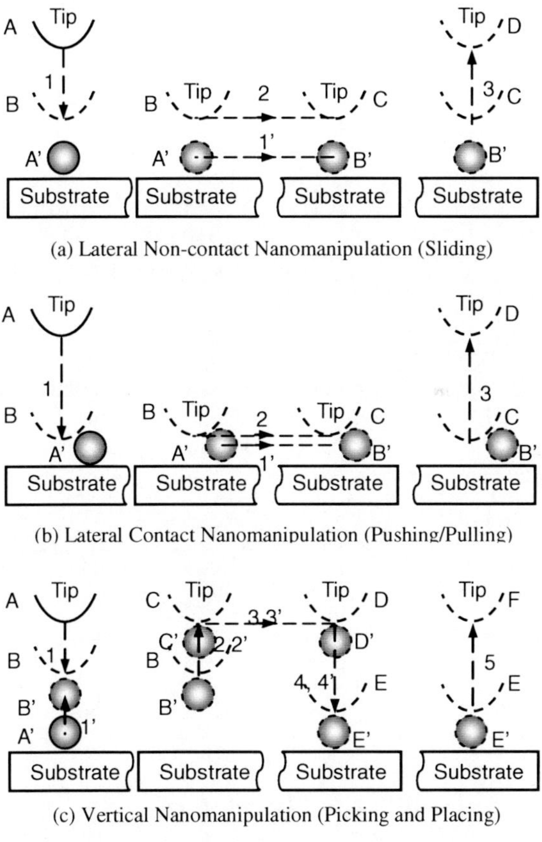

Fig. 1.7 Basic Strategies for Nanomanipulations

Laser trapping (optical tweezers) and magnetic tweezers are possible for non-contact manipulations of nano-order bio-samples, e.g. DNAs [20, 21]. Non-contact manipulations combined with STMs have revealed many important strategies to manipulate atoms and molecules. However, for the manipulation of CNTs, there were no demonstrations shown. A main difficulty is the relatively strong interaction of a nanotube and a surface because of the relatively large size of a nanotube as comparing with an atom.

Table 1.2 Non-contact Manipulations

Manipulator	Object	Substrate	Temp.	Environment	Strategy of Manipulation	Result /Application	Refs.
STM	Xe atom	Metal Ni(110)	4K	UHV	Van der Waals force	Logo "IBM"	[55]
STM	Fe atom	Cu(111)	4K	UHV	Van der Waals force	Quantum corrals: wave-particle nature	[13]
STM	Cs	GaAs(110) InSb(110)	RT	UHV	Electric-field-induced electrostatic force	RT bonding dissociation	[14]
STM	Si	Si(111)-(7x7)	RT	UHV	Chemical interaction and electrostatic force	Bonding the tip to selected atom	[15]
STM	$B_{10}H_{14}$	Si(111)	RT	UHV	Tunneling current induced local heating	Molecule dissociation	[16]
STM	H	H-terminated Si(100)	RT	UHV	Inelastic tunneling caused vibration	Atom/molecule adsorption	[17]
STM	C_{60}	Step Cu(111)	RT	UHV	Repulsive pushing	Abacus: monoatomic step of molecules	[18]
STM	CO-Fe, Fe(CO)-CO	Ag(110)	13K	UHV	Electric-field-induced electrostatic force	Synthesis of molecules	[19]
OT	DNA	None	RT	Liquid	Laser trapping	DNA cutting	[20]
MT	DNA	None	RT	Liquid	Magnetic trapping	DNA cutting	[21]

Lateral Contact Manipulation Strategies

Pushing or pulling nanometer objects on a surface with an AFM is a typical manipulation using this method as shown in Figure 1.7 (b). Early work showed the effectiveness of this method for the manipulations of nano-particles [22-25]. This method has also been shown in nano-constructing [26] and biomanipulations [27]. A virtual reality interface facilitates such manipulations [28, 29] and may provide for other kinds of manipulations. This technique has been used in the manipulations of nanotubes on a surface. The first demonstration was given by Lieber and co-workers by bending a 4.4nm-diameter multiwalled carbon nanotube on an oxidized silicon substrate with an AFM tip. The images (0.9 μm by 0.9μm) show a 4.4nm-diameter multi-walled carbon nanotube (MWNT) before and after bending on an oxidized silicon substrate. The nanotube was then bent by scanning with the tip in contact with the surface. The right image shows bumps along the nanotube that correspond to positions at which buckling occurs. Both images were recorded in tapping mode to minimize the applied lateral forces. This manipulation has been used for the investigation of the elasticity, strength, and toughness of nanotubes.

The similar strategy was used by Falvo et al. for the investigation of the behaviours of nanotubes under large strain [30]. The white scale bars at the bottom of each panel *a–d* represent 500 nm. Panal *a* shows the original adsorbed shape of the tube, 10.5 nm in diameter and 850 nm long. The tube is bent in steps, first upwards as shown in panel *b*, until it bends all the way back onto itself as shown in panels *c* and *d*.

To investigate the dynamics of rolling at the atomic level, sliding and rolling of a nanotube were performed by Falvo et al. on graphite surfaces by using an AFM [31]. Panel *a* shows the tube lies in its original position. Grid lines are overlain so that one of the grid axes corresponds to the original orientation of the tube axis. The absolute position of the grid relative the fiducial feature indicated by the arrow was adjusted to be consistent with that in b so that the point of rotation could be determined. Panel *b* shows the tube's orientation after AFM manipulation. The pivot point and push point are indicated by the bottom and top arrows, respectively. The white dashes to the right of the push point are markers indicating the trajectory of the AFM tip during manipulation. Inset shows the lateral force trace during a sliding manipulation. It was shown the rolling motion of a nanotube. The tube is 500 nm long and has an average radius at the ends of 13.3 nm and at mid-length of 8.5 nm. Panels *a-f* shows the tube as it is manipulated from left to right. The tube is imaged before and after each of the 5 manipulations. The insets between each topographical image show the lateral force during each manipulation. The tube is moving from left to right not gradually, but in sudden slips in a stick-slip type rolling motion.

Vertical Manipulation Strategies

The pick-and-place task as shown in Figure 1.7 (c) is especially significant for 3D nanomanipulations since the main purpose of such manipulations is to assemble pre-fabricated building blocks into devices. The main problem is how to achieve the control of the interactions between the tool and the object, and between the object and the substrate. Two strategies have been presented earlier for micromanipulation [32] and have proven to be also effective for nanomanipulation recently [54, 33]. One is to apply a dielectrophoretic force between the tool and object as an additional external force, and another is to modify the van der Waals and other intermolecular and surface forces between the object and substrate. For the former one, an AFM cantilever is ideal to be applied as one electrode to generate a non-uniform electrical field between it and the substrate.

1.2.4 Related Works of Micromanipulations

Micromanipulation is manipulations for the micro-scale objects or micro-scale resolution. In this section, we focus on the micromanipulation system under optical microscope. The classifications of micromanipulations are summarized in Table 1.3.

The micromanipulation is mainly divided into two manipulation types, contact or non-contact types. Contact manipulation type uses an approach to the target object in a contact manner. This configuration is simple, reliable, and traditional way, however some limitations make them difficult to apply in some situation, for example space, environment, numbers of target objects and so on. On the other hand, the non-contact type realizes an approach to the object in a non-contact manner, by changing the environmental fields, the target objects can be manipulated. In the following section, main methods are introduced for micromanipulation.

Table 1.3 Classification of micromanipulations

Types	Mechanism	Features
Contact	Serial mechanism	Simple mechanism enable simple modeling and low-cost.
	Parallel mechanism	Compact, multi-DOFs, and High-rigidity, however complex structure.
Non-Contact	Electric force	Simple set-up, easy to make in small scale
	Magnetic force	Large operation force, high response time
	Optical force	High accuracy and good operability, however low manipulation force
	Ultrasonic force	Compact and Large operation force, however low resolution and operatability
	Liquid force	Good transportation and operatability however low response time
	Gas force	Large operation force however difficulty on the resolution and operability

Contact Micromanipulations

The basic micromanipulator is a mechanical manipulator by operating manually. It was developed by a rack and pinion mechanism which is one of the ordinal linear mechanisms, as an example. When we operate the mechanical manipulator directly by our hand, the vibration of our hand is easily transmitting to manipulator and end-effector caused for failure in work.

For this reason, a master-slave micromanipulator was widely used. Until now, the hydraulic (water/oil) pressure delivering mechanism is still used for the operation in a microscopic field. It can prevent transmitting the vibration directly. However, the hydraulic pressure types have some limitations on the moving velocity of a slave system and being hard to carry out operation to a master.

Recently, a micromanipulator is developed using the ultrasonic linear motor with light weight, high torque, and high stillness holding power compared with the electromagnetism motor. Moreover, the micromanipulator using the Piezo drive mechanism is extremely used with high precision, and low vibration.

For example, the laminated type piezoelectric device is used for the parallel link manipulator which the mountain was developed (Figure 1.8) [34]. This manipulator has a micro hand of two at a tip, and there is the feature of being manufactured by the parallel linkage mechanism. Although the flexible mechanism is useful that can realize comparatively compactly and high positioning accuracy in the free positioning within a microscopic field. As an example of serial micromanipulator, Figure 1.9 shows the serial micromanipulator with a coarse and fine movable mechanism with seven degrees of freedoms.

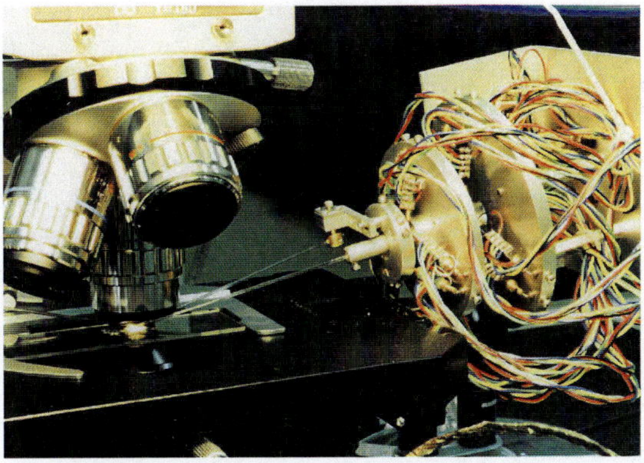

Fig. 1.8 Piezo electric parallel micromanipulator (with permission)

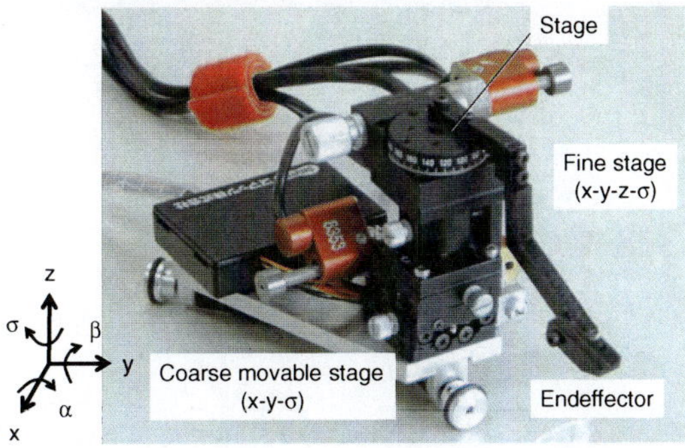

Fig. 1.9 Example of serial micromanipulator with coarse and fine movable mechanism

Non-contact Micromanipulations

The local environmental control technique has been mainly studied by chip type devices and probe type devices. The chip type devices can control and measure the local environment without disturbance because of the operation in the closed space. For example, the microchips with the microchannels were developed for such as sensing and controlling the temperature in the microchannel [35, 36], generating the concentration gradient [37], generating the water/oil droplets [38, 39]. The microchip with the confined space in which some biological sample was captured was developed to control and measure the environmental conditions of the confined space [40]. These devices control and measure the local environmental conditions in the nearly equal or bigger scale compared to the dimensions of the target microbes. Therefore, the microchip devices with the microchannel or the confined space do not have enough resolution to reveal the local information of the single cell.

The microchip devices with the nanochannel were also developed as shown in Figure1. 10 [41, 42]. These microchip devices with the nanochannel have enough resolution to reveal the local information of the single cell. However, because the channels are built-in inside the microchip, the chip type devices have no facultativity of the positioning to operate, though they have the advantage of the stable operation space. The probe type devices; such as glass micropipettes, have difficulty to operate in the closed space. On the other hand, the applications of the probe type devices are various, for example, injection of the reagents into the cells/microbes [43].

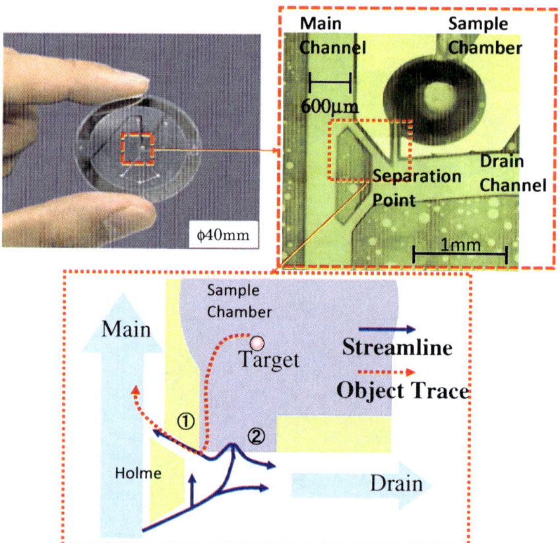

Fig. 1.10 Separation of target cells by fluidic micro-channel

Recently, reconfigurable micro-channels were proposed to realize flexible channel configuration [44]. This mechanism is the optoelectronic phenomenon created by di-electrophoresis (DEP) [45]. The local thawing of solidified compound was achieved by optoelectronic effect with a low power. The micro-channel was created by annihilation and reconfiguration of frozen liquid depending on the light pattern. As shown in Figure 1.11, arbitral shape of micro-channel was patented through the proposed method.

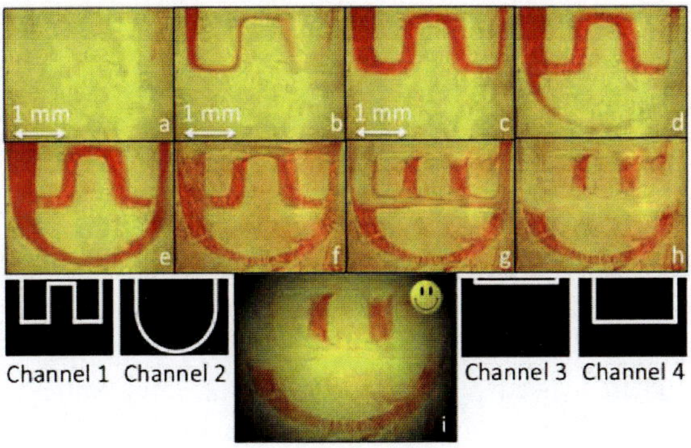

Fig. 1.11 Red-dyed DMSO to form a specific pattern. (with permission)

The optical manipulation is also widely applied in micro-scale objects. At first, A. Ashkin developed a manipulation by optical force and this manipulation method [46]. While electrical force works in filed, optical tweezers works at laser focal point. Multiple objects can be trapped simultaneously by scanning the laser focal point. The laser scanning approach is suitable for the attitude control of target and the trajectory control of the object since it is easy to create and change multiple laser trap potentials [47]. As shown in Figure 1. 12, we developed synchronized laser micromanipulation (TSS) for an independent trajectory control of multi-objects by using single laser with a scanning system of its focal point [48], [49].

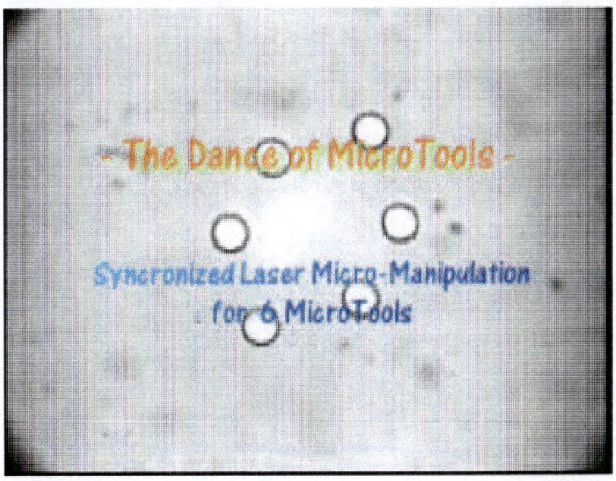

Fig. 1.12 Example of laser micromanipulation

1.2.5 Related Works of Nanomanipulations

Previously, the nanomanipulations were enabled by the inventions of STMs [11], AFMs [50], and other types of scanning probe microscopes (SPMs), which can used as both microscopes and manipulators. Besides these, optical tweezers (laser trapping) [51] and magnetic tweezers [52] are also possible nanomanipulators. Characterized by the ability of 3D positioning, position and orientation control, independently actuated multiple end-effectors, separate real-time observation system, and possibility to combine SPMs inside them, NRMs [53, 54] largely extend the complexity of nanomanipulations.

The first practice on nanomanipulation was performed by Eigler and Schweizer in 1990 [55]. They applied a STM at low temperature (4K) to position individual xenon atoms on a single-crystal nickel surface with atomic precision. The manipulation enabled them to fabricate rudimentary structures of their own design, atom by atom. The result is the famous set of images showing how 35 atoms were moved to form the three-letter logo "IBM", which showed that people could indeed maneuver things atom by atom as was dreamed of by Feynman [1].Strategies for nanomanipulation are basically determined by the environments—air, liquid or vacuum, which is further decided by the property and size of the objects, and observation methods [56] There are lots of research approaches on the nanomanipulations and typical examples related of nanomanipulations are listed in Table 1.4.

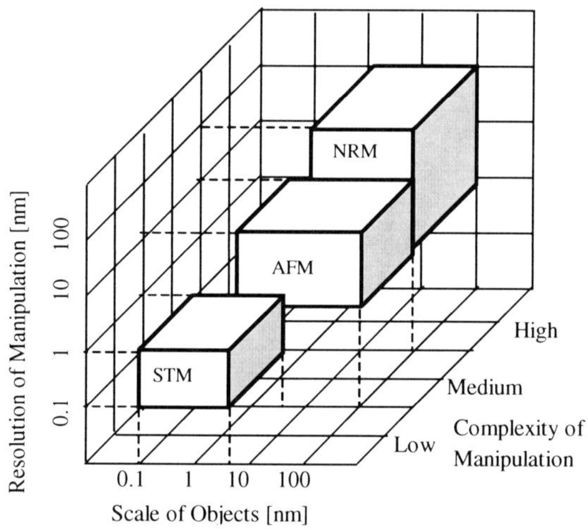

Fig. 1.13 Comparison of Nanomanipulators

Table 1.4 Related works of nanomanipulations

Microscopes	**Papers**	**Objects**	**Methodology of Nanomanipulation**	**Dimensions**
SPM(STM)	[62]	Xenon atoms	van der Walls forces	2D
SPM(STM)	[63]	MoS_2 atoms	Electric-field-evaporation	2D
SPM(STM)	[57]	Supermolecules	Contact pushing	2D
SPM(AFM)	[64]	CNTs	Contact pushing	2D
SPM(AFM)	[65, 66]	Nanoparticles, DNA	Contact pushing	2D
SPM(AFM)	[58]	DNA	Contact pushing	2D
SEM	[69]	FAB materials	Contact pushing	3D
SEM	[70]	CNTs, Nanowires,...	Contacts	2D/3D
SEM	[71, 72]	Nanoparticles,...	Contacts	2D/3D
SEM	[59]	CNTs	Contacts	3D
TEM	[74]	CNTs, Nanowires,...	Contacts	2D/3D
TEM	[75]	Au wire	Contacts	2D
TEM	[60]	Nanotubes, Nanobelts	Contacts	2D/3D
TEM	[61]	Clacks of ceramics	Contacts	2D

A simple comparison of a STM, an AFM, and a NRM is shown in Figure 1. 13. With its incomparable imaging resolution, STMs can be applied to particles as small as atoms with atomic resolution. But limited by its 2D positioning and available strategies for manipulations, standard STMs can hardly be used for complex manipulations and cannot be used in 3D space. AFMs can get nearly as high resolution as STMs for either conductive or insulate objects, and can be applied to manipulate relatively large nanometer objects mainly by mechanical pushing in a 2D surface. The major limitation of SPM-type manipulators is that they do not have rotation degrees-of-freedom (DOFs) and hence cannot control its orientation. Another important demerit is that their observation and manipulation abilities cannot be used simultaneously. By separating the imaging and manipulation functions, nanorobotic manipulators can perform nanomanipulations under the real observations of electronic microscopes. They generally have much more DOFs with orientation control, and hence can be used for the manipulations of 0D (symmetric spheres) to 3D objects in 3D free space. Limited by the relative lower resolutions of electron microscopes, it is still hardly to be used for the manipulations of atoms, but its capability of 3D positioning, orientation control, independently actuated multi-end-effector, separated real-time observation system, and possibility to include SPMs inside it makes it the most promising way for nanomanipulations.

In 1990, Eigler and Schweize demonstrated that the first atom practice nanomanipulation with STM [62]. They applied a STM at low temperature (4K) to position individual xenon atoms on a single-crystal nickel surface with atomic precision. The manipulation enabled them to fabricate rudimentary structures of their own design, atom by atom. The result images atoms were moved to form the three-letter "IBM". This work shows the impact feasibility of the control the atoms through nanomanipulation. Hosoki et. al also demonstrated the atomic letter by STM [63]. The MoS_2 atoms were evaporated by electric field at room temperature.

Avouris et al. applied an AFM to bend and translate carbon nanotubes (CNTs) on a substrate [64].The objects are pushed and deformed by the tip of an atomic force microscope (AFM). They combined the techniques with an inverse process, namely straightening, by pushing along a bent tube, and realized the translation of a tube to another place. They also showed the assembly of two bent tubes and a straight one to form a "θ" letter. Ning Xi et. al at Michigan Stagte University, developed AFM based nanomanipulation system with interactive operation system [65,66]. The system realized a real-time visual feedback during AFM based nanomanipulation. They also proposed some physical models of the interaction among the AFM tip, substrate and objects to feel the real-time interactive forces through haptic system for operator. The "msu" pattern was re-presented by pushed nanoparticles on a glass surface (Figure 1. 14). A. Requicha et al. was proposed fully automatic AFM system for arraignment of nanoparticle with diameters ~10 nm (Figure 1. 15) [67]. This technique is important because of reducing the require time in a matter of minutes whereas the past interactive systems would require several hours by a skilled operator to construct similar, but usually less accurate patterns. Metin Sitti et. al also proposed the tele-nanorobotic system based on AFM probe [68]. It consists of a piezoresistive AFM probe as mechanical manipulation probe and topology sensor. A haptic device and virtual reality display are embedded for a force-feedback and a visual feedback system.

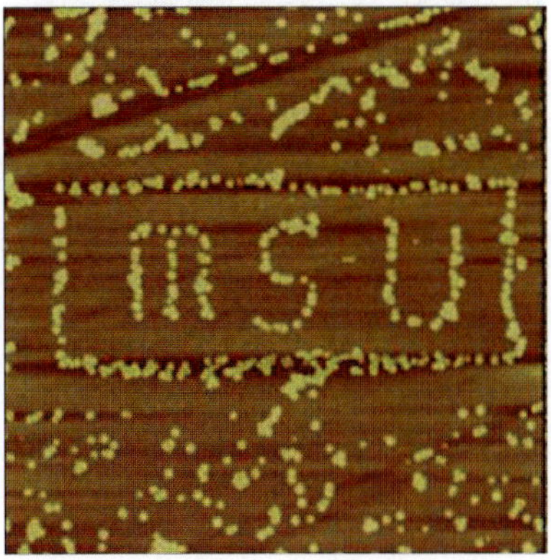

Fig. 1.14 "msu" letter with pushed 100 nm latex particles (Area: 10 μm × 10 μm) (with permission)

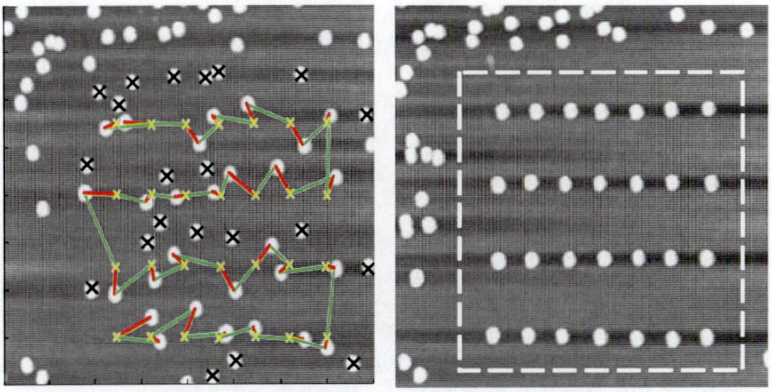

Fig. 1.15 Arrangement of nano-particles in line array of 16 nm Au particles on mica coated with ply-L-lysine (with permission)

Tsuchiya et al at Tokyo Univesity, Nakao Lab. have been proposed nano-manufacturing world (NMW) [69]. NMW is table top type micro-nano manipulation with assembly and fabrication simultaneously. The system has two chambers as assembly and marching chambers. In the machining chamber, fast-atom-beam (FAB) source is effectively used for the micro-machining. In the assembly chamber, two SEMs are embedded for enough 3-D observation area. The 2-Unit remote manipulators are also embedded for the micro assembly with two probes. As commercially available products, Zyvex Corp. produced the nanomanipultors inside a SEM [70]. The over-view of the system is as shown in. The manipulator can be realized the X-Y-Z positioning inside SEM vacuum chamber with real-time observation. The size of manipulator is ~100×100 cm×50 mm and the working area is 12 mm×12 mm×12 mm. The system can be introduced the vacuum rate is ~ 10^{-6} Torr order. The manipulated objects can be considered to be micro to nano meter scale objects. NanoEffector™ is used as the actuators. The 4-probe manipulation can be operated independently. The system applied to measurement of the mechanical and electrical properties and assembly of nanotube or nanomaterials.

S. Saito proposed SEM manipulation system for autonomous manipulation system [71,72]. The system has following features, (1) an optical microscope is equipped with orthogonal to the SEM to obtain full 3D information, (2) five DOFs including rotation, (3) a configuration that the probe tip is always located the center of the field of view to guarantee the consistent acquisition of well-focused images. As a result of the automatic task execution, the polymer particles, whose sizes are about 30 μm in diameter, were arranged into a linear pattern.

V. Eichhorn and S. Fatikow proposed the automatic SEM manipulation system for nanotubes by specially designed micro-gripper based on the SEM imaging analysis [73]. They mainly investigated on the nanorobotic strategies such as mechanical and electrical characterization of nanowires, automatic pick-and-place handling of individual nanowires CNTs and so on. The electro-thermal micro-grippers or micro-four-point probes were developed by their corroborator. As shown in Figure 1.20, the pick and place of single CNT was demonstrated in the 3D space based on the AFM-based handling inside a SEM.

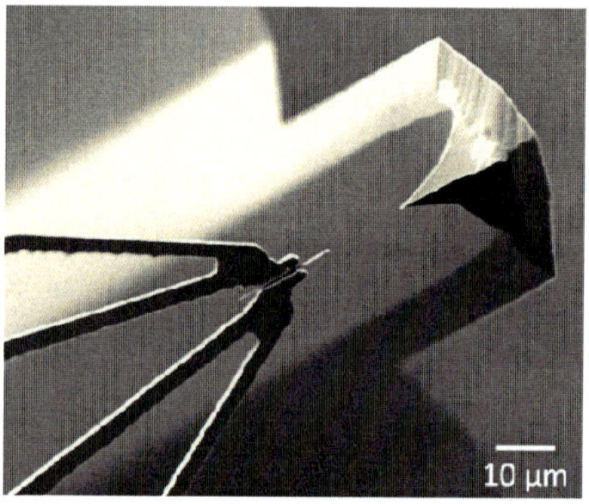

Fig. 1.16 Grasping and attaching a carbon nanotube at the tip of AFM cantilever by SEM nanomanipulator (accepted by author)

Instruments Corp. manufactured the nanomanipulator inside transmission electron microscope (TEM), called as Nanofactory™. It has three types; the scanning tunneling microscope (STM) probe type, atomic force microscope (AFM) probe type, and nanoidentation type [74]. The manipulator actuated with Pb(Zr,Ti)O_3 (PZT) tube elemental devices. These manipulators have been applied for some investigations of nano-material property such as inter-layer resistance of multi-walled carbon nanotube, conductively change of nanowire applied forces, or contact resistance with Au nanowire.

Kizuka et.al proposed the manipulation holder inside high-resolution transmission electron microscope (HR-TEM). The manipulator was specially designed with atomic level positioning resolution [75]. For rough positioning, the motor and micro-gear mechanisms are used the actuation. For the precise positioning, a piezo-tube is used for 3 degrees of freedom (x-y-z) actuations. The maximum working area of rough positioning is ±1 mm. One of precise positioning is 1.2 μm in x direction, 11.4 μm in y direction, and ~11 μm in z direction. *In situ* fabrication of 2.6 nm single Au atomic nanowire is fabricated at the contact point of Au layers inside a HR-TEM as shown in Figure 1. 17 [76].

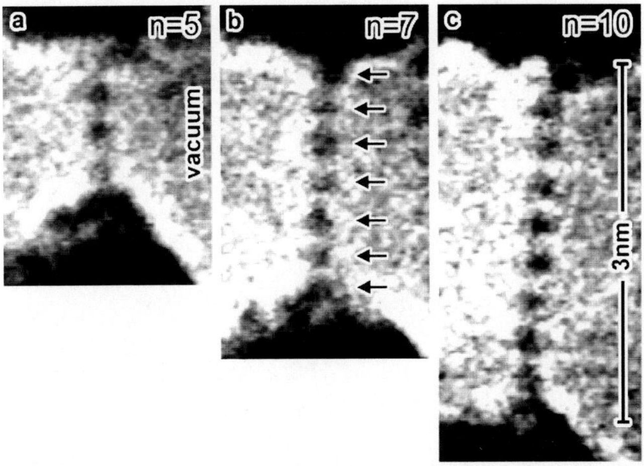

Fig. 1.17 HR-TEM image of stable gold single atom wire (with permission)

1.3 Application Fields of Micro-Nanorobotic Manipulation Systems

1.3.1 For Micro-Nano Mechatronics

"Microtechnology" is commonly used in our daily life, such as applications to the automobiles, computer peripheries, printers, cameras, amusements, robotics, automation, environmental monitoring, energy resource, biological/medical treatments, and so on. In particularly, "Micromechatronics" is the one of the important technology especially in sensors and actuators to miniaturize elemental devices in micro-scale. Recently, "Nanotechnology" has more important role in the advanced field of mechatronics as "Nanomechatronics". To realize various applications for industry, some techniques are important, especially micro-nano fabrication, assembly, control, material, and evaluation techniques (Figure 1. 18).

Historically, the mechatronics was born in Japan around 1969. At that moment, a lot of machines were changed their actuation mechanism from the hydraulic actuators to electric motors. The mechatronics was grown the research and development to combine mechanical and electrical system efficiently. Normally, the mechatronics contribute the development for Robotics and Mechatronics, System Design, Sensor and Actuator, Assembly and Control, Design Goal and Knowledge , Design Principle and so on.

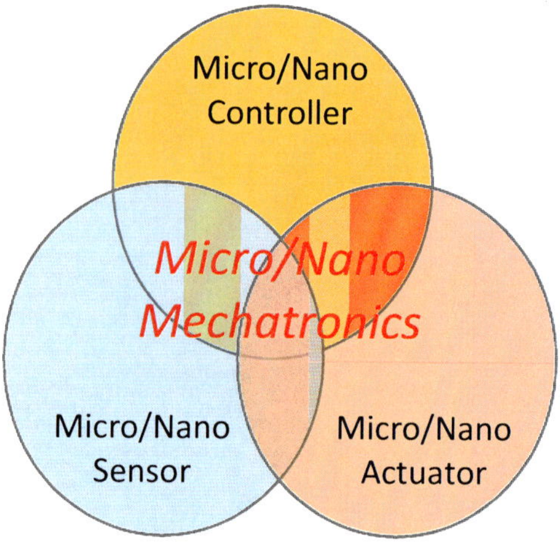

Fig. 1.18 Micro-Nano Mechatronics

For industrial applications, various devices are developed and available. Those devices are investigated based on the micro and nanomechatronics technologies to realize high-efficiency, high-integration, high-functionality, low-energy consumption, low-cost, miniature, and so on (Figure 1. 19). The applications are basically categorized into the "Mechanical", "Electrical", and "Biological/Medical" applications as shown in Table 1.2. "Mechanical" applications are relatively based

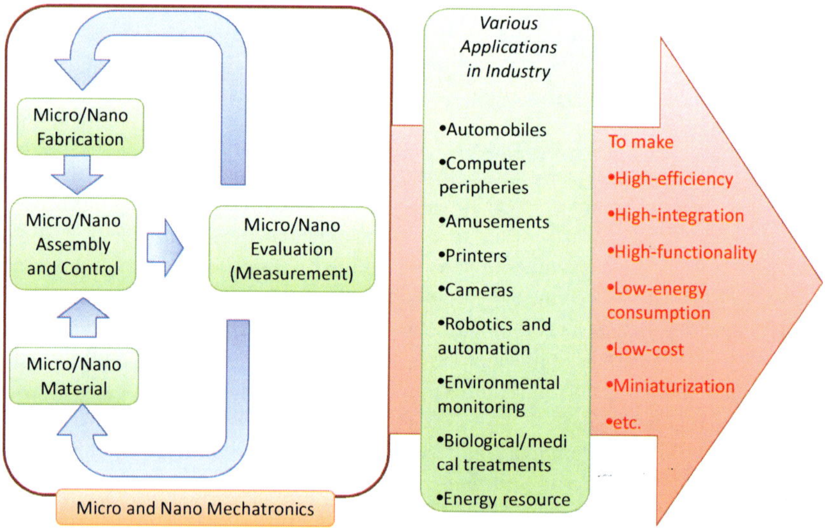

Fig. 1.19 Micro-Nano Mechatronics for industrial applications

on the inorganic materials or technologies, such as lithography technique. On the other hand, "Biological/Medical" applications, the organic materials or technologies are used, such as self-assembly technique. In between them, "Electrical" applications are placed for delivering or calculating information and so on. Table 1.3 shows the important technologies and examples of micro and nano mechatronics categorized application fields.

Table 1.2 Categories of Micro and Nano Mechatronics

	Category	Basics	Components	Applications
Inorganic (Dry)	Mechanical	Nanocarbon, Battery,...	Fuel Cells, Micro-Nano Actuators, Inkjet System...	MEMS, Photonic Devices, Storage,...
↕	Electrical	Photonic Crystals,...	Semi-conductor Devices,...	IC, Memory, Molecular Computing, ...
Organic (Wet)	Biological/Medical	Biomaterials, Artificial Cells,...	Bio-sensors, Bio-catalyst,	Bio-chips, DNA Computing,...

Table 1.3 Industrial application of micro and nano mechatronics

Application Fields	Important Technology	Examples
Automobiles	Si etching, Micromachining, Ceramics processing, Metal coating, Photolithography...	Temperature sensor, Gas sensor (O_2, NO_X), Knock sensor, Pressure meter(Fuel), Acceleration meter (Air-bag), Yaw rate sensor, ...
Computer peripheries	Photolithography, Precision machining, Micromachining, Nano-imprinting...	CPU, Hard disk drive, Memory device, Electric circuits, ...
Printers	Micromachining, Ceramics processing...	Ink-jet head, Ink solution delivery line, Surface potential sensor,
Cameras	Micromachining, Ceramics processing...	Photo-sensitive elements, Lens, Lens actuator, Shutter, ...
Amusements	Photolithography, Micromachining...	Acceleration meter (Gravity/Angle sensor), Display, Touch panel (Display), Wireless communicator, ...
Robotics and Automation	Photolithography, Precision machining, Micromachining, ...	Force sensor, Pressure sensor, Displacement sensor, Gyro sensor, IR sensor, Switch, ...
Environmental monitoring	Photolithography, Micromachining, Database...	Temperature sensor, Humility sensor, Solar irradiation sensor, CO_2 sensor, ...
Energy resource	DNA delivery, Micro/Nano materials, ...	Transgenic plants, Energy conversion system (Solar panel)...
Biological/medical treatments	Pharmacy, Cell analysis, Cell treatments, ...	Single cell injection, Nuclear transplantation, Tissue engineering, Regeneration medicine, Surgical system, Surgical simulator...

From the view point of future direction of micro-nanomechatronics, we proposed to construct "super simulator" by mimicking the biological system (Figure 1. 20). To realize the super simulator, the micro-nano fabrication/manipulation/sensing/auction/assembly techniques are required. Their materials need to be used biocompatible, biodegradable for using real biological cells in vitro environment. The key technology should be minimized the gap between in vivo and in vitro systems (Figure 1. 21). This challenging issue may be possible to solve to corroborate different science/engineering fields cautiously, such as stem cell engineering, tissue engineering, micro-nano manipulation technologies and so on.

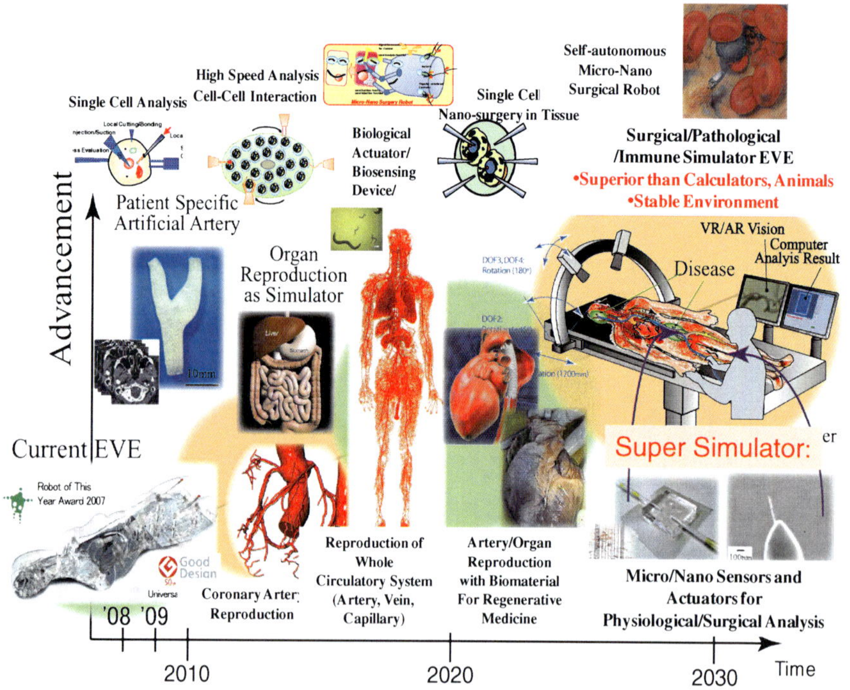

Fig. 1.20 Example of goal of Micro-Nano Mechatronics "Super Simulator"

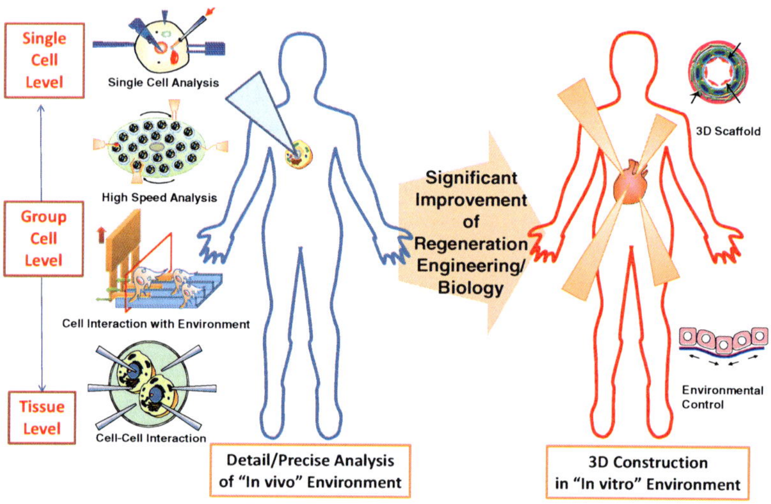

Fig. 1.21 Research direction of 3D construction of biological system in vitro from the functions of biological cells in vivo

1.3.2 For Micro-Nanorobotics

The requirements for the micro-nanorobots are various, such as sensing, actuation, data transmission, remote control, power supply subsystems etc. and their embedment or integration to the micro-nanorobots. Hence, those all components need to be downsized by improving the manufacturing technique or using novel materials.

Recently, micro-nanorobots have been actively studied. If the highly functional micro-nanorobots are realized, it is expected to lead the developments of novel technologies especially in the medical field such as drug delivery system (DDS) and micro-nano surgery inside the human body [77,78,79,80] as shown in Figure 1. 22 [81].

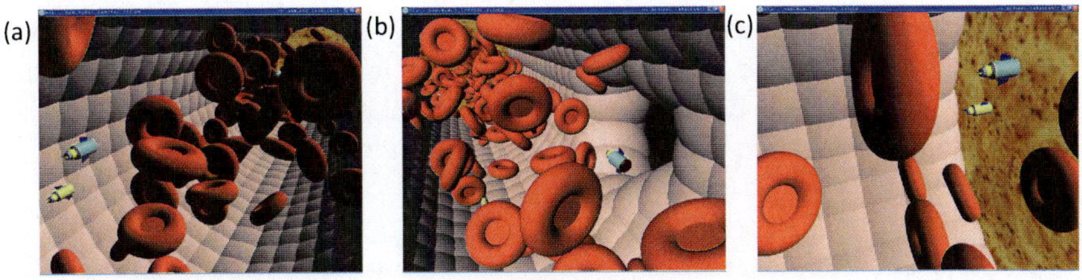

Fig. 1.22 Nanorobot for the detection of aneurysm (a) nanorobot enter the vessel, (b) nanorobot detects aneurysm from chemical gradient change, (c) nanorobot flow with bloodstream until they leave the vessel. (with permission)

The micro-nano transportation systems were also proposed and demonstrated by using the biomolecular motors. Microobjects, such as microbeads and micromachined structures, modified with the kinesin were transported on the microtubule [82,83]. The autonomous transportation system with the microtubules on the kinesin array was shown for the delivery of the lipid vesicles as a reactor connected to the microtubules [84]. These transportation systems and afore mentioned actuation systems with the biomolecular motors basically move on the specific "rail" structures, because Brownian motion inhibits the free directional motion. So, these systems are not suitable for the versatile micro-nanorobots because of the difficulty and unavailability to built-in such rail structures in the operation environment. The rotational actuation of the nanostructure was also demonstrated by combining the manufactured nanopropeller with F_1-ATPase biomolecular motor [85]. To utilize these biomolecular motors, it is needed to purify and reconstitute them after extracting from the living cells. It is relatively difficult and long in doing.

The actuations of the micro structures using intact tissues and cells extracted from the living organism or cultured from the preserved cells were reported. For example, the manufactured micromechanical structure was moved as a consequence of collective cooperative contractions with the cardiomyocytes muscle bundles [86]. And, the micropillars were contracted [87] and the inverted two-row micropillar array based microrobtic structure was moved [88] by the insect dorsal vessel tissue (DVT; muscle like tissue). Such bio actuators in tissue level require long time in day order to culture and relatively severe environmental control (though the insect dorsal vessel tissue is more environmentally robust than the cardiomyocytes).

1.3.3 For System Cell Engineering

The "System Cell Engineering" is a new academic field aiming at synthetic understanding of cell function using engineering (Figure 1. 23). The biological cell is a unit of our organism in micro-meter scale (cell level). It is consisted of component of nano-meter scale structures, such as DNA, ribo nucleic acid (RNA) and so on (molecular level). On the other hand, the multiple cells get together as some another functional component (tissue level). To understand the cells' function and mechanism, it is need to analysis/integration in multiple levels. It means that it is need to understand the communication and control principle of bare function and integration function of the cell. This academic field also aims to reconstruct the cell function and functional control of tissue.

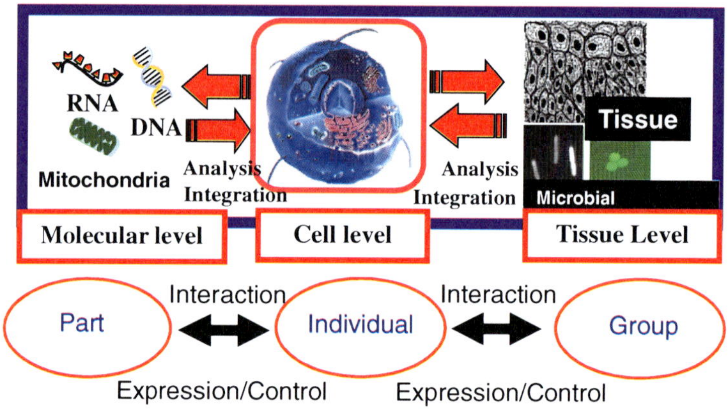

Fig. 1.23 Schematic of research area of system cell engineering

To achieve the system cell engineering, it is need to analysis in wide range from nano-meter to milli-or centi- meter scale. One of the key technologies is the "multi-scale manipulation" as shown in Figure 1. 24. For analysis of biological cells, a lot of manipulation techniques are required such as isolation, separation, immobilization, stimulation and so on. Micro-nano manipulation techniques can be applied to measure or control individual component freely in more complex and detail conditions.

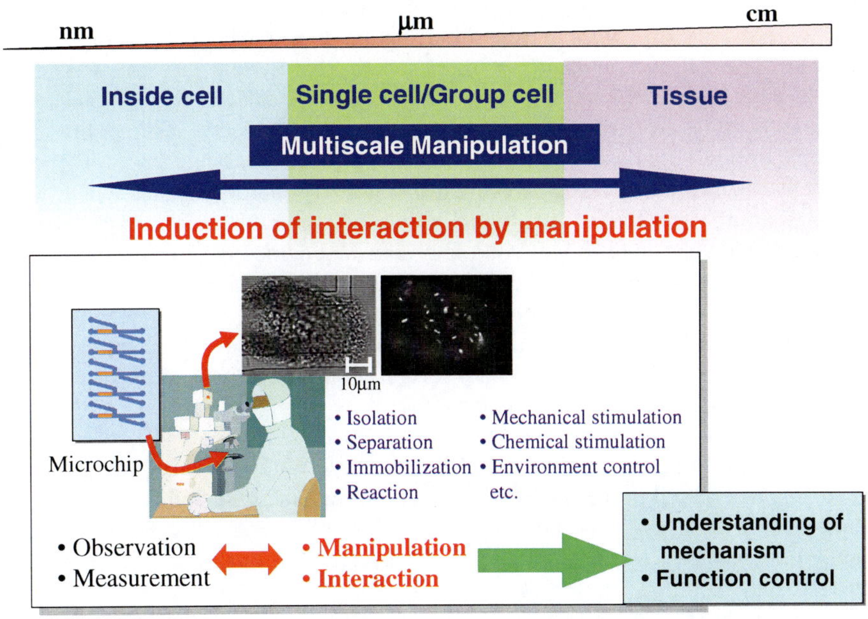

Fig. 1.24 Schematic of research area of multi-scale manipulation

1.3.4 For Single Cell Analysis

In recent years, micromanipulation of bio-microparticles such as microorganisms and cells has become essential in various industries such as food industry, medical industry, and environmental purification and so on. Many microorganisms such as oil-degrading bacterium basophilic bacterium and thermopile bacterium have been already discovered. By employing the useful characters of these microorganisms, the development of new medicine, the production of useful materials, and environmental purification were achieved. For example, recombination microorganisms used for production of biological plastic (polyhydroxyalkanpoate (PHA)) [89] or degradation of biodegradable plastics (poly-butylene succinate-co-adipate (PBSA)) [90].

As ordinary technique, a flow cytometer is apparatus which can measure the antigen level on a cell surface quantitatively by measuring the fluorescence which puts the cell dyed by the fluorescent antibody in the style of liquid, passes it, and is made to pass the focus of laser light, and each cell emits. Not only the fluorescence intensity of the inside and outside of a cell but biological character, such as a size (forward-scattered light forward scatter and FSC) of a cell and an internal structure (side scattering light side scatter and SSC), can be measured. In addition, it is also possible to isolate only a specific cell individually an aseptic technique with keeping in a living condition. General classification of a flow cytometer will divide it into the apparatus (analyzer) in which only the analysis of each parameter is possible, and the apparatus (cell sorter) which has a preparative isolation function. There are mainly two types of cell sorter, one is droplet charged method and the other one is a cell capture method.

The advantage of plastic production by microorganism is that oil, which is conventionally used for plastic, is not necessary to produce plastic. This is an example of microorganism application to post-petrochemical industry amid energy problems such as an oil resource decrease. However, microorganisms which were discovered are estimated as less than 10% of the whole microorganisms. For the remarkable advancement of microbiology and bioengineering, it is important to develop a new technology of manipulation and separation and discover the unknown microorganism and their characters.

Many research works on cell analysis have been done for investigation of unknown properties of them. There are several different methods to understand the biological properties of cells. For example, group cell analysis and single cell analysis based on the direct observation of each specimen. Of course, both methods are important, but each method has characteristics. With the former method, the time – course of cell properties such as a cell volume increase and DNA synthesis during the division cycle has been studied. Group cell analysis is very important to apply to industry. However, we can get only averaged information on each cell. In the latter method, we can acquire detailed information on the individual cell by controlling the spatial distribution of each cell as well as the environmental conditions.

Recently, a demand of single cell level analysis has been increasing. In the actual microbial community, it is heterogeneous population as shown in Figure 1. 25. After stable phase through the exponential growth phase, this population divides into three phases such as viable but not culturable, apoptosis, and adaptation to a new environment.

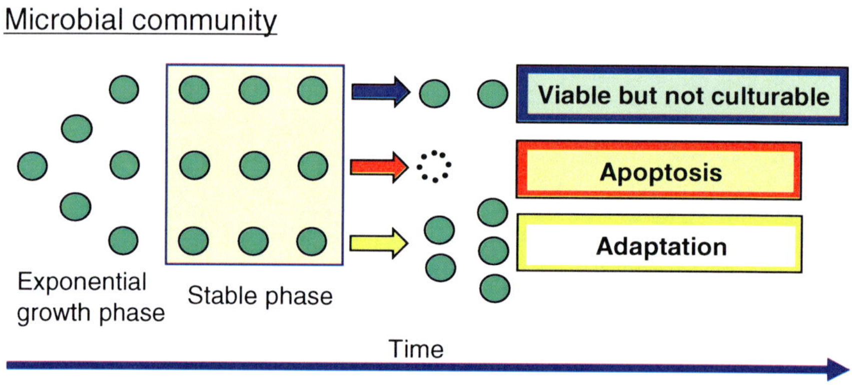

Fig. 1.25 Mechanism of microbial community

There are other reasons for significance of single cell analysis as shown in Figure 1.26. As mentioned in above, cell population is not homogeneous and the mechanism of differentiation to the heterogeneous is not known. Microbial activity is not uniform and the mechanism of gene expression is not known. In addition, gene expression is dynamically changing and its property is unknown. For single cell analysis, an extraction of single target cell from a large number of samples and continuous monitoring of cell are necessary. For these reasons, bio micro manipulation such as single cell manipulation and observation of single cell reaction is very important. The configuration of microchip for cell manipulation is shown in Figure 1. 27.

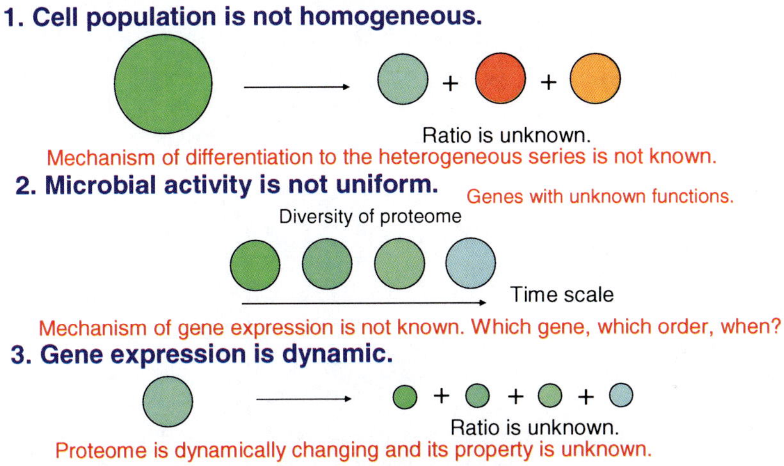

Fig. 1.26 Significance of single cell tracking and observation

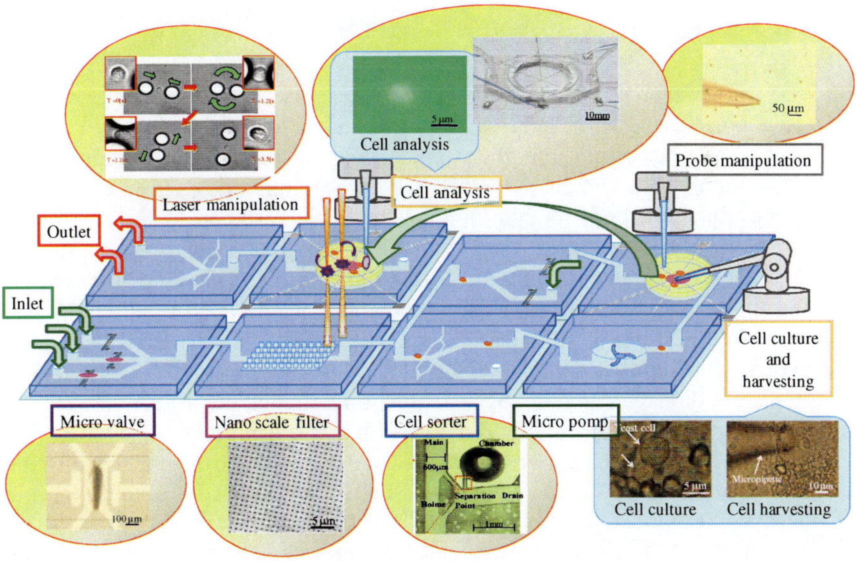

Fig. 1.27 Configuration of Microchip for Cell Manipulations

The cell manipulation system needs to have functions to control the local environment surrounding the cells as shown in Table 1.5.

Table 1.5 Role and Issue of Cell Manipulation

Role	Conceptual figure	Application	Issue
1. Mechanical interaction Biomechanics Differentiation, growth	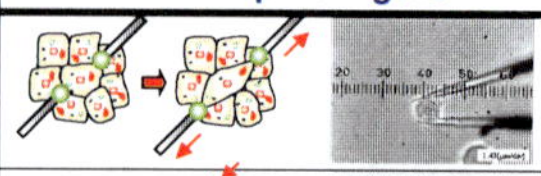	Micro-hand Manipulation μTAS, BioMEMS	Local control, 3D Force measurement - nN (Force) - 100nm (Position)
2. Gene manipulation, injection Stability in the cell Transportation, expression		Nano-probe Injection μTAS, BioMEMS	High speed, mass production High efficiency Gene expression
3. Separation of single cell On-chip cell analysis Gene expression		Immobilization Optical tweezers DEP, Electrophoresis	Separation Single cell High speed, in-situ, less damage
4. Environment control Chemical, mechanical, and electrical stimulation	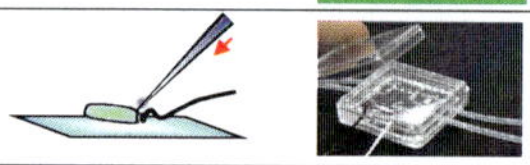	Nano-probe Injection μTAS, BioMEMS	Limited space Density gradient Real time Probes
5. Measurement, imaging Real time measurement Gene expression	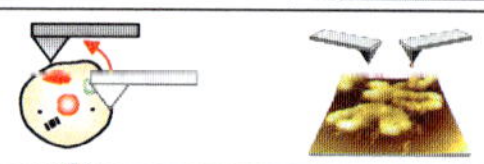	AFM Nano-probe Optical tweezers	Real time imaging Force, torque, location measurement
6. Molecular manipulation Cell model reconstruction Functional protein synthesis	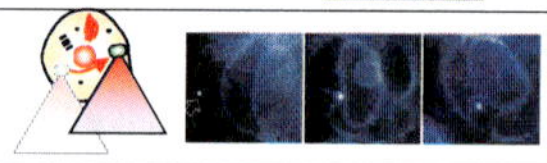	Optical tweezers Nano-probe DEP, Electrophoresis	Protein synthesis Mutation Function control
7. Shape control of cell / tissue Chemical control of surface Shape and pattern control Cell sheet manipulation	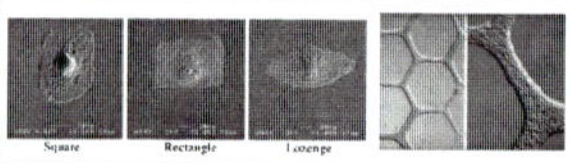	Micro-nano fabrication SPM μTAS, BioMEMS	Surface and shape Pattern generation 2D, 3D organization Function control

1.3.5 *For 3D Cell Assembly*

The cells inside real tissues and organs are usually arranged with large amount according to certain patterns and shapes, such as neural cells, skin and blood vessel [91]. In tissue engineering, important issues are cell encapsulating inside certain structures and high-throughput assembly of these structures [92].

The convenient methods for cell immobilization are based on such as aspiration, pressure of solution and fluidic structure [93]. The advantage of aspiration and pressure is that fixing force is large, while the disadvantage is damaging the cells [94]. By special fluidic structures, cells are immobilized inside microfluidic chip. The immobilized cells are difficult to be further analysis [95]. On-chip fabrication based on photo-crosslinkable resin via UV illumination is a creative way for immobilizing cells. Cells are directly immobilized inside the microfluidic chip. There are several advantages such as high-speed, low-cost and arbitrary shaped [96].

One of the important issue to be solved is the deficiency of oxygen and nutrients and the accumulation of wastes, the survival rates of cells inside scaffolds are low. This also applies to stacks made of cell-sheet. The conservation limit of the thickness of stacks of cell-sheet by diffusion of oxygen, nutrients and wastes is less than 200 μm [97]. Hence, blood vessel structure is important to construct with three-dimensional tissue.

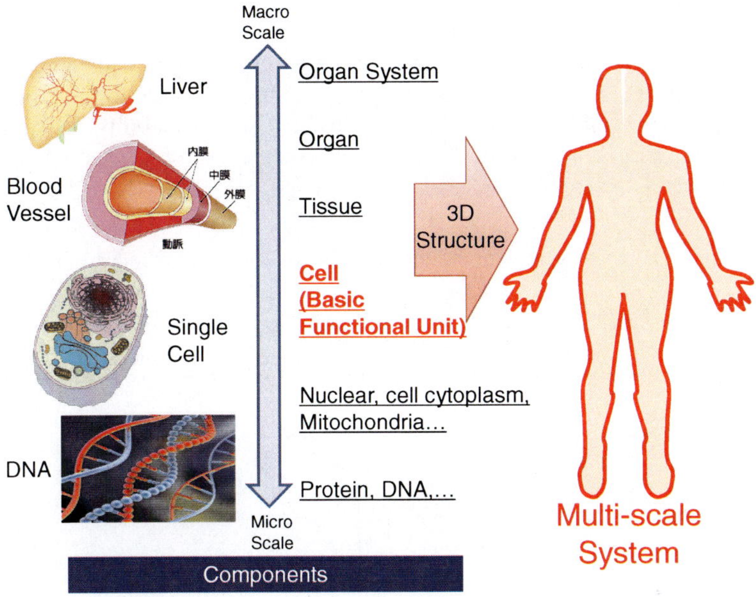

Fig. 1.28 A schematic drawing of the cell assembly method

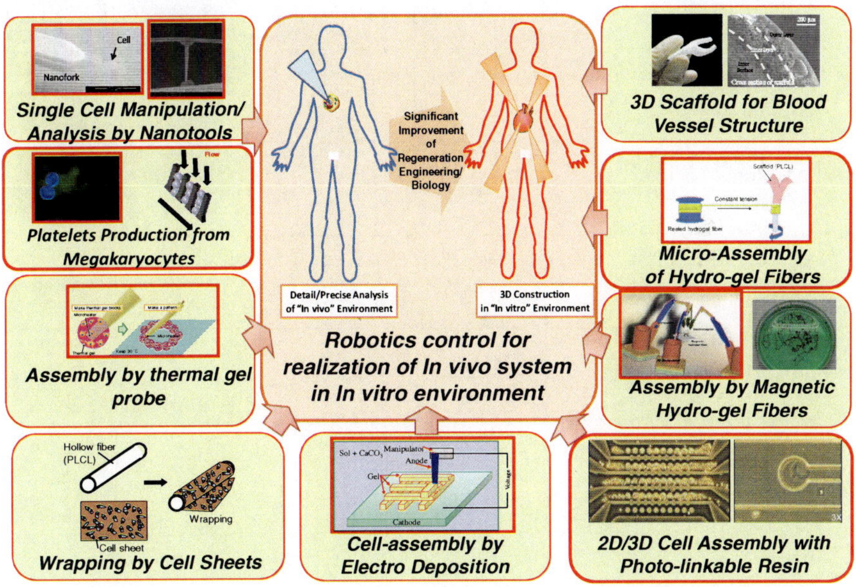

Fig. 1.29 A schematic of Bio Analysis/Assembly/Synthesis based on micro-nano manipulation

The micro-nano manipulation is a key technology to realize cell assembly (Figure 1. 29). Because it is effectively used for achieve arbitral shape construction and local analysis. Table 1.6 shows the summary of the assembly methods of cellar structures. From the components in certain dimension, the three-dimensional structures are assembled. Their important features are "Scalability", "Fabrication Speed", and "Bio-compatibility". These are subjectively scored by authors from the current research progress.

Table 1.6 3D cellar structural assembly methods (all figures with permission)

Methods	Component Dimension	Scalability	Fabrication Speed	Bio-compatibility	Example
Optical-polymerization	2D	++	++	+	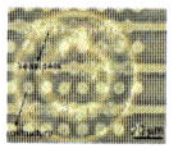
Scaffold	2D	+++	++	++	
Hydro-gel Fibers	1D	++	+	++	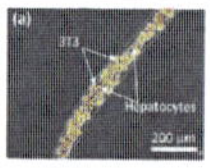
Bioprinting	0D	++	+	++	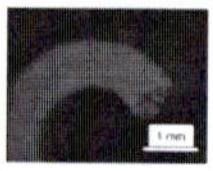
Hydro-gel Beads	2D	++	+	++	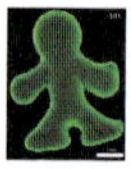
Gravity force/ Spheroid	0D	+	+	+++	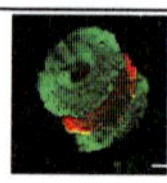
Electro-Deposition	2D	+	++	+++	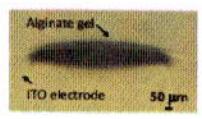
Magnetic force	0D	++	++	+	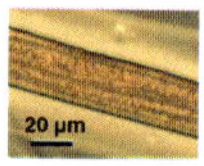
Origami	2D	+	+	+	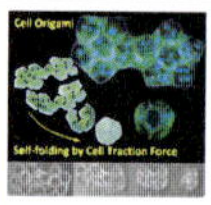

Optical Based Assembly Strategies

Optical fabrication has an advantage to fabricate the structure with non-contact ways. More over the optical pass can be well controlled especially using a laser system. To apply this method for cell assembly, the laser writing system was proposed to fabricate the micro-structures with cells [98]. The photo-linkable resin was used to encapsulate the cells. The laser system was scanned to fabricate the structure in three-dimensional structure. By changing the mixing cell types, the co-assembly of different type of cells were also achieved.

For tissue engineering, it is required to construct cell patterns and immobilize patterned cells inside certain structures. In this paper, we present a novel method of forming cell pattern by dielectrophoresis (DEP) and immobilizing cells by photo-crosslinkable resin inside microfluidic chips. The whole research procedure is shown in Figure 1. 30. We have been proposed a cell assembly method based on cell immobilization by on-chip fabrication and microfluidic self-assembly inside microfluidic devices. There are 3 advantages in this research. First, the arbitrary shaped microstructures are able to be efficiently on-chip fabricated inside microfluidic chips. Second, the large amounts of cells are able to be immobilized inside microstructures. Third, the assembly of microstructures is with high efficiency because of the microfluidic self-assembly.

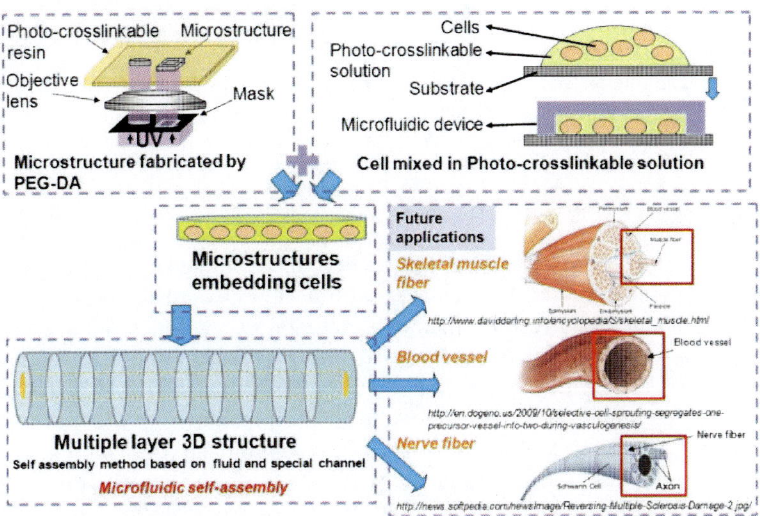

Fig. 1.30 A schematic drawing of the cell assembly method by fluidics chip using DEP and photo-resistive materials

For applying DEP force, several microelectrodes are fabricated by Indium Tin Oxides (ITO) and Cr/Au which are coated on the glass. The two kinds of DEP responses of yeast cell (W303) and other particles are experimentally confirmed. Based on negative DEP phenomenon, cell traps generated by microelectrode are demonstrated (Figure 1. 31). Position control and transportation of yeast cells is performed by using cell traps. Besides, a cell trap matrix is fabricated and high speed cell pattering is performed. The experimental results show that the cell line patterns which contain hundreds of yeast cells can be formed by DEP within 1 second.

The on-chip fabrication for arbitrary shapes of microstructures based on Poly Ethylene Glycol Diacrylate (PEG-DA) is reported. With the cell patterning by DEP and immobilizing by on-chip fabrication, microstructure which contains 3 lines of yeast cells is fabricated in the microfluidic channel, inside PEG-DA and NaCl solution. We propose a novel design of microfluidic device to separated patterning and fabrication areas. It is designed for movable microstructures embedding microbeads of which the concentration is controllable.

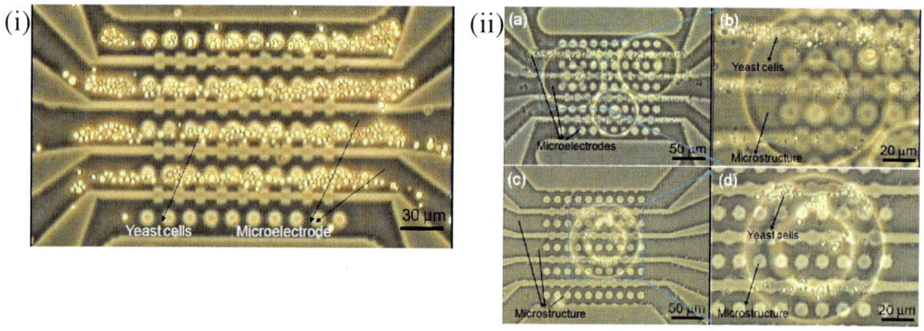

Fig. 1.31 (i) The line pattern of yeast cells formed by n-DEP based on ITO microelectrode. (ii)The images of each steps of fabrication of microstructureswith 3 lines of yeast cells (a)/(b): before cleaning away the cells outside the microstructure,. (c)/(d): after cleaning.

Scaffold Strategies

In the field of tissue engineering, researchers have studied aspects of polymer chemistry, engineering and life sciences with a view toward the development of biological substitutes to restore, maintain, or improve malfunctioning tissue and organs. Because isolated cells cannot grow into functional three-dimensional cell constructs that are similar to the native tissues, scaffolds serving as extracellular matrices are commonly used to provide a viable microenvironment for host cell attachment, infiltration, proliferation and differentiation and as a guiding template for cells to form new tissues. Typically, scaffolds for tissue engineering are made using biocompatible, biodegradable and bioresorbable materials, they have a controlled degradation rate that matches tissue in growth, and have high porosity, pore density and interconnectivity for the transport of oxygen and nutrients and excretion of metabolic waste. Moreover, scaffolds should possess sufficient mechanical strength to provide physical support for cell growth. We proposed a 3D porous biodegradable scaffold by salt/sugar reaching method [99]. This method uses a mold, which is fabricated by three dimensional printing technology as arbitrary shapes. The mold is coated by a biodegradable material (PLCL: poly(l-lactide-co-ε-caprolactone)) through dip coating method. By mixing PLCL with salt/sugar particles in certain sizes, these particles can be washed out after solidification of PLCL. Recently, the direct laser writhing method is investigated to solidify and fabricate three-dimensional structure with high accuracy [100].

Fluidics/Gel Based Assembly Strategies

The micro-chip device has advantages to control the fluidics environment effectively with minimal usage of materials [101]. A microfluidic system was proposed to fabricate the hydrogel fibers contentiously with animal cells [102]. In this paper, the authors proposed multi-channels to insert into a main channel as a laminar flow to realize the co-culture of different type of cells.

On the other hand, the direct fabrication of three-dimensional structures as bio-printing technology [103]. The biological cells are encapsulated with hydrogel solution and collagen to keep in survival condition. By printing the solution with cells from the inkjet nozzle continuously, the three-dimensional hollow pipe structure was fabricated.

To combine with self-assembly technique, the hydrogel beads with cells were patterned in molding technique [104]. The nutrients are diffused between each hydrogel beads, the cells were grown to be a three-dimensional tissue.

Gravitational/Electric/Magnetic Assembly Strategies

The gravitational/electric/magnetic fields are also non-contact method to control the position of cells. The gravitational force is one of the traditional ways to pattern cell on particular substrate. By combining with micro-fabricated well substrate, the ring type of structure was well assembled [105]. The cells were formed by themselves to aggregate near the pillar at the center of micro-well.

The electric/magnetic fields are more controllable of positioning forces. The electrical [106]/magnetic[107] assembly strategies were investigated. The hydrogel structures were fabricated by electrolyzation. The hydrogel structures are fabricated on the electrode part only. This method

Spheroid Assembly Strategies

The spheroid is a cell aggregated structure in three-dimensional shape. It is need to combine tougher the multiple cells. The advantage to construct spheroid is pure structure form cells, hence this is one of the confident technique as cell assembly strategy.

From this fabrication view point, the micro-chip device or gravitational/ electric/ magnetic forces and other techniques are applied to construct spheroid structure. For example, the spheroid structure was fabricated by vortex flow inside the micro-chip device [108]. One of the challenging issues of spheroid is to make them in large scale or size. The bio-printing technique was applied to construct spheroid in three-dimensional space [109]. This technique can realize relatively large-size of spheroid structures in arbitrary shapes.

Self-assembly Strategies

The self-assembly strategies are basically a method to construct our natural system. One of the interesting strategies of self-assembly in three-dimensional structure is "origami". The "origami" is a folding paper art and comes from a Japanese word. The "origami" is a method to extend in three-dimensional space from two-dimensional paper by folding. This method is one of the candidate to apply the cell assembly propose. The "cell origami" was investigated for three-dimensional cubic shape of structure [110]. The cell was placed on the parylene micro-plates. When the cell is grown on the micro-plate, it is formed in cubic shape by self-folding by cell traction forces which is around several tens nN [111].

References

[1] Feynman, R.P.: There's Plenty of Room at the Bottom. Caltech's Engineering and Science 23, 22–36 (1960)

[2] Nanotechnology Research Directions: Interagency Working Group on Nanoscience, Engineering and Technology (IWGN) Workshop Report (Vision for Nanotechnology R&D in the Next Decade), Washington D.C. (September 1999)

[3] National Nanotechnology Initiative: Leading to the Next Industrial Revolution, A Report by the Interagency Working Group on Nanoscience, Engineering and Technology, Committee on Technology, National Science and Technology Council, Washington D.C. (February 2000)

[4] Drexler, K.: Nanosystems: Molecular Machinery, Manufacturing and Computation. Wiley Inter Science (1992)

[5] Iijima, S.: Helical Microtubules of Graphitic Carbon. Nature 354, 56–58 (1991)

[6] Kawai, T.: Introduction to Nanotechnology, Ohmsha (2002) (in Japanese)

[7] Degrado, W.F.: Design of Peptides and Proteins. Adv. Protein Chem. 39, 51–124 (1998)

[8] Lehn, J.M.: Supramolecular Chemistry: Concepts and Perspectives. VCH Verlagesellschaft (1995) (Japanese Translation)

[9] Whitesides, G.M., Grzybowski, B.: Self-Assembly at All Scales. Science 295, 2418–2421 (2002)

[10] Fujita, M., Fujita, N., Ogura, K., Yamaguchi, K.: Spontaneous Assembling of Ten Small Components into a Three-Dimensionally Interlocked Compound Consistingof the Same Two Cage Frameworks. Nature 400, 52–55 (1999)

[11] Binnig, G., Rohrer, H., Gerber, C., Weibel, E.: Surface Studies by Scanning Tunneling Microscopy. Phys. Rev. Lett. 49, 57–61 (1982)

[12] Avouris, P.: Manipulation of Matter at the Atomic and Molecular Levels. Acc. Chem. Res. 28, 95–102 (1995)

[13] Crommie, M.F., Lutz, C.P., Eigler, D.M.: Confinement of Electrons to Quantum Corrals on a Metal Surface. Science 262, 218–220 (1993)

[14] Whitman, L.J., Stroscio, J.A., Dragoset, R.A., Cellota, R.J.: Manipulation of Adsorbed Atoms and Creation of New Structures on Room-Temperature Surfaces with a Scanning Tunneling Microscope. Science 251, 1206–1210 (1991)

[15] Lyo, I.-W., Avouris, P.: Field-Induced Nanometer-Scale to Atomic-Scale Manipulation of Silicon Surfaces with the STM. Science 253, 173–176 (1991)

[16] Dujardin, G., Walkup, R.E., Avouris, P.: Dissociation of Individual Molecules with Electrons From the Tip of a Scanning Tunneling Microscope. Science 255, 1232–1235 (1992)

[17] Shen, T.C., Wang, C., Abeln, G.C., Tucker, J.R., Lyding, J.W., Avouris, P., Walkup, R.E.: Atomic-Scale Desorption Through Electronic and Vibrational-Excitation Mechanisms. Science 268, 1590–1592 (1995)

[18] Cuberes, M.T., Schittler, R.R., Gimzewski, J.K.: Room-Temperature Repositioning of Individual C60Molecules at Cu Steps: Operation of a Molecular Counting Device. Appl. Phys. Lett. 69, 3016–3018 (1996)

[19] Lee, H.J., Ho, W.: Single-Bond Formation and Characterization with a Scanning Tunneling Microscope. Science 286, 1719–1722 (1999)

[20] Yamamoto, T., Kurosawa, O., Kabata, H., Shimamoto, N., Washizu, M.: Molecular Surgery of DNA Based on Electrostatic Micromanipulation. In: Proc. of the France-Japan Workshop from Nano to Micro Science and Technology Through Microsystems, pp. 38–43 (1998)

[21] Haber, C., Wirtz, D.: Magnetic Tweezers for DNA Micromanipulation. Rev. Sci. Instrum. 71, 4561–4570 (2000)

[22] Schaefer, D.M., Reifenberger, R., Patil, A., Andres, R.P.: Fabrication of Two-Dimensional Arrays of Nanometer-Size Clusters with the Atomic force Microscope. Appl. Phys. Lett. 66, 1012–1014 (1995)

[23] Junno, T., Deppert, K., Montelius, L., Samuelson, L.: Controlled Manipulation of Nanoparticles with an Atomic force Microscope. Appl. Phys. Lett. 66, 3627–3629 (1995)

[24] Sheehan, P.E., Lieber, C.M.: Nanomachining, Manipulation and Fabrication by Force Microscopy. Nanotechnology 7, 236–240 (1996)

[25] Baur, C., Gazen, B.C., Koel, B., Ramachandran, T.R., Requicha, A.A.G., Zini, L.: Robotic Nanomanipulation with a Scanning Probe Microscope in a Networked Computing Environment. J. Vac. Sci. & Tech. B 15, 1577–1580 (1997)

[26] Resch, R., Baur, C., Bugacov, A., Koel, B.E., Madhukar, A., Requicha, A.A.G., Will, P.: Building and Manipulating 3-D and Linked 2D Structures of Nanoparticles Using Scanning Force Microscopy. Langmuir 14, 6613–6616 (1998)

[27] Hu, J., Zhang, Z.H., Ouyang, Z.Q., Chen, S.-F., Li, M.Q., Yang, F.J.: Stretch and Align Virus in Nanometer Scale on an Atomically Flat Surface. J. Vac. Sci. & Tech., B 16, 2841–2843 (1998)

[28] Sitti, M., Horiguchi, S., Hashimoto, H.: Controlled Pushing of Nanoparticles: Modeling and Experiments. IEEE/ASME Trans. on Mechatronics 5(2), 199–211 (2000)

[29] Guthold, M., Falvo, M.R., Matthews, W.G., Paulson, S., Washburn, S., Erie, D.A., Superfine, R., Brooks Jr., F.P., Taylor II, R.M.: Controlled Manipulation of Molecular Samples with the Nanomanipulator. IEEE/ASME Transactions on Mechatronics 5, 189–198 (2000)

[30] Falvo, M.R., Clary, G.J., Taylor, R.M., Chi, V., Brooks, F.P., Washburn, S., Superfine, R.: Bending and Buckling of Carbon Nanotubes Under Large Strain. Nature 389, 582–584 (1997)

[31] Falvo, M.R., Taylor II, R.M., Helser, A., Chi, V., Brooks Jr., F.P., Washburn, S., Superfine, R.: Nanometre-Scale Rolling and Sliding of Carbon Nanotubes. Nature 397, 236–238 (1999)

[32] Arai, F., Andou, D., Fukuda, T.: Micro Manipulation Based on Micro Physics–Strategy Based on Attractive Force Reduction and Stress Measurement. In: Proc. of IEEE/RSJ Int. Conf. on Intelligent Robotics and Systems, vol. 2, pp. 236–241 (1995)

[33] Dong, L.X., Arai, F., Fukuda, T.: 3-D Nanorobotic Manipulations of Nanometer Scale Objects. J. of Robotics and Mechatronics 13(2), 146–153 (2001)

[34] Tanikawa, T., Arai, T., Saeki, M.: Two-Finger Micro Hand. In: Proc. IEEE International Conference on Robotics and Automation, May 21-27, pp. 1674–1679 (1995)

[35] Lin, Y.C., Yamanishi, Y., Fukuda, T., Arai, F.: On-chip cell manipulation systems -Part 3: Local temperature measurement and cell immobilization with thermosensitive gel-. In: Proc. of the 2007 JSME Conf. on Robotics and Mechatronics, 2A1-O07 (2007)

[36] Yamanishi, Y., Teramoto, J., Magariyama, Y., Ishihama, A., Fukuda, T., Arai, F.: On-chip Cell Immobilization and Monitoring System Using Thermosensitive Gel Controlled by Suspended Polymeric Microbridge. IEEE Trans. on NanoBiosci. 8(4), 312–317 (2009)

[37] Dertinger, S.K.W., Chiu, D.T., Jeon, N.L., Whitesides, G.M.: Generation of Gradients Having Complex Shapes Using Microfluidic Networks. Anal. Chem. 73, 1240–1246 (2001)

[38] Huang, S.H., Tan, W.H., Tseng, F.G., Takeuchi, S.: A monolithically three-dimensional flow-focusing device for formation of single/double emulsions in closed/open microfluidic systems. J. Micromech. Microeng. 16, 2336–2344 (2006)

[39] Nisisako, T., Torii, T., Higuchi, T.: Droplet Formation in a Microchannel Network. Lab on a Chip 2, 24–26 (2002)

[40] Suzuki, H., Tabata, K.V., Noji, H., Takeuchi, S.: Development of an Assay System for ABC Transporters Reconstituted in an Artificial Lipid Bilayer. In: Proc. of 10th Int. Conf. on Miniaturized Systems for Chem. and Life Sci., pp. 1363–1365 (2006)

[41] Harnack, O., Hospach, I., Yasuda, A.: Fabrication of PDMS-Based Nanochannels for Enzymatic Processing and Detection of Biomolecules. In: Proc. of 11th Int. Conf. on Miniaturized Systems for Chem. and Life Sci., pp. 1601–1603 (2007)

[42] Hibara, A., Saito, T., Kim, H.B., Tokeshi, M., Ooi, T., Nakao, M., Kitamori, T.: Nanochannels on a Fused-Silica Microchip and Liquid Properties Investigation by Timed-Resolved Fluorescence Measurements. Anal. Chem. 74, 6170–6176 (2002)

[43] Kimura, Y., Yanagimachi, R.: Intracytoplasmic Sperm Injection in the Mouse. Biology of Reproduction 52, 709–720 (1995)

[44] Haulot, G., Benahmed, A.J., Ho, C.: Optoelectronic Reconfigurable Microchannels. In: Proc. of MEMS 2011, pp. 53–56 (2011)

[45] Yu Chiou, P., Ohta, A.T., Wu, M.C.: Massively parallel manipulation of single cells and microparticles using optical images. Nature 436, 370–372 (2005)

[46] Ashkin, A., Dziedzic, J.M., Bjorkholm, J.E., Chu, S.: Observation of a single-beam gradient force optical trap for dielectric particles. Opt. Lett. 11, 288–291 (1986)

[47] Mio, C., Marr, D.W.M.: Optical Trapping for the Manipulation of Colloidal Particles. Adv. Mater. 12, 917–920 (2000)

[48] Arai, F., Sakami, T., Yoshikawa, K., Maruyama, H., Fukuda, T.: Synchronized Laser Micro-manipulation of Microtools for Assembly of Microbeads and Indirect Manipulation of Microbe. In: Proceedings of the 2003 IEEE/RSJ Int'l Conf. on Intelligent Robot and Systems, pp. 2121–2126 (2003)

[49] Arai, F., Yoshikawa, K., Sakami, T., Fukuda, T.: Synchronized laser micromanipulation of multiple targets along each trajectory by single laser. Appl. Phys. Lett. 85(19), 4301–4303 (2004)

[50] Binnig, G., Quate, C.F., Gerber, C.: Atomic force Microscope. Phys. Rev. Lett. 56, 93–96 (1986)

[51] Ashkin, A., Dziedzic, J.M.: Optical Trapping and Manipulation of Viruses and Bacteria. Science 235, 1517–1520 (1987)

[52] Crick, F.H.C., Hughes, A.F.W.: Exp. Cell Res. 1, 37 (1950)
[53] Yu, M.F., Dyer, M.J., Skidmore, G.D., Rohrs, H.W., Lu, X.K., Ausman, K.D., Von Ehr, J.R., Ruoff, R.S.: Three-Dimensional Manipulation of Carbon Nanotubes Under a Scanning Electron Microscope. Nanotechnology 10, 244–252 (1999)
[54] Dong, L.X., Arai, F., Fukuda, T.: 3D Nanorobotic Manipulation of Nano-Order Objects Inside SEM. In: Proc. of the 2000 International Symposium on Micromechatronics and Human Science, Nagoya, Japan, pp. 151–156 (2000)
[55] Eigler, D.M., Schweizer, E.K.: Positioning Single Atoms with a Scanning Tunneling Microscope. Nature 344, 524–526 (1990)
[56] Fukuda, T., Arai, F.: Prototyping Design and Automation of Micro/Nano Manipulation System. In: Proc. of IEEE Int'l Conf. on Robotics and Automation (ICRA 2000), San Francisco, U.S.A., vol. 1, pp. 192–197 (2000)
[57] Jung, T.A., Schlittler, R.R., Gimzewski, J.K., Tang, H., Joachim, C.: Controlled Room-Temperature Positioning of Individual Molecules: Molecular Flexure and Motion. Science 271, 181–1846 (1996)
[58] Lee, G.U., Chrisey, L.A., Colton, R.J.: Direct Measurement of the Forces Between Complementary Stands of DNA. Science 266, 771–773 (1994)
[59] Yu, M.F., Lourie, O., Dyer, M.J., Moloni, K., Kelly, T.F., Ruoff, R.S.: Strength and Breaking Mechanism of Multiwalled Carbon Nanotubes. Science 287, 637–640 (2000)
[60] Wang, Z.L.: Properties of Nanobeletsnad Nanotubes Measured by *In Situ* TEM. Microscopy and Microanalysis 10, 158–166 (2004)
[61] Ikuhara, Y., Wang, L.: Properties of Nanobeletsand Nanotubes Measured by *In Situ* TEM. Microscopy and Microanalysis 10, 158–166 (2004)
[62] Eigler, D.M., Schweizer, E.K.: Positioning Single Atoms with a Scanning Electron Microscope. Nature 344, 524–526 (1990)
[63] Hosoki, S., Hosaka, S., Hasegawa, T.: Surface modification of MoS2 using STM. Appl. Surf. Sci. 60/61, 643 (1992)
[64] Hertel, T., Martel, R., Avouris, P.: Manipulation of Individual Carbon Nanotubes and Their Interaction with Surfaces. J. Phys. Chem. B 102(6), 910–915 (1998)
[65] Li, G., Xi, N., Yu, M., Fung, W.-K.: Development of Augmented Reality System for AFM based Nanomanipualiton. IEEE Trans. on Mechatronics 9, 358–365 (2004)
[66] Li, G., Xi, N., Chen, H., Pomeroy, C.: Videolized" Atomic Force Microscopy for Interactive Nanomanipulation and Nanoassembly. IEEE Tran. on Nanotech. 4, 605–615 (2005)
[67] Requicha, A.A.G., Arbuckle, D.J., Mokaberi, B., Yun, J.: lgorithms and Software for Nanomanipulation with Atomic Force Microscopes. Int'l J. Robotics Research 28 (2009)
[68] Sitti, M., Hashimoto, H.: Teleoperated Touch Feedback from the Surfaces at the Nanoscale: Modeling and Experiments. IEEE Trans. on Mechatronics 8, 287–298 (2003)
[69] Nakao, M., Tsuchiya, K., Matsumoto, K., Hatamura, Y.: Micro Handling with Rotational Needle-Type Tools under Realtime Observation. Annals of the CIRP 50, 9–12 (2001)
[70] Zyvex Corporation, `http://www.zyvex.com/`
[71] Saito, S., Miyazaki, H.T., Sato, T., Takahashi, K.: Kinematics of Mechanical and AdhesionalMicromanipulation under a Scanning Electron Microscope. J. of Applied Physics 92, 5140–5149 (2002)

[72] Kasaya, T., Miyazaki, H.T., Saito, S., Koyano, K., Yamaura, T., Sato, T.: Image-based Autonomous Micromanipulation System for Arrangement of Spheres in a Scanning Electron Microscope. Rev. of Scientific Instruments 75, 2033–2042 (2004)

[73] Sardan, O., Eichhorn, V., Petersen, D.H., Fatikow, S., Sigmund, O., Boggild, P.: Rapid prototyping of nanotube-based devices using topology-optimized microgrippers. Nanotechnology 19(495503), 9p. (2008)

[74] Nanofactory Instruments, `http://www.nanofactory.com/`

[75] Kizuka, T., Yamada, K., Deguchi, S., Naruse, M., Tanaka, N.: Cross-sectional Time-resolved High-resolution Transmission Electron Microscopy of Atomic-Scale Contact and Noncontact-Type Scanning on Gold Surfaces. Phys. Rev. B 55, R7398–R7401 (1997)

[76] TokushiKizuka: Atomic configuration and mechanical and electrical properties of stable gold wires of single-atom width. Phys. Rev. B 77(155401), 11 (2008)

[77] Cavalcanti, A., Shirinzadeh, B., Kretly, L.C.: Medical Nanorobotics for Diabetes Control. Nanomedicine: Nanotechnology, Biology, and Medicine 4, 127–138 (2008)

[78] Martel, S., Mohammadi, M., Felfoul, O., Lu, Z., Pouponneau, P.: Flagellated Magnetotactic Bacteria as Controlled MRI-trackable Propulsion and Steering Systems for Medical Nanorobots Operating in the Human Microvasculature. Intl. J. Robotics Research 28(4), 571–582 (2009)

[79] Martel, S., Felfoul, O., Mathieu, J.-B., Chanu, A., Tamaz, S., Mohammadi, M., Mankiewicz, M., Tabatabaei, N.: MRI-based Medical Nanorobotic Platform for the Control of Magnetic Nanoparticles and Flagellated Bacteria for Target Interventions in Human Capillaries. Intl. J. Robotics Research 28(9), 1169–1182 (2009)

[80] Patel, G.M., Patel, G.C., Patel, R.B., Patel, J.K., Patel, M.: Nanorobot: A Versatile Tool in Nanomedicine. J. Drug Targeting 14(2), 63–67 (2006)

[81] Cavalcanti, A., Shirinzadeh, B., Fukuda, T., Ikeda, S.: Nanorobot for Brain Aneurysm. Intl. J. Robotics Research 28(4), 558–570 (2009)

[82] Yokokawa, R., Takeuchi, S., Kon, T., Nishiura, M., Ohkura, R., Edamatsu, M., Sutoh, K., Fujita, H.: Hybrid Nanotransport System by Biomolecular Linear Motors. J. Microelectromechanical Systems 13(4), 612–619 (2004)

[83] Knoblaucha, M., Peters, W.S.: Biomimetic Actuators: where Technology and Cell Biology Merge. CMLS, Cell. Mol. Life Sci. 61, 2497–2509 (2004)

[84] Hiyama, S., Moritani, Y., Gojo, R., Takeuchi, S., Sutoh, K.: Biomolecular-motor-based Autonomous Delivery of Lipid Vesicles as Nano- or Microscale Reactors on a Chip. Lab on a Chip 10, 2741–2748 (2010)

[85] Soong, R.K., Bachand, G.D., Neves, H.P., Olkhovets, A.G., Craighead, H.G., Montemagno, C.D.: Powering an Inorganic Nanodevice with a Biomolecular Motor. Science 290, 1555–1558 (2000)

[86] Xi, J., Schmidt, J.J., Montemagno, C.D.: Self-assembled Microdevices Driven by Muscle. Nature Mater. 4, 180–184 (2005)

[87] Akiyama, Y., Iwabuchi, K., Furukawa, Y., Morishima, K.: Long-term and room temperature operable bioactuator powered by insect dorsal vessel tissue. Lab on a Chip 9, 140–144 (2009)

[88] Akiyama, Y., Hoshino, T., Iwabuchi, K., Morishima, K.: Design and Fabrication of Temperature-Tolerant Micro Bio-Robot Driven by Insect Heart Tissue. In: Proc. of 2010 Intl. Symposium on Micro-Nano Mechatronics and Human Sci., pp. 115–120 (2010)

[89] Chee, J.-Y., Yoga, S.-S., Lau, N.-S., Ling, S.-C., Abed, R.M.M., Sudesh, K.: Bacterially Produced Polyhydroxyalkanoate (PHA): Converting Renewable Resources into Bioplastics. Current Research, Technology and Education Topics in Applied Microbiology and Microbial Biotechnology 2, 1395–1404 (2010)

[90] Kitamoto, H.K., Shinozaki, Y., Cao, X., Morita, T., Konishi, M., Tago, K., Kajiwara, H., Koitabashi, M., Yoshida, S., Watanabe, T., Sameshima-Yamashita, Y., Nakajima-Kambe, T., Tsushima, S.: Phyllosphere yeasts rapidly break downbiodegradable plastics. AMB Express 1:44, 11 Pages (2011)

[91] Suuronen, E.J., Sheardown, H., Newman, K.D., et al.: Building In Vitro Models of Organs. In: Kwang, W.J. (ed.) International Review of Cytology, pp. 137–173. Academic Press (2005)

[92] Langer, R., Vacanti, J.P.: Tissue Engineering. Science 260, 920–926 (1993)

[93] Tsutsui, H., et al.: Efficient Dielectrophoretic Patterning of Embryonic Stem Cells in Energy Landscapes Defined by Hydrogel Geometries. Annals of Biomedical Engineering 38, 3777–3788 (2010)

[94] Tixier-Mita, A., et al.: A silicon micro-system for parallel gene transfection into arrayed cells. In: Proceedings of the uTAS 2004 Symposium. The Royal Society of Chemistry, pp. 180–182 (2004) ISBN 0-85404-896-0

[95] Di Carlo, D., Aghdam, N., Lee, L.P.: Single-cell enzyme concentrations, kinetics, and inhibition analysis using high-density hydrodynamic cell isolation arrays. Analytical Chemistry 78(14), 4925–4930 (2006)

[96] Yue, T., Nakajima, M., Ito, M., Kojima, M., Fukuda, T.: High speed laser manipulation of on-chip fabricated microstructures by replacing solution inside microfluidic channel. In: Proc. of the 2011 IEEE/RSJ Int. Conf. on Intelligent Robots and Systems (IROS 2011), pp. 433–438 (2011)

[97] Derda, R., Laromaine, A., Mammoto, A., Tang, S., Mammoto, T., Ingber, D., Whitesides, G.: Paper-supported 3Dcell culture for tissue-based bioassays. Proc. Nat. Acad. of Sci. 106, 18457–18462 (2009)

[98] Chan, V., Zorlutuna, P., Jeong, J.H., et al.: Three-dimensional photopatterning of hydrogels using stereolithography for long-term cell encapsulation. Lab on a Chip 10(16), 2062–2070 (2010)

[99] Uchida, T., Ikeda, S., Oura, H., Tada, M., Nakano, T., Fukuda, T., Matsuda, T., Negoro, M., Arai, F.: Development of biodegradable scaffolds basedon patient-specific arterial configuration. Journal of Biotechnology 133, 213–218 (2008)

[100] Yamada, A., Niikura, F., Ikuta, K.: A three-dimensional microfabrication system for biodegradable polymers with high resolution and biocompatibility. J. Micromech. Microeng. 18, 025035 (2008)

[101] Mironov, V., Visconti, R.P., Kasyanov, V., Forgacs, G., Drake, C.J., Markwald, R.R.: Organ printing: tissue spheroids as building blocks 30, 2164–2174 (2009)

[102] Yamada, M., Utoh, R., Ohashi, K., Yamato, M., Okano, T., Seki, M.: Formation of Complex Hepatic Organoids Using Microfabricated Anisotropic Hydrogel Fibers. In: Proc. of the 15th International Conf. on Miniaturized Systems for Chemistry and Life Sciences, pp. 1385–1387 (2011)

[103] Henmi, C., Nakamura, M., Nishiyama, Y., Yamaguchi, K., Mochizuki, S., Takiura, K., Nakagawa, H.: Development of an effective three dimensional fabrication technique using inkjet technology for tissue model samples. In: Proc. 6th World Congress on Alternatives & Animal Use in the Life Sciences, pp. 689–692 (2007)

[104] Matsunaga, Y., Morimoto, Y., Takeuchi, S.: Molding cell beads for rapid construction of macroscopic 3D tissue architecture. Advanced Materials 23, H90–H94 (2011)

[105] Masuda, T., Takei, N., Nakano, T., Anada, T., Suzuki, O., Arai, F.: A microfabricated platform to form three-dimensional toroidal multicellular aggregate. Biomed. Microdevices 14, 1085–1093 (2012)

[106] Cheng, Y., Luo, X., Tsao, C.Y., Wu, H.C., Betz, J., Payne, G.F., Bentley, W.E., Rubloff, G.W.: Biocompatible multi-address 3D cell assembly in microfluidic devices using spatially programmable gel formation. Lab Chip 11, 2316–2318 (2011)

[107] Xu, F., Max Wu, C., Rengarajan, V., Finley, T.D., Keles, H.O., Sung, Y., Li, B., Gurkan, U.A., Demirci, U.: Three-Dimensional Magnetic Assembly of Microscale Hydrogels. Adv. Mater. 23, 4254–4260 (2011)

[108] Ota, H., Miki, N.: Microfluidic experimental platform for producing size-controlled three-dimensional spheroids. Sensors and Actuators A 169, 266–273 (2011)

[109] Mironov, V., Visconti, R.P., Kasyanov, V., Forgacs, G., Drake, C.J., Markwald, R.R.: Organ printing: Tissue spheroids as building blocks. Biomaterials 30, 2164–2174 (2009)

[110] Kuribayashi-Shigetomi, K., Onoe, H., Takeuch, S.: Cell Origami: Self-Folding of Three-Dimensional Cell-Laden Microstructures Driven by Cell Traction Force. Plos One 7(12), e51085, 8 (2012)

[111] Tan, J.L., Tien, J., Pirone, D.M., Gray, D.S., Bhadriraju, K., Chen, C.S.: Cells lying on a bed of microneedles: an approach to isolate mechanical force. Proc. of Natl. Acad. Sci. U. S. A. 100(4), 1484–1489 (2003)

Chapter 2
Physics in Micro-Nano Scale

2.1 Scaling Effects in Micro-Nano Scale

At the micro-nano meter scale, the mechanics is dramatically changed from our living world. The micro-nano mechanics is intuitively understood by the scaling law (Table 2.1). In the less than nano meter scale, quantum effects appear.

Table 2.1 Scaling law

Physical Parameters	**Symbols**	**Equations**	**Size Effects**
Length	L	L	L
Area	S	$\propto L^2$	L^2
Volume	V	$\propto L^3$	L^3
Mass	m	ρV	L^3
Pressure	f_p	SP	L^2
Gravity		mg	L^3
Inertia Force	f_g	md^2x/dt^2	L^4
Viscous (Kinetic Friction) Force		$uS/d \cdot dx/dt$	L^2
Elastic Force	f_i	$eS\Delta L/L$	L^2
Spring Constant	K	$2UV/(\Delta L2)$	L
Resonant Frequency	ω	$\sqrt{K/m}$	L^{-1}
Moment of inertia	I		L^5
Deflection	D	$\alpha m r_I^2$	L^2
Reynolds number	R_e	M/K	L^2
		f_i/f_f	

Table 2.1 (*continued*)

Electrostatic Force	F_e	$\varepsilon SE^2/2$	L^2
van der Waals Force	F_{vdw}	$F_{vdw} = \dfrac{Hd}{12z^2}$	L
Dielectric Force	F_d	$2\pi L^3 \varepsilon_1 \dfrac{\varepsilon_2 - \varepsilon_1}{\varepsilon_2 + 2\varepsilon_1} \nabla(E^2)$	L^3

Remarks: ρ: density, P: pressure, g: Acceleration of gravity, x: displacement, t: time, U: d: density, e: Young's modulus, a: constant, r_l: radius of body of rotation, ε: dielectric constant, E: electric field, d: conversed radius, C: constant of the atom-atom potential.

The phenomenon is not completely revealed. In this section, van der Waals forces and electric static forces are described as the important role on the mechanics of nano manipulation and self-assembly

2.2 Mechanics in Micro-Nano Scale

2.2.1 van der Waals Forces

The van der Waals force always acts on the sample in the dominative region of intermolecular forces. This effect is exactly considered like to be a gravity forces for atoms or molecules [1].

When two materials approach each other, attracting forces generated with each other. Then the distance is enough small, the repulsive force is higher than attracting forces. On the region, the Lennard-Jones potential equation can be expressed with radius of molecule r.

$$\phi(r) = 4\varepsilon\left\{\left(\frac{\sigma}{r}\right)^{12} - \left(\frac{\sigma}{r}\right)^{6}\right\} \tag{2.1}$$

The van der Waals force is the attracting force on the Lennard-Jones potential.

The dispersal interactive energy between same two atoms (molecules) is expressed as London's equation,

$$e(r) \approx -\frac{3}{4}\frac{\alpha_0{}^2 h\gamma}{(4\pi\varepsilon_0)^2 r^6} = -\frac{3}{4}\frac{\alpha_0{}^2 I}{(4\pi\varepsilon_0)^2 r^6} = -\frac{C}{z^6} \tag{2.2}$$

where C is constant of the atom-atom potential. In the case of the different two atoms (molecules), the dispersal interactive energy is expressed is as follows,

$$e(r) \approx -\frac{3}{2}\frac{\alpha_{01}\alpha_{02}}{(4\pi\varepsilon_0)^2 r^6}\frac{h\gamma_1\gamma_2}{(\gamma_1 + \gamma_2)} = -\frac{3}{2}\frac{\alpha_{01}\alpha_{02}}{(4\pi\varepsilon_0)^2 r^6}\frac{I_1 I_2}{(I_1 + I_2)} \tag{2.3}$$

where ρ_1, ρ_2 are number density of each materials To lead van der Waals energy, (2.3) is integrated with the object shapes.

For example, the van der Waals energy E_{vdw} between a particle and a plate is expressed as following equation.

$$E_{vdw} = -\frac{H}{6}\left\{\frac{d}{2z} + \frac{d}{2(z+d)} - \ln\frac{z}{z+d}\right\} \tag{2.4}$$

H is the Hamaker constant defined as following equation.

$$H = \pi^2 C\rho_1\rho_2 \tag{2.5}$$

where z is gap between particle and plate, d is the diameter of particle. From differentiation (2.4) with respect to z, the van der Waals forces F_{vdw} is as follows.

$$F_{vdw} = \frac{H}{6}\left\{\frac{d}{2z^2} + \frac{d}{2(z+d)^2} - \frac{1}{z} + \frac{1}{z+d}\right\} \tag{2.6}$$

From $z >> d$ condition, this equation is approximated as follows.

$$F_{vdw} = \frac{Hd}{12z^2} \tag{2.7}$$

2.2.2 Elastic Properties in Micro-Nanometer Scale

By the progress of the MEMS and NEMS technology, the sensor and actuators are fabricated in smaller and smaller sizes. It is important to understand the mechanical property, especially elastic property. For example, fracture behavior is depended on the elastic or plastic defamation mechanism. In atomic scale, the defects, such as dislocation are affected for the mechanical properties.

One of the major materials for MEMS and NEMS is the silicon material, which can be fabricated in unique 3D shapes by the photolithography process depending the crystal orientation. The silicon material is normally brittle, however it has a ductile fracture involved by plastic deformation under high temperature (higher than 500 degrees C) or nanometer scale size as same as metallic material [2].

2.3 Electronics in Micro-Nano Scale

2.3.1 Electrostatic Force

One of the most important characteristics of electrostatic forces, it is one of remote forces, which can be reached each other without intermediates. Hence the movement of charged objects can be controlled by them with noncontact condition [3]. In the real world, a lot of types of electrostatic forces have relationships each other in complex manner.

The charged two objects have coulomb forces. Assuming the central distance between two objects: *r*:, each charge of objects: q_1, q_2, and the dielectric constant under vacuum: ε_0: the coulomb force is expressed as follows.

$$F_e = \frac{1}{4\pi\varepsilon_0}\frac{q_1 q_2}{r^2} \tag{2.8}$$

If the charge is same sign, it is repulsive force, the other case, it is attracting force. When the distance between two objects are quite small compared to the sizes of objects, (2.8) is approximated as follows,

$$F_e \approx \frac{\pi\sigma_1\sigma_2}{\varepsilon_0} d^2 \tag{2.9}$$

where the σ_1, σ_2 are surface density of charges defined as $\sigma_i = q_i / \pi d_i^2$ (i=1,2), and *d* is conversed radius. However, if the electric filed, the shapes of objects, or charge distribution are complex, it is not easy to lead the electrostatic force from Coulomb equations. Hence, the electric-tensile force f_e (Maxwell's strain force) can be expressed as following equation using the electric field E [4].

$$f_e = \frac{1}{2}\varepsilon_0 E^2 \tag{2.10}$$

2.3.2 Dielectric Force

As shown in Figure 2.1, the neutral particle has dielectric polarity in the medium. When the particle has shape with one-dimensional direction, orientation force is generated. In the un-uniformed electric field (Figure 2.1 (b)) the objects is attracted in intense electric filed. If the polarity is changed counter part, the direction of force is not changed. The dielectric force F_d is expressed as follows with the radius of particle *r*,

$$F_d = 2\pi r^3 \varepsilon_1 \frac{\varepsilon_2 - \varepsilon_1}{\varepsilon_2 + 2\varepsilon_1} \nabla\left(E^2\right) \tag{2.11}$$

From this equation, when the dielectric constant of particle is much higher than one of medium ($\varepsilon_2 > \varepsilon_1$), the dielectric force is directed to the in intense electric field. On the other hand, when the dielectric constant of particle is much lower than one of medium ($\varepsilon_2 < \varepsilon_1$), the dielectric force is directed to the weak electric field.

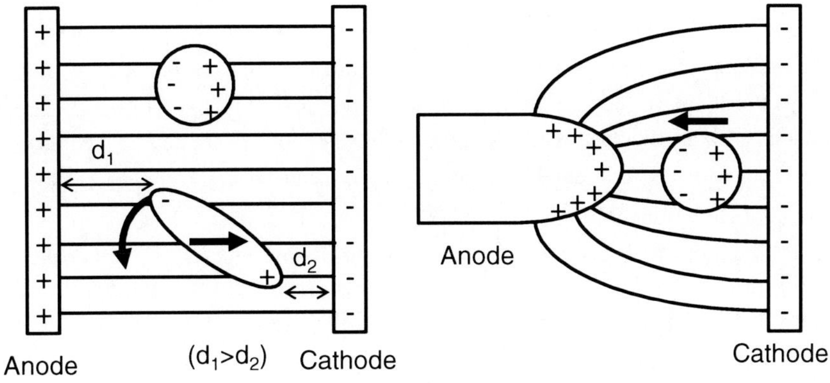

(a) (b)

Fig. 2.1 Schematic diagram of the principle of dielectrophoresis

2.3.3 *Electrophoretic Force*

The electrophoretic force F_v under electric field E is expressed as follow,

$$F_v = q_v E \tag{2.12}$$

where q_v is electric charge. This force have typical characteristics as follows, (1) The direction of forces is determined by the direction of electric field and positive or negative charge, (2) It is generated uniform and un-uniformed electric field, (3) At the low frequency electric field, vibration behavior has been observed, on the other hand, at high frequency electric field, the force is decreased.

2.3.4 *Field Emission Mechanism*

The theory of field emission from metals was first derived by Fowler and Nordheim in 1928 [5], who showed that if the temperature is not too high, most of the emitted electrons originate from a small energy interval around the Fermi level of the metal. The Fowler-Nordheim law gives a relation between the emitted current I (A) and the local field at the emitter surface F (V/m). F is usually related to the applied voltage V as $F = \gamma V / d$, where d is the inter-electrode distance and γ quantifies the ability of the emitter to amplify the applied field V/d and is named the field enhancement factor. In this frame, the Fowler-Nordheim law is written as [6]

$$I = A\frac{1.05\times10^{-6}}{\phi}\left(\frac{V}{d}\right)^2\gamma^2\exp\left(\frac{10.4}{\sqrt{\phi}}\right)\exp\left(-\frac{6.44\times10^{9}\phi^{1.5}d}{\gamma V}\right) \tag{2.13}$$

where A has the dimension of an area (m^2) and ϕ is the work function (eV).

The easiest way to determine A and γ from the measurements is to trade a Fowler-Nordheim plot, i.e., to plot $ln(I/V^2)$ versus $1/V$. The I-V curve should appear as a linear function with a slope that depends on d, ϕ, and γ. Since d is known from the experimental configuration, γ can be determined by taking a reasonable value for ϕ.

The strength of the field emission current is influenced by some factors including the inter-electrode distance, the work function of the emitters, the field enhancement factor being determined by emitter geometries, and the applied voltage.

$$I = f(\phi,\ g,\ \gamma,\ V) \tag{2.14}$$

where, ϕ is the work function of nanotube emitter, g the inter-electrode distance, γ the field enhancement factor, and V the applied voltage.

2.3.5 *Optical Dielectrophoresis*

Optical dielectrophoresis (DEP) was proposed to manipulate particles or biological cells under liquid environment as shown in Figure 2.2 [7]. This phenomenon was well established, and basically an attractive forces are created by light patterning.

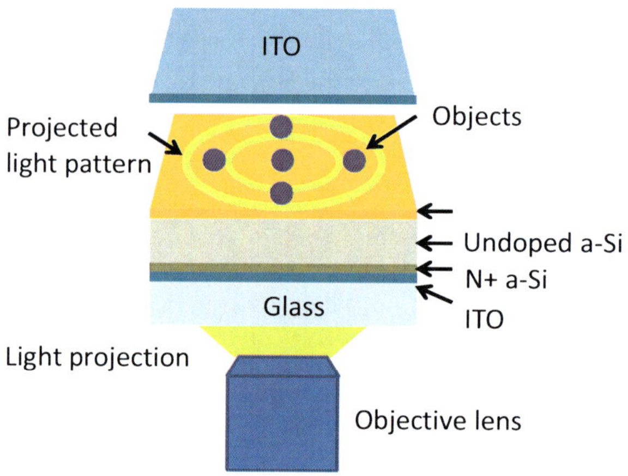

Fig. 2.2 Experimental configuration of optical DEP

2.4 Fluidics in Micro-Nano Scale

2.4.1 Brownian Motion

The Brownian motion is the random motion of small objects under thermal condition. It is considered to be a motion caused by pushing from the atoms of suspended environment (fluid or gas). This motion was found the pollens under water by Robert Brawn in 1827. In the micro-nano scale, the micro-nano objects are basically always affected by Brownian motion.

The theoretical model of Brownian motion is described by the Wiener process. T The Wiener process W_t is characterized by the following factors:

1. $W_0 = 0$
2. W_t is almost surely continuous
3. W_t has independent increments
4. W_t - W_s ~N (0, t-s) (for 0<s<t)).

where N is denotes the normal distribution.

2.4.2 Diffusion Phenomenon

Diffusion is a physical phenomenon to spread the particle, energy to the environmental naturally. In micro-nano scale, the small particles or ions move in a media by this phenomenon. Basically, the phenomenon is described by the Fick's laws of diffusion.

$$J = -D\frac{dc}{dx} \tag{2.15}$$

The fundamental solution of 1-D diffusion equation is

$$C(x,t) = \frac{1}{\sqrt{4\pi Dt}}\exp\left[-\frac{x^2}{4Dt}\right] \tag{2.16}$$

where C is the concentration, D is the diffusion constant, and t is the time. The time until the solution reaches the cell is given as the division of the distance between the nanopipette and the cell by the flow velocity.

2.4.3 Navier-Stokes Equations

The fluids dynamics is expressed as a Navier-Srokes equation for fluid substance delivering from Newton's second law. Under the assumption of incompressible flow (ρ is constant), the Navier-Srokes equation is as follows,

$$\frac{\partial u}{\partial t} + u \cdot \nabla u = -\frac{1}{\rho} \nabla p + \gamma \nabla^2 u \tag{2.17}$$

where, u is the flow velocity, ρ is the fluid density, γ is the dynamic viscosity, and p is the pressure. The left-hand side is the summation of unsteady acceleration and convective accelerations terms. The right-hand side is the summation of pressure gradient and viscosity terms. This equation includes the unknown parameters "u" and "p". Hence, it is needed another equation to resolve. Under the in compressive flow, the equation of continuity

$$\nabla u = 0 \tag{2.18}$$

is used to resolve, however it is not easy to obtain exact solution of Navier-Srokes equation. Normally, the fluidics dynamics is analyzed by the numerical simulations using the above equations.

2.4.4 Reynolds Number

The Reynolds number is defined as following equation,

$$\mathrm{Re} = \frac{\rho \mu L}{\eta} = \frac{\mu L}{v} \tag{2.19}$$

where v is the mean velocity of the object relative to the fluid (SI units: m/s), L is a characteristic linear dimension, (travelled length of the fluid; hydraulic diameter when dealing with river systems), μ is the dynamic viscosity of the fluid, υ is the kinematic viscosity, and ρ is the density of the fluid (kg/m^3).

2.4.5 Knudsen Number

Flow field is strongly influenced by the interaction of molecules with a solid boundary rather than intermolecular collisions.

Knudsen number is defined as following equation.

$$K_n = \frac{1}{L} \tag{2.20}$$

where, l is mean free path, and L is a characteristic dimension. Continuum approach invalid when the K_n is higher than 0.01. This high K_n Number low is in the micro-nano scale, such as large l or small L.

As the value of the Knudsen number increases, non-continuum effects became more important.

$Kn < 0.01$	: Continuum flow regime
$0.01 < Kn < 0.1$	: Slip flow regime
$0.1 < Kn < 10$	: Transition regime
$10 < Kn$	: Free molecular flow regime

2.4.6 Capillary Force

For example of glass capillary, a water being sucked up by a hydrophilic capillary as a result of positive capillary (Figure). This phenomenon is described as following equation,

$$\gamma_{SL} + \gamma_{LG} \cos q_C = \gamma_{SG} \tag{2.21}$$

where, γ: surface tension, q_c: contact angle. The Energy Reduction by rising over distance l in a capillary as shown in following equation.

$$(\gamma_{SG} - \gamma_{SL})\, 2\pi r l \tag{2.22}$$

The work performed in raising the liquid column

$$D_p = \pi r^2 l \tag{2.23}$$

where, r: the capillary radius, D_p: the pressure difference between the top and bottom of raised column.

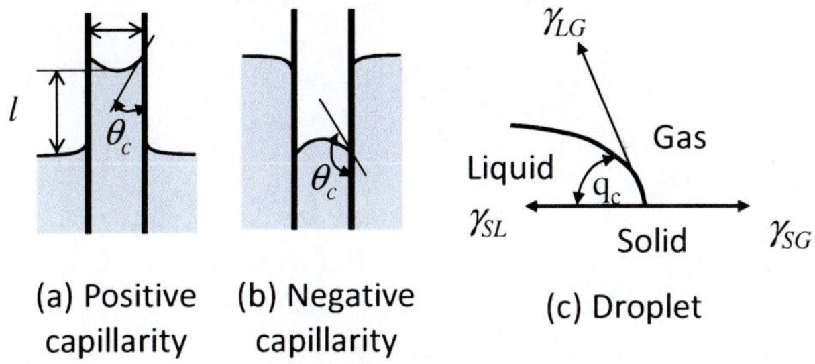

Fig. 2.3 Capillary phenomenon

2.5 Surface Interaction in Micro-Nano Scale

Pick-and-place is the most fundamental 3D nanomanipulation, but in most case is also the most difficult one for nanometer scale objects that is caused by lacking of proper tools. The micro-electro-mechanical system (MEMS) based micro gripper can work with nanometer resolution, but the size of the gripper itself is still too large for nanometer order objects. Nano tweezers have been demonstrated to be promising for nanomanipulation. For example, nanotube tweezers have been applied for the manipulations of micro-order objects [8].

Presently, it is more practice to adopt a sharp probe, e.g., STM probe or AFM cantilever, to realize the pick-and-place manipulation. To do this, it is necessary to control the intermolecular and surface forces between the tool and the object, and between the object and a substrate. Here we show two strategies. The first one is to increase the adhesive force between the tool and the object by applying dielectrophoresis/electrophoresis, and the second one is to decrease the adhesive force between the object and the substrate by using "rougher" substrate. It is more effective to combine the two way together.

2.5.1 Analysis of Intermolecular and Surface Forces

As dealing with micro-nanometer scale objects, the interactions caused by quantum and electromagnetic effects can no longer be ignored, which is different from that in the macro world. For instance, the interactions between a sphere with diameter of d and an infinite plate include:

The van der Waals force:

$$F_{vdW} = \frac{H}{6}\left\{\frac{d}{2z^2} + \frac{d}{2(z+d)^2} - \frac{1}{z} + \frac{1}{z+d}\right\} \tag{2.24}$$

The electrostatic forces between two charged objects

$$F_e = \frac{\pi\sigma_1\sigma_2}{\varepsilon_0} d^2 \tag{2.25}$$

The electric field imaging

$$F_{ei} = \frac{\pi}{4\varepsilon_0}\frac{\varepsilon - \varepsilon_0}{\varepsilon + \varepsilon_0} d^2\sigma^2 \tag{2.26}$$

and the contacting charging

$$F_{ce} = -\frac{1}{2}\pi\varepsilon_0 \frac{V_c^2}{z^2}\left\{\frac{Hkd^2}{32z^2}\left(1 + \frac{H^2k^2d}{108z^7}\right)\right\}^{2/3} \tag{2.27}$$

where, F_{vdW} is the van der Waals force between the sphere and the surface, H is a Hamaker constant, d is a diameter of the sphere, z is a distance between the sphere and the surface, F_e is an electrostatic force, σ_1 and σ_2 are the surface charge densities by $\sigma_i = q_i / \pi d_i^2$ ($i = 1,2$, q_i is the charges), ε_0 is dielectric coefficient of circumstance, F_{ei} is the electrostatic force by electric imaging, ε the dielectric coefficient of the sphere, F_{ce} the electrostatic force by contacting charging, V_c the potential difference, and k the specific spring constant.

An example is shown in Figure 2.4. It can be found that for micrometer/nanometer order objects, adhesive forces rather than gravity governs the interactions. In air, the surface tension forces are the largest one, while van der Waals forces will be the second important one. In a vacuum circumstance, the largest force for a neutral particle will be van der Waals force as shown in Figure 2.5.

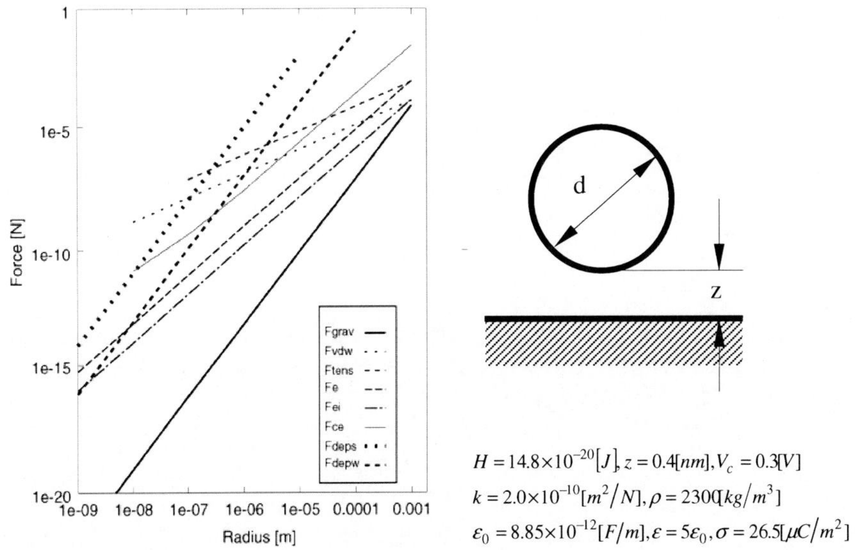

Fig. 2.4 Comparison of Typical Adhesive Forces and Dielectrophoretic Force (Fdepw and Fdeps: Dielectrophoretic force at 100V and at 1000V)

2.5.2 Picking Up by Applying Dielectrophoresis under Surface Forces

To pick up an object from a surface, a probe with sharp tip will be readily used. However, the adhesive force between the probe and the object is generally smaller than that between the object and the substrate. Two strategies have been tried to change the situation. The first is to increase the force between the probe and the object by using dielectrophoresis. Another way is to decrease the adhesive force between the object and the substrate.

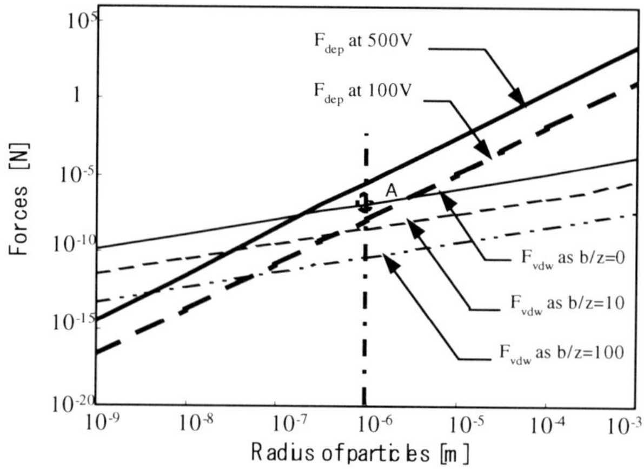

Parameters used in computations:

$$H = 14.8\times10^{-20}[J],\ z = 0.4[nm],\ \rho = 2300[kg/m^3],$$
$$\varepsilon_0 = 8.85\times10^{-12}[F/m],\quad \varepsilon = 5\varepsilon_0,\ \sigma = 26.5[\mu C/m^2]$$

Fig. 2.5 Control Dielectrophoretic Force and/or van der Waals Force

The dielectrophoretic force F_{dep} is a function of the radius of objective bead r, the dielectric coefficient of circumstance ε_0, the dielectric coefficient of the objective bead ε and the electric field intensity E_0.

$$F_{dep} = 2\pi r^3 \frac{\varepsilon_0(\varepsilon-\varepsilon_0)}{\varepsilon+2\varepsilon_0}\nabla\left(E_0^2\right) \tag{2.28}$$

Because the dielectrophoretic force is a function of the gradient of the electric field intensity, it is easy to control it by changing the applied voltage. It is also easier to be realized than to control other kinds of adhesive forces. For generating a nonuniform electric field gradient, it is effective to apply a sharp pin such as an AFM cantilever vs. to a plate as two electrodes. Hence, it is possible to pick up an object placed in a plate such as that shown in if the dielectrophoretic force is greater than van der Waals forces acted on a particle. In order to realize such manipulations, two ways can be effectively used. One of them is to decrease the van der Waals forces as described in 1) and 2). Another way is to control the intensity and gradient of the electric field.

2.5.3 *Adhesion Force of Micro-Nano Fibers*

The adhesive micro-nano fibers have unique properties to stick on a surface as biomimetic phenomenon. For example, geckos have a remarkable ability to adhere on smooth or rough surface in wet or dry, clean or dirty surface [9]. They are in millions of micron-scale setae on each toe of the gecko form a self-cleaning dry adhesive.

As natural gecko setae arrays are made of beta-keratin which is relatively hard protein [10]. Single seta of gecko is approximately in 100 μm in length and 4 μm in diameter. Each seta consists of hundreds of nanoarray of spatula structures which are approximately 0.2 μm in length and width at the tip. The elastic properties of gecko setae are approximately 1.5 GPa [11].

As artificial micro-nano fivers adhesive various materials are used for example, polyimide, polypropylene, parylene, polyurethane, carbon nanotubes, Poly(methyl methacrylate) (PMMA), and poly(dimethylsiloxane) (PDMS) [12]. Gecko adhesion can be mechanical switching. The adhesive condition is released by relaxing sliding tensional forces. Thus, it is depended the geometry of fibers.

One of the uniqueness of adhesion of gecko fiber is directional adhesion. The adhesion forces are totally changed in along and against directions [13]. In the along direction, an adhesive force is generated (at the steady-state sliding F_s=~9 N). On the other hand, in the against direction, the adhesive force is quite small (F_s=~0.2 N).

The anisotropic adhesion of angled fiber arrays was modeled by elastic bending model of angled beam [13]. The angle θ(s) from the surface normal due to the shear force applied at the free end of a fiber ΔFs is expressed as the following differential equation,

$$EI\frac{d^2\theta}{d^2s} + \Delta F_S \cos\theta = 0 \tag{2.29}$$

where s is the distance from the fixed end of the beam, I(=$\pi r^4/4$) is the moment of inertial, r is the fiber radius, θ(=θ_0+Δθ) is the angle from the surface normal, and Δθ is the deviation angle from the initial angle $\tilde{\theta}$due to the shear force ΔF_s (Figure 2.6).

As experimental results, the steady-state shear force ΔF_s ~ 298 nN at sliding along the fiber direction, and ΔF_s ~ 119 nN at sliding against the fiber direction [13]. From this result, the sliding along the fiber direction produce higher shear force by ~2.5 times.

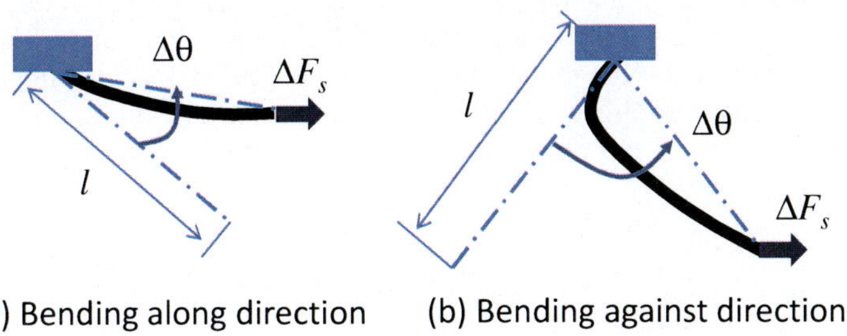

Fig. 2.6 Modeling of directional microfiber in along and against directions

2.6 Laser Trapping Mechanism

Generally, the laser trapping or optical trapping is the applying forces by the light pressure of laser to the object. [14,15]. This phenomenon is effectively explained to understand the physics of objects of electromagnetic diffraction in small scale [16]. Hence, the micro-nano particles have a lot of effects, such as electrostatic force, thermal forces, gravity forces, frictional forces, Brownian motion, van der Waals force and so on. These effects are totally changed by the size of objects and environmental conditions as explained scaling effects. It was demonstrated to trap biological material or cells using this technique [17,18].

The effect of light pressure is explained by the dipole of dielectric material under electromagnetic field. When the wavelength of light is larger than the size of object, the kinetic momentum of dielectric material P under electromagnetic field is described as the following equation,

$$P = S / v^2 \tag{2.30}$$

where, S is the pointing vector, v is the speed of light inside object. The trapping force is generated by the kinetic momentum change, hence the trapping force F is described by the momentum change ΔP during time Δt as follows,

$$F = \Delta P / \Delta t \tag{2.31}$$

here the intensity of light is defined as I, the trapping force F is described as follows.

$$F = QIn / c \tag{2.32}$$

where, n is refraction index of outer martial of object, c is the light speed in vacuum. Q is a trapping efficiency and normally 0-2 constant value depending on the momentum change. The momentum change is affected by the interactions of reflection, refraction, dispersion, absorption and so on. Hence, two important factors on the laser trapping are the wavelength of light and the size of object.

The laser trapping in the Mie scattering region is depicted in Figure 2.7. This phenomenon is used to trap the object to focus point by the momentum change of light on the object. The A of Figure 2.7 in shows the focus of laser light at the center of object. The B and C are the images when the focus of laser is out of alignment in horizontal or vertical axes. The lines of a and b show the light axes. These lines reflected by the passing through the object. The F_a and F_b are the applying forces by the momentum change of light axes a and b. When the each light pressure change is proportional to the energy of light axes, the trapping force is described as following equation.

$$F = Q\left(n_1 \frac{P_L}{c} \right) \tag{2.33}$$

where, P_L is the laser power.

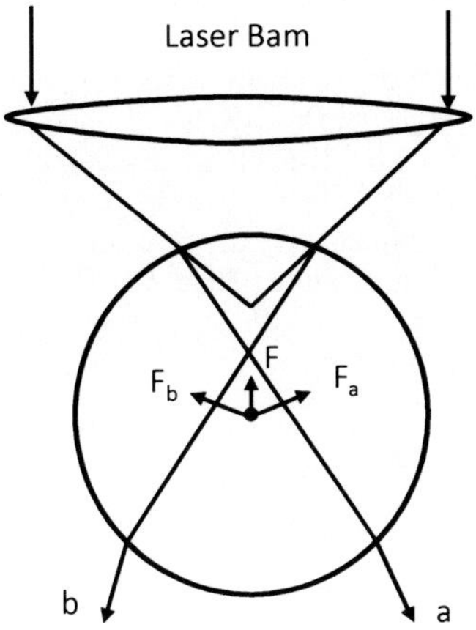

Fig. 2.7 Laser trapping model on geometrical optics

References

[1] Israelachvili, J.N.: Intermolecular And Surface Forces. Academic Pr. (1996)

[2] Han, X., Zheng, K., Zhang, Y., Zhang, X., Zhang, Z., Wang, Z.L.: Low-Temperature In Situ Large-Strain Plasticity of Silicon Nanowires. Adv. Master. 19, 2112–2118 (2007)

[3] Institute of Electrostatics Japan, Electro-static Handbook, Ohmsha (1981) (in Japanese)

[4] Sano, M.: New ・Basics Electromagnetics, Saiensu-sha (2005) (in Japanese)

[5] Nordheim, L.W.: ZurTheorie der thermischen Emission und der Reflexion von Elektronen an Metallen. Zeitschrift fur Physik. 46(11-12), 833–855 (1928)

[6] Gadzuk, J.W., Plummer, E.W.: Field emission energy distribution. Rev. Mod. Phys. 45, 487–548 (1973)

[7] Chiou, P.Y., Ohta, A.T., Wu, M.C.: Massively parallel manipulation of single cells and microparticles using optical images. Nature 436, 370–372 (2005)

[8] Kim, P., Lieber, C.M.: Nanotube Nanotweezers. Science 286, 2148–2150 (1999)

[9] `http://robotics.eecs.berkeley.edu/~ronf/Gecko/index.html`

[10] Autumn, K.: Gecko Adhesion: Structure, Function, and Applications. MRS Bulletin 32(06), 473–478 (2007)

[11] Peattie, A.M., Majidi, C., Corder, A., Full, R.J.: Ancestrally high elastic modulus of gecko setal β-keratin. R. Soc. Interface 4(17), 1071–1076 (2007)

[12] `http://robotics.eecs.berkeley.edu/~ronf/Gecko/gecko-compare.html#Hard_Polymer_Arrays_`

[13] Lee, J., Fearing, R.S., Komvopolous, K.: Directional adhesion of gecko-inspired angled microfiber arrays. Applied Physics Letters 93(191910) (2008)

[14] Ashkin, A.: Acceleration and trapping of particles by radiation pressure. Phys. Rev. Lett. 24, 156–159 (1970)

[15] Ashkin, A.: Trapping of atoms by resonance radiation pressure. Phys. Rev. Lett. 40, 729–732 (1978)
[16] Richard, B., Wolf, E.: Electromagnetic diffraction in optical systems II Structure of the image field in an aplanatic system. Proc. R. Soc. London A 253, 358–379 (1959)
[17] Ashkin, A., Dziedzic, J.M.: Optical trapping and manipulation of viruses and bacteria. Science 235, 1517–1520 (1987)
[18] Ashkin, A., Dziedzic, J.M., Yamane, T.: Optical trapping and manipulation of single cells using infrared laser beams. Nature 330, 769–771 (1987)

Chapter 3
Related Technologies on Micro-Nanorobotic Manipulation Systems

3.1 Materials and Science in Micro-Nano Scale

Historically, the micro-nano materials are investigated in chemistry and biology fields. For example, biological components are basically tiny size ranging in micro-nano meter scale, such as DNA with diameter ~2 nm. The material is one of the basic component of science In 1991, carbon nanotube was investigated by Prof. Sumio Iijima, then nano-technology was widely paid attention and changed this field [1].

3.1.1 Carbon Nanomaterials

Carbon material is one of the key material in micro-nano meter size, because it has different configuration in 0D fullerens, 1D nanotube, 2Dgraphene, 3D graphite/diamond as shown in Figure 3.1 [2]. The grapheme has a lot of attention because of the novel prize in 2010. For example, the grahene has an excellent electric properties as a transparent electrode.

3.1.2 Carbon Nanotubes

The explorations of carbon nanotubes over the past decade have revealed many exceptional properties of nanotubes and proposed broad potential applications of them.

Structures of Carbon Nanotubes

Carbon nanotubes have well-defined geometries. They can have only one layer, or as many as tens of layers. Single-walled nanotubes (SWNTs) [3,4] consist of a single graphite sheet seamlessly wrapped into a cylindrical tube, whereas multiwalled nanotubes (MWNTs) [1] comprise an array of concentrically nested tubes. The diameters of SWNTs range from ~0.4 to >3nm, whereas those of

MWNTs can be ~1.4 to at least 100nm [5]. Typically, nanotubes are several μm long, but very long MWNTs can reach up to several mm [6]. Centimeter long SWNT strands have also been synthesized [7].

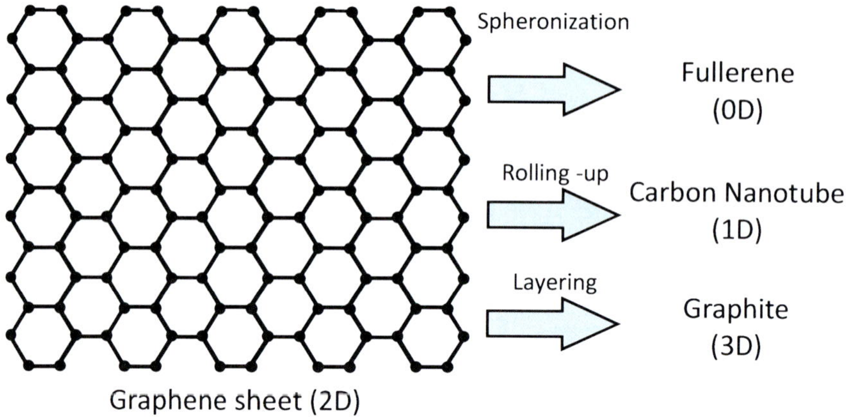

Fig. 3.1 Structure of carbon family from graphne sheet

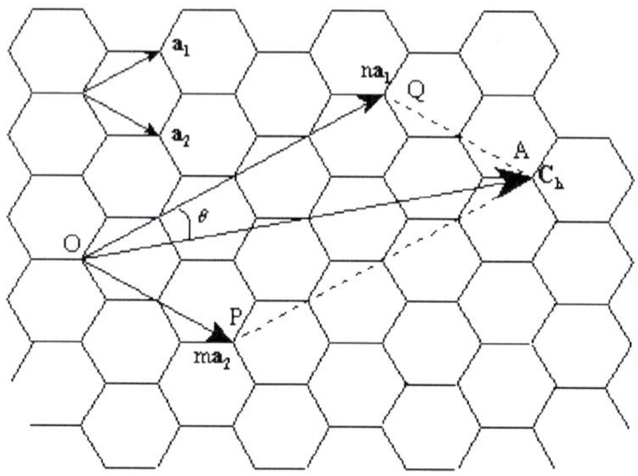

Fig. 3.2 Structure of a Carbon Nanotube

The structure of a SWNT or a layer of a MWNT is specified by the chiral vector $C = (n, m) = na_1 + ma_2$, where n and m are integers and, $0 \le |m| \le n$, and a_1 and a_2 are real space unit vectors as shown in the unrolled honeycomb lattice of a nanotube in Figure 3.2 [8]. The chiral vector corresponds to a section of the nanotube perpendicular to the tube axis and denotes the nanotube type. Depending on the appearance of a belt of carbon bonds around the nanotube diameter, the nanotube is either of the armchair (n = m), zigzag (n = 0 or m = 0), or chiral (any other n and m) variety (Figure 3.3 [9]). All armchair SWNTs are metals; those with n – m =3k, where k is a nonzero integer, are semi-conductors with a tiny band gap; and all others are semiconductors with a band gap that inversely depends on the nanotube diameter. Given the chiral vector *(n,m)*, the diameter of a carbon nanotube can be determined using the relationship $d = 0.0783\,(n^2 + m^2 + nm)^{1/2}$ [nm].

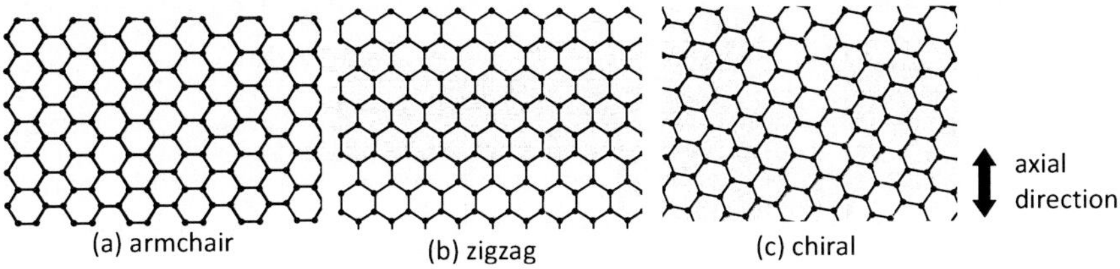

Fig. 3.3 Chiral types of carbon nanotube

Properties of Carbon Nanotubes

The mechanical properties of nanotubes have been intensively investigated. The measured Young's modulus of a MWNT can be 0.4 to 4.15TPa (mean 1.8TPa) [10], 1.3TPa [11], 0.1~1TPa [12] and 0.27~0.95TPa [13], and that of a SWNT can be 1.25TPa [14], 1TPa [15] and 0.32~1.47TPa [16]. It seems that the diversity comes from the different definition of cross-section of nanotubes, the different synthesis methods (rates of defects), and the measurement errors. However, most measured values suggest a Young's modulus as high as about 1TPa. Tensile strength has the similar tendency. Measured values are 11~63GPa for MWNTs [32], 13~52GPa (mean 30GPa) [35], and 45±7GPa[17] for ropes of SWNTs. Although data for mechanical properties of nanotubes are quite diverse, it seems that they are one of the strongest materials known so far.

Depending on the chirality, SWNTs can be either metallic or semiconducting, but all MWNTs are metallic [18-20]. Metallic SWNTs and MWNTs are ballistic conductors, enabling them to carry high currents with essentially no heating [21, 22]. Phonons also propagate easily along the nanotube: The measured room temperature thermal conductivity for an individual MWNT (>3kW/mK) is greater than that of natural diamond and the basal plane of graphite (both 2kW/mK) [23]. Nanotubes also have excellent capability for field emission. The main properties of CNTs is shown in Table 3.1.

Table 3.1 Properties of Carbon Nanotubes

Property	Item	Data
Geometrical	Layers	Single/Multiple
	Aspect Ratio	10-1000
	Diameter	~0.4 nm to >3 nm (SWNTs) ~1.4 to >100 nm (MWNTs)
	Length	Several μm (Rope up to cm)
Mechanical	Young's Modulus	~1 TPa (steel: 0.2 TPa)
	Tensile Strength	45 GPa (steel: 2 GPa)
	Density	1.33~1.4 g/cm^3 (Al: 2.7 g/cm^3)
Electronic	Conductivity	Metallic/Semi-conductivity
	Current Carrying Capacity	~1TA/cm^3 (Cu: 1 GA/cm^3)
	Field Emission	Activate Phosphorus at 1~3 V
Thermal	Heat Transmission	>3 kW/mK (Diamond. 2 kW/mK)

Applications of Carbon Nanotubes

All of these remarkable properties promise nanotubes many applications [28]. The Table 3.2 summarizes the applications of nanotubes. In the bulk state, nanotubes can be used to synthesize conductive and high-strength composites, to fabricate field emission devices, to save and convert electrochemical energy, to store hydrogen, and so on. However, the most promising applications of nanotubes that have much deeper implications for molecular nanotechnology need to maneuver the tubes individually to build complex nanodevices. Such devices include nanoelectronics and nano-electro-mechanical systems (NEMS).

The first example of such devices, a nanotube probe for an atomic force microscope (AFM) was demonstrated by Dai et al. for improving the spatial resolution of an AFM and protecting the tip from "tip crash" [24]. In the device, a MWNT was manually assembled onto a commercially available silicon cantilever. Further development improved the construction technique through direct chemical vapor deposition (CVD) [25], controlled assembly [26], and picking up a tube from vertically aligned SWNTs grown from planar substrate surfaces [27].

Nanotube tweezers have been constructed with two bundles of nanotubes on a glass fiber and driven by the electrostatic interaction [28]. A nanotube linear bearing [29] has been demonstrated with an opened MWNT [30]. Chemical sensors based on individual semiconducting SWNTs have also been demonstrated [31].

Table 3.2 Applications of Carbon Nanotubes

State	Device	Main Properties Applied
Bulk /Array	Composite	High strength, conductivity, etc.
	Field emission devices: flat display, lamp, gas discharge tube, x-ray source, microwave generator, etc.	Field emission: stable emission, long lifetimes, and low emission threshold potentials, high current densities
	Electrochemical devices: super-capacitor, battery cathode, electromechanical actuator, etc.	Large surface area, conductivity, high strength, high reversible component of storage capacity
	Fuel cell, hydrogen storage, etc.	Large surface area
Individual	Nanoelectronics: wire, diode, transistor, switch, memory, etc.	Small sizes, semiconducting /metallic
	NEMS: probe, tweezers, scissors, sensor, actuator, bearing, gear, etc.	Well defined geometries, exceptional mechanical and electronic properties

For nanoelectronics, individual SWNTs can function as room-temperature (RT) field-effect transistors (FETs) by placing a semiconducting SWNT between two electrodes (as source and drain), in which gating has been achieved by applying a voltage to a submerged gate beneath the SWNT [32]. Integrated nanotube devices involving two nanotube transistors have been reported [33]. However, these nanotube transistors still involved lithographically fabricated electrodes as a part of them, and their overall size is not smaller than silicon-based FETs.

Pure nanotube circuits [34-36] by interconnecting nanotubes of different diameters and chirality might lead to further shrinkage of the integration. Nanotube intermolecular and intramolecular junctions are basic elements for such systems. An intramolecular kink junction behaving like a rectifying diode has been reported [37]. RT single electron transistor (SETs) [38] has been shown with a short (~20 nm) nanotube section that is created by inducing local barriers into the tube with an AFM and Coulomb charging has been observed. With a cross-junction of two SWNTs (semiconducting/metallic), three- and four-terminal electronic devices have been made [39]. A suspended cross-junction can function as electromechanical nonvolatile memory [40].

Almost all of the applications of nanotubes for nanoelectronics and NEMS involve characterizing, placing, deforming, modifying, and/or connecting nanotubes. Although chemical synthesis may provide a way for patterned structure of nanotubes in large-scale [41], and self-assembly may generate better regular structures, we still lack the capability to construct irregular complex nanotube devices. Nanomanipulation, with its "bottom up" nature, is the most promising way for this purpose.

Field Emission Applications Using Carbon Nanotubes

Why are nanotubes such exceptional field emitters? The very low turn-on fields measured for nearly all emitters originate certainly from the small diameter and elongated shape of the tubes that lead to a high geometrical field enhancement. In fact, the local electric field just above the emitter surface needed for field emission is around 2-3 V/nm as for metallic emitters, as can be estimated from the applied field and the field amplification factor [42]. On the other hand, nanotubes do not behave like very sharp metallic tips. Furthermore, the influence of the structural properties of the tube on the field emission is not clearly assessed at present. This issue is complicated by the fact that the physical and chemical properties are deeply modified by preparation and purification steps [43, 44]. Despite these difficulties we will summarize here the principal models that have been proposed for the field emission from carbon nanotubes. Some authors recorded I-V characteristics that follow the Fowler-Nordheim law, from which they concluded that carbon nanotubes behave as metallic emitters [45,46,47].

Rinzler et al. interpreted the higher efficiency of open MWNTs and their behavior under laser irradiation as emission from a single carbon chain that unravels from the tube edge and produces very high field amplification [43]. This model could not be confirmed yet experimentally, and would probably be valid only for open tubes. Again, this model would be valid only for open MWNT.

Luminescence induced by the electron emission could arise from radiative transitions between two levels participating to the field emission [48]. Actually, theoretical calculations predict the presence of localized states at the tube cap, with a density of electronic states that differs markedly to that of the tube body [49,50]. This was recently confirmed experimentally by STM measurements on MWNTs [12] as well as on SWNTs [51]. The full width at half maximum (FWHM) of these states and their separation is in good agreement with the values measured for the light emission and with field emission energy distribution observations carried out at 20 °C and at 600 °C [52].

A complete study by Dean et al. suggests complementary mechanisms, and shows clearly that the emission behavior of nanotubes is far more complex than the one expected from a very sharp metallic tip with a workfunction of ~5 eV [53,54]. Different emission regimes on single SWNTs were identified and depended on applied field and temperature. A first regime corresponding to resonant tunneling through an adsorbate was found under "usual" experimental conditions at low temperatures and applied fields. The involved molecule has been identified as water and it appears that this adsorbate-assisted tunneling is the stable field emission mode at room temperature. These molecules desorb either at high fields and emitted currents or at temperatures higher than 400°C. The other regimes correspond to the intrinsic emission from the cleaned tube and show a far lower emitted current for comparable voltages with strongly reduced current fluctuations. The origin of these (at least two) instrinsic regimes is not clear yet but the emission mechanism involves probably non-metallic electronic states, such as enhanced field emission states above the Fermi level or a non-metallic density of electronic states.

Obraztsov et al. proposed that the graphene layers form sharp bends at the open end of the nanotube, where the carbon atoms show sp^3-like atomic bonds instead of the sp^2 configurateion typical for graphene [55]. This change in coordination would decrease the height of the potential barrier and could explain the very low workfunction that the authors estimated from the slope of Fowler-Nordheim plots. J.-M. Bonard et al. concluded from their observations that the electrons are emitted from sharp energy levels due to localized states at the tube cap [6]. Luminescence induced by the electron emission could arise from radiative transitions between two levels participating to the field emission [56]. Actually, theoretical calculations predict the presence of localized states at the tube cap, with a density of electronic states that differs markedly to that of the tube body [57,58].

Two points have to be mentioned concerning this hypothesis. If several energy levels participate in the emission, the occupied level nearest to the Fermi energy will supply nearly all the emitted electrons. Since the position of this level would depend strongly on the local atomic configuration (tube diameter, chirality, presence of pentagons and other defects), significant differences of the emitted currents can be expected from one tube to another. Second, these localized states often show far higher carrier densities as compared to the tube body at the Fermi level [12]. As the field emission current depends directly on this carrier density, we speculate that the emitted current would be far lower for a nanotube without such states.

The present understanding is that the emission involves a non-metallic density of electronic states and/or adsorbate-resonant tunneling. Supplementary informations on the electronic structure of the nanotube cap and on the influence of absorbates or bonded groups are required for a better comprehension of the emission. It seems us important to mention again the work of Lovall et al. [59]. They show the way for a complete characterization of nanotubes, where the electronic density of electronic states is measured by field emission energy distribution and the exact structure of the tube cap is determined by field ion microscope (FIM). This simultaneous determination of the electronic and structural properties of individual nanotubes would without doubt yield valuable insights on the emission mechanism.

The work function corresponds to the energy difference between the Fermi level and the maximum of the surface barrier. It depends on materials of the emitter originally. Figure 3.4 shows the properties of field emission through various emitters at different work function (4.6eV, 4.9eV, 5.0eV). The material with small work function has better field emission characteristics. The work function of an emitter can be measured according to Fowler-Nordheim theory. Another method based on the mechanical resonance of the CNT has been demonstrated by Wang and co-workers [60]. They found that of a part of MWNTs is ~0.2-0.4 eV lower than that of carbon likely to be metallic, and that of the other MWNTs is ~0.6 eV higher than that of carbon (5 eV) likely to be semiconductive. To improve the surface work function of an emitter, it is effective to decorate it with metal or oxide metal.

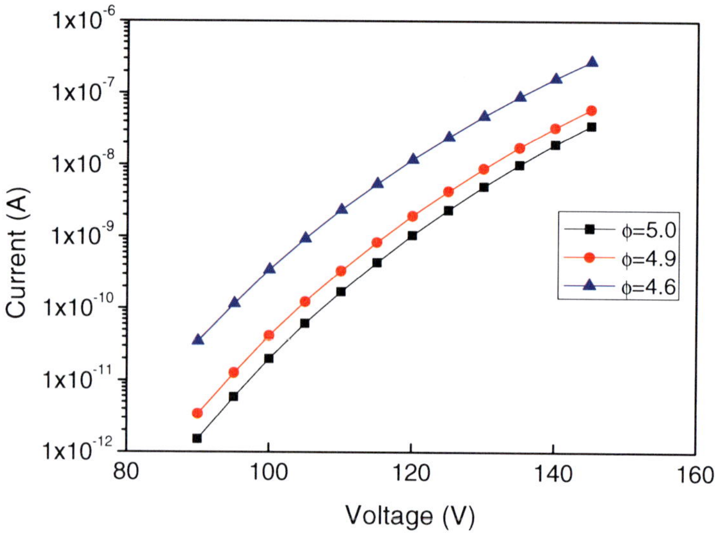

Fig. 3.4 Field emission properties of different work function emitters

The inter-electrode distance is adjusted with nanorobotic manipulators within several nanometers of resolution. When the distance is set to 0.5, 1, 3, and 5 μm, the onset voltages (as emission current becomes larger than 25 pA) are shown in Figure 3.5. The onset voltage is reducing when the emitter approaches to the substrate. The field enhancement factor depends on the geometries of an emitter such as diameters, lengths and helical angles of the emitter.

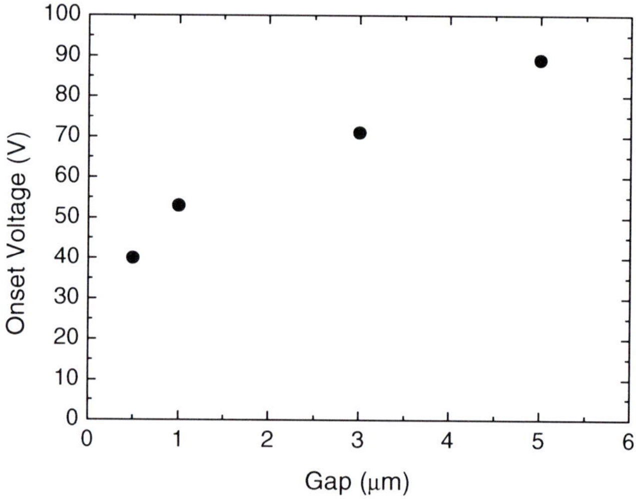

Fig. 3.5. Onset voltages at difference inter-electrode distances

3.1.3 *Sol-Gel Material*

We developed a novel technique to investigate individual properties of microorganisms on a chip using in-situ microfablication of permeation membrane with photo-crosslinkable resin. A cage having vertical permeation membrane was fabricated in situ in the microchannel by providing the photo-crosslinkable resin and curing it in situ by the focused UV irradiation. An arbitrary selected individual cell was transported inside the cage by the newly developed laser micromanipulation method employing a microtool made by the local thermal gelation. The cage was freely and properly constructed at arbitrary position in the microchannel. Since the cage is made of the permeable membrane, we can culture the cells in it and observe their chemical reactions microscopically by providing growth media or reagents. In experiment, we confirmed repeatability of on-demand analysis of isolated individual yeast cells.

3.1.4 *Hydrophilic/Hydrophobic Material*

The formation of a macromolecule thin film and etching of a silicon wafer by plasma have already carried out the big contribution to simplification of drying of an electronic industry material or a chipmaking process, and a process, etc.

Compared with this, the plasma processing techniques of a polymer material by the grade which there is the plasma Drais tripping method in a broad sense, and was well known mainly as photoresist removal technique in a wafer process, For example, the adhesive improvement of cleaning of a printed circuit board, a polymer material, parts, etc., the other improvement in the printing performance of a high polymer film of a material list side, or canal-ized processing etc., Becoming a leading applicable field of the plasma in the electronic industry material field in the future is expected, and energetic research is expected.

Plasma Treatment of Solid Surface

When plasma classifies the situation about the solid surface and the typical action which plasma treatment especially does to a polymeric surface, a main action and effect of plasma. It is classified into (1) physical change of the surface by etching etc., (2) surface bridge construction, (3) chemical modification, (4) bulk material list side polymerization, polymerization by (5) plasma, etc.

Although these effects are related also to the time of processing, ESCA (Electron Spectroscopy ChemicalAnalysis), It is specified as the range about several tens to thousands from the material list side from the result of the surface observation by a scanning electron microscope, and there is no internal change as bulk of material itself, i.e., the material depths. The action of the plasma in the polymeric surface described here, A mere thermal effect and ion, It is not the sputtering action by the collision to material list sides, such as an electron, The kinetic energy which the chemical functioning which plasma, such as chemical modification by the activated species in a polymeric surface, bridge construction, and polymerization, has rather is in the mainstream, and is loose compared with sputtering, etching, etc. as for the reaction, and contributes to a reaction 10 eV remainder and sputtering, Compared with it of etching, a thing very small in energy is mentioned especially.

Hydrophilic Surface of Polymer Surface by Oxygen Plasma

Oxygen plasma is reactive gas which has played an active part in the surface treatment most.

When the PDMS surface is processed using [6-9]. oxygen plasma it becomes possible to carry out property modification of the surface of polymer materials, such as PDMS, for example like by using oxygen plasma as shown in Figure 3.6 It is about 105 before [10]. which shows the time variation of the angle of contact of water droplets, as shown in Figure 3.7. It turns out that the angle of contact which existed becomes immediately after processing by oxygen plasma, and the PDMS surface is hydrophilic treatment.

Moreover, the effect of processing by oxygen plasma is lost mostly in after-processing about 3 hours, and it also turns out that it has returned to the state before processing. Although it is by performing processing by oxygen plasma temporarily at a target, it is possible to carry out property modification of the surface to hydrophilic property.

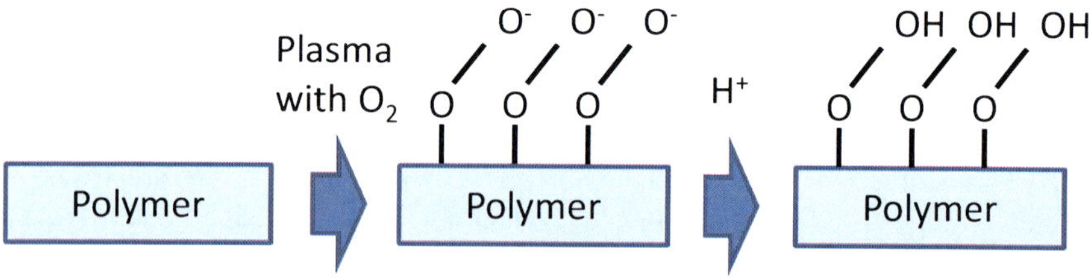

Fig. 3.6 Mechanism of surface treatment by oxgen plasma on PDMS surface

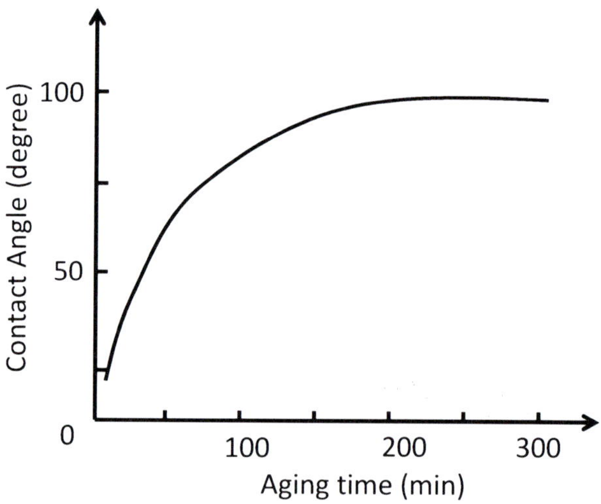

Fig. 3.7 Example of contact angle of water droplet by treatment of hydrophilic surface (PDMS)

The plasma treatment has some advantages that plasma treatment is afflicted by neither in many cases processing time being very short and treating a dangerous oxidization medical fluid, nor waste contamination, although these are the good methods which do not require cost, The worries about a bend processing uniform with the material which is not a flat side, for example, a thing with complicated relief structure, is difficult, and according to heat etc. existing ., however the present place plasma treatment, It is not carrying out till the place taken and replaced with processing a cheap being [it / a plastic bottle] mass production type material by the sandblast or the economical and efficient practical use method like flame processing.

Excimer Exposure with VUV/O_3 Washing

An ultraviolet-rays ozone wash ultraviolet rays Use and active oxygen The organic matter which made it generate and adhered to the substrate face simultaneously It decomposes, and the light of the short wavelength not more than 200nm which is a dry washing method which carries out oxidization volatilization removal is called VUV (Vacuum Ultra Violet), and washing (VUV/O_3 washing) using 172nm excimer light is distinguished from washing (UV/O_3 washing) of the low-pressure mercury lamp using 254 nm.

3.1.5 Biocompatible Material

The method of manufacturing micro structure by molding made from the elastomer of a silicon system which is represented by PDMS came to be tried from the second half of the 1980s to the first half of the 1990s. The silicone elastomer is used for uses, such as manufacture of micro channel structure, and surface patterning by micro contact printing, in those examples. This method comes to be widely used as a method which will manufacture a micro chemical chip from the second half of the 90s, and is named "soft lithography" generically in recent years as contrasted with the conventional lithography which used silicon and glass.

Photo-Linkable Resin

Photo-curing resin adds a photosensitizer to the monomer and oligomer (prepolymer) which are generally the macromolecules of an unbuilt bridge, and it is possible to make bridge construction induce and polymerize by irradiating with the light of a specific wavelength. In photo-curing resin, diazo, azide, vinyl, compounds are mainly used as photosensitive group. In many cases, the resin shows the photosensitivity for ultraviolet region. On the other hand, some resins are developed for the light in the visible or infrared regions. The insoluble resin is mainly used a diazo compound. The diazo compound decomposes and polymerizes by the exposure of light. One of soluble resin is mainly used a diacyl compound. The photo-polymerized part is partially soluble and un-photo-polymerized part is remained.

The photodimerization type resin has a mechanism to cleave the photosensitive photodimerization bond to be dimer of four membered ring. One of the photosensitive compound is cinnamylidene group.

The cinnamyl group is not sensitive for humility or temperature, and is good to keep in stable condition. Various researches have been achieved for photosensitive material, and it is possible to have a photosensitivity for the light in the visible or infrared regions with control the photosensitive speed.

The photo polymerization materials are activated by the photosensitive group with a monomer or an oligomer to change the larger size as a polymer or the other type of monomer and oligomer. The advantages of these polymers are 1) unlimited combination using various resin and photosensitizer, 2) steric-photo-fabrication, 3) transparency. The examples of equations are shown as Figure 3.8 for some polymerization processes using photosensitive resins.

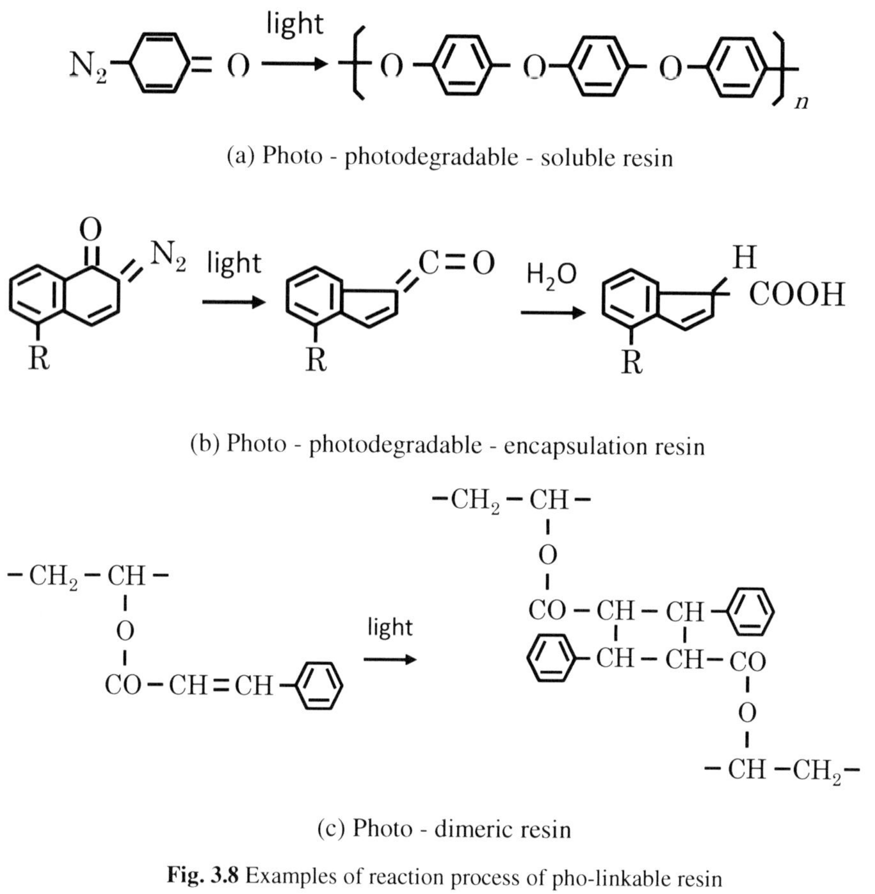

Fig. 3.8 Examples of reaction process of pho-linkable resin

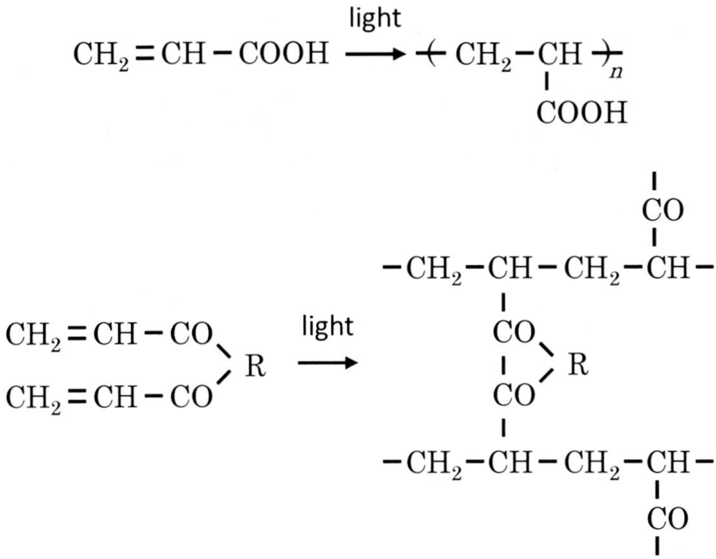

(d) Photo - polymerisation resin

Fig. 3.8 *(continued)*

3.2 Microscopes in Micro-Nano Scale

Nanomanipulation has been received much more attention ever since, because it is an effective strategy for the property characterizations of individual nano-scale objects, and the construction of nano-scale devices to realize high-integration, multi-function, low-energy consumption, and so on. To manipulate nano-scale objects, it is needed to observe them with a resolution higher than nano-scale. Hence, a manipulation system and an observation system, microscope in general, are necessary for nanomanipulations.

3.2.1 Optical Microscopes

The optical microscope has an old history in a microscope, and research and development are going on further. As compared with an electron microscope or scanning probe microscopy, the advantages of optical microscope are 1) real-time observation, 2) inner structural observation, 3) light emitted observation and so on.

On the other hand, the imaging resolution is limited in ~200 nm in principal. The imaging resolution of an optical microscope is proportional to the wavelength of light, so that it is given by the following equation.

$$\delta = \frac{0.61\lambda}{NA} \qquad (NA = n\sin\alpha) \tag{3.1}$$

where, δ is resolution, λ is wavelength of light, n is refraction index of medium (n in air: 1.0, n in oil: ~1.5, n in water: ~1.3), α is aperture angle, NA is a numerical aperture of microscope as shown Figure 3.9. When the wavelength of light is 400 nm and the NA is 1.5 as normal microscope value, the imaging resolution is calculated as ~200 nm. The optical microscope is basically divided into several categories depending on 1) Applications, 2) Observation Methods, and 3) Imaging Principals.

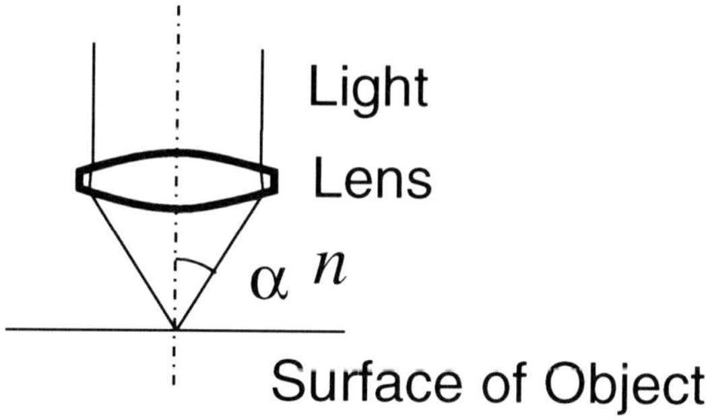

Fig. 3.9 Condenser lens of optical microscope

Applications of Optical Microscope

· **Biological Microscope**

Since the depth of focus is shallow, a cover glass is used fundamentally. It uses, when observing by the transmitted light and observing the thin fervent piece of a microbe or animals and plants, etc. by one 40 times as many as about 1000 times the magnification of this.

· **Metallurgical Microscope**

To a sample, from the object lens side, it irradiates with light and observes the physical properties of a sample by the light and darkness of the catoptric light, or the contrast of a color.

Although resolution is good, since the depth of focus is shallow, problems, like it is near also have observation distance, and it needs to process a subject, and needs to put it on a stage.

· **Stereoscopic Microscope**

It is observable in the state where there is no distortion perpendicularly and horizontally in three dimensions. An image looks being the same as that of the image seen with the naked eye. The depth of focus is comparatively deep and it is possible to observe operating the magnification of a lens continuously with a zoom lens etc.

Observation Methods of Optical Microscope

Most popular microscope to use It is placed on a stage, and it observes from the upper part. A sample is arranged on a stage and it observes from the lower part. It can observe without being restricted to the height of a subject. When manipulation equipment is installed, since it does not become an obstacle of observation, it is suitable also for . cultured cell observation. It is efficient to obtain enough space for extensions such as micro-manipulation purposes.

Imaging Principles Optical Microscope

- **Bright Field Microscope**

When a sample is illuminated by uniform incident radiation, since the absorption spectra of light differ in each portion of a sample the microscope. using contrast being attached to the image of the transmitted light -- it is necessary to devise that various pigment dyes an object by the sample (for example, transparent objects, such as a cell and a crystal) whose absorptivity it is thin and is small since contrast is low and a clear image is not obtained etc.

Usually, a light source for a bright field microscope is mainly from the illumination of an object (or the lens and reflector for collecting sunlight). The eyepiece for making eyes and a camera image the light which passed the object lens for condensing the stage for placing the capacitor for condensing on slide glass, and the slide glass which carried the sample to observe, and the light from a sample, and the object lens, It consists of parts, such as a body tube used as the optical path from an object lens to an eyepiece.

Being used generally has a capacitor under a stage, and although it is a erected microscope removed from a top, there is also a formal inverted microscope removed from the bottom for observation of a cultured cell, etc.

- **Dark-Field Microscope**

It is the method of observing dispersion and reflection of the light with which it irradiated from across to the sample unlike the bright field microscope observed by the transmitted light. By this method, the background of a view is black, and a sample shines and it looks contrary to a bright field microscope. This method is realizable only by inserting a dark field capacitor in the usual optical microscope. Although it is unsuitable for observation of a detailed structure of the object surface or an inside, it is possible to observe existence of an object smaller than the wavelength of visible light by high contrast.

- **Polarization Microscope**

According to an internal structure or a crystal structure, an object has the polarization nature which changes the oscillating direction of light. It is the method of observing this polarization nature (polarizing plate) is put on the place of the capacitor of an optical microscope, a delay board and an analyzer (polarizing plate) are placed behind an object lens, and the polarization nature and form birefringence of a sample are observed as a difference in light and darkness or a color.

According to rotation of a polarizing plate, a crystal is observed in a vivid color. It is used for observation of the substance of the crystalline material contained in the crystal and living thing sample of rock etc.

- **Phase-Contrast Microscope**

The light passing through the inside of a medium with a big microscope, the refractive index for observing samples which consist of the portion from which a refractive index differs although it is water-white, such as a living thing sample. It is delayed for the light passing through the inside of a medium with a small refractive index in that phase. capacitor and the object lens which are the microscopes using the diffraction ray in connection with this phase difference It is possible to observe the internal structure of an almost transparent living thing cell by the . this gentleman method for making the diffraction rays from which the phase shifted interfere, changing phase difference into light and darkness, and observing it.

- **Fluorescent Microscope**

It is a microscope which observes the fluorescence shown from a sample. Fluorescence is the light emitted from a sample, when light is applied to a sample to observe, and since it is very weak in many cases, the device is given so that the light applied to a sample may interfere with fluorescence and may not disappear. This restricts the wavelength of the light which comes out of a light source beforehand by some methods (excitation filter etc.), and is trying only for a certain specific wavelength to hit a sample in a fluorescence microscope unlike the usual bright field microscope.

If the light of the wavelength is received in an observation sample and the substance which shows a fluorescence is contained, the fluorescence which comes out of a sample is observable under a microscope. Usually, the wavelength of the fluorescence which a substance shows is the light of a wavelength longer than the wavelength of the light (excitation light) shed in order to make fluorescence come out, and the wavelength was decided by the substance which gives off fluorescence.

Since the light which the light applied to the sample has only reflected on the other hand is the light of the same wavelength as the wavelength of a basis, the fluorescence shown from the sample can distinguish it with the wavelength. Usually, a wavelength enables it to observe only fluorescence by the filter in which transmissivity differs, and using a prism etc.

- **Confocal Microscope**

After scanning a pinhole pass for the fluorescence (or light reflected from the sample) which scanned laser from the object lens and was emitted from the excited sample, a detecting device is used and a picture is reconstructed on detection and a computer. Since the fluorescence from other than the same focal (confocal point) side can be shut out by using a pinhole, the optical section image of thickness depending on a numerical aperture can be obtained.

For example, although the optical section of thickness about 200 nm will be obtained and it is greatly inferior to a transmission electron microscope when the lens of the numerical aperture 1.33 is used by Ar laser (wavelength 488 nm), space resolution conventionally higher than an optical microscope can be obtained easily. Moreover, compared with the case of a transmission electron microscope, specimen preparation is easy.

- **Near-Field Optical Microscope**

The atomic electrical dipole is happened for the small object which has a size smaller than the wave length of light. The electromagnetic filed (evanescent field) is used as a light for observation of object by scattering light. This light is called as evanescent wave.

The evanescent wave is generated by a light through a quite-small hole which has a size smaller than the wavelength of light. By using the evanescent light, the high resolution imaging is achieved as near-field optical microscope. This microscope achieves the high resolution imaging higher than the limitation caused by the wave length of optical light without any coating .or setting inside vacuum environment.

- **Total Internal Reflection Fluorescence Microscope**

This microscope use the evanescent wave generated by illuminating at the glass-water interface. The light is totally internally reflected at the interface. Usually, the observation region is less than 200 nm height. But, this microscope is effective to observe biological sample with especially fluorescent information with quite low noise for biophysics or molecular biology fields.

3.2.2 Scanning Probe Microscopes

The strategy of nanomanipulations with various kinds of microscopes is shown in Figure 3.10. Scanning probe microscopes (SPMs), like scanning tunneling microscopes (STMs) or atomic force microscopes (AFMs), have functions of both observation and manipulation. Their high resolution makes them capable of atomic manipulation. However their observation space is too narrow and constrained in a plane. Optical microscope (OM) is one of the most historical and basic microscope. However, its resolution is limited to ~100 nm because of the diffraction limit of optical wavelength (: ~400 - ~800 nm) [61]. Hence, the special techniques (using ex. evanescent light or fluorescent light) are needed for the observation under nanometer scale objects [62]. Recently, the high-speed AFM system has a lot of attentions [63].

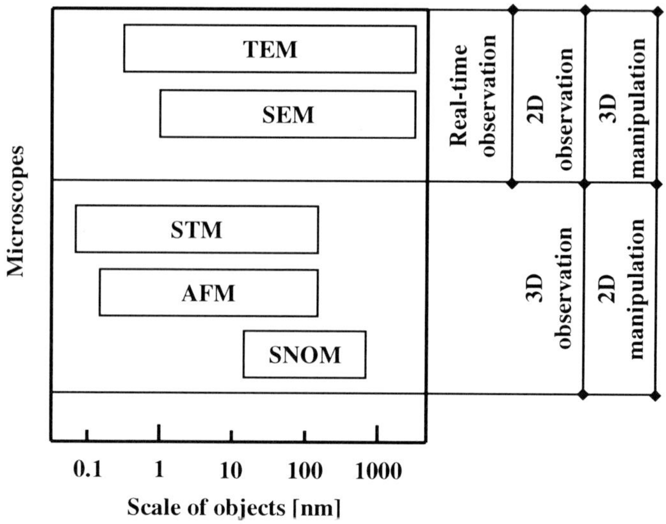

Fig. 3.10 Strategy of nanomanipulations with microscopes

3.2.3 Electron Microscope

On the other hand, electron microscopes (EMs) provide atomic scale resolution with the electron beam which wave length is less than ~0.1Å. EMs are divided mainly two types as scanning electron microscopes (SEMs) and transmission electron microscopes (TEMs). SEM images provide us surface images with ~1 nm resolution. However the resolution is generally one order in magnitude lower than that of a TEM. High resolution and transmission image of TEMs are useful for measurement and evaluation of nano-scale objects. However, the specimen chamber and observation area of TEM are too narrow to contain manipulators with complex functions. Hence, special sample preparation techniques are also needed. Before an ideal microscope being invented, it is a practical strategy to combine different microscopes so as to take both the resolution and complexities into advantage.

The history of electron microscopy began with the development of electron optics. The theoretical foundation of electron optics was mainly contributed by two scientists, Hans Busch and Louis de Broglie, in mid 1920s [50]. In a simple word, Busch showed that electrons could be focused by magnetic fields, and de Broglie proposed that electron has wave nature. From these two notions, it is possible to use an electron as an optics microscopy source.

In 1931, the first transmission electron microscope (TEM) was developed by Ruska and his research group in Berlin. By using TEM, the resolution limit has improved to 0.2 nm. Few years later, Max Knoll built the first scanning electron microscope (SEM) in 1935. He was not using demagnifying lenses to produce a fine probe. The resolution limit of SEM was around 100 mm because of the diameter of the focused beam on the specimen [51]. The first true SEM was described and developed in 1942 by Zworykin, who showed that secondary electrons provided topographic contrast by biasing the collector positively relative to the specimen. He reached a resolution of 50 nm, which was still considered low in comparison with the performance obtainable in TEM [64]. Although the resolution of SEM is still lower than TEM, the main advantage of SEM over TEM is to yield three dimensional information from the surfaces of bulk specimens over a large range of length-scales. Further improvements of SEM were contributed by Charles Oatley (1956) who developed a stigmator to correct lens cylindrical imperfections and Everhart-Thornley (1960) who improved the secondary electron detection.

In 1965, the first commercial prototype of SEM was developed by the Cambridge Scientific Instruments company as ''Stereoscan'' [65]. The SEMs used currently are not very different from this first instrument.

Development of environmental scanning electron microscope (E-SEM) was originally driven by a growing need to overcome the limitations of conventional SEM. A fundamental limitation of conventional SEM is the need of high vacuum conditions ($\sim 1.33 \times 10^{-3\ \mathrm{to}\ -5}$ Pa) throughout the system, in order to prevent unacceptable scattering of the primary electron beam [66,67]. However, many biological specimens and other non-conducting materials contain water, oil or other volatile substances that evaporate under high vacuum. Preparation of such specimens may therefore involve dehydration, chemical fixing, and freeze-drying, which can be very sophisticated or time consuming and may change the structural features [68]. Furthermore, unless the specimen is electrically conductive, negative charges can quickly build up. Hence insulators must be subjected to further treatment of a metallic coating. Coated specimens give only topographic contrast, due to the short escape depths of secondary electrons from metals, and therefore valuable compositional contrast from the underlying specimen is lost [69].

Realizing the needs to operate the SEM at various pressure levels, scientists have incorporated several features on a standard SEM which make the operations of SEM under vacuum and non-vacuum conditions become possible. Gerry Danilatos has played a major role in the development of E-SEM. The ability to observe samples under high pressure level ranging from 10 up to 2700 Pa and humidity level up to 100 % makes E-SEM suitable for observing biological samples without any prior sample treatment. By the late 1980s, the first commercial environmental scanning electron microscopes (E-SEMs by Electroscan, a company created for the purpose of gaseous microscope manufacturing) had been produced.

In ~1930, E. Ruska et. al have firstly developed Transmission Electron Microscope (TEM) [70]. In 1939, the first commercially available TEM was developed in the world by Siemens Company. Recently, TEM is one of the highest resolution images with improvement of the system. [71]. On the other hand, Scanning Electron Microscope (SEM) is experimentally manufactured in 1938 by V. Ardenne.

When the sample is irradiated by the electron beam, various electromagnetic waves and secondly electrons are generated from interaction with sample (Figure 3.12). TEM use the transmission electron for getting the transmission image from thin sample. Figure 3.13 (a) shows the configuration of a TEM. The TEM has optional various analysis functions as exceptional tools of local area characterizations, for example electronic diffraction method, Electron Energy-Loss Spectroscopy (EELS), and Energy Dispersive X-ray Spectroscopy (EDS) [72]. The samples of TEM are normally fabricated as thin layer, less than ~100 nm.

SEM detects the secondly electrons for the observations. The SEM images are basically surface image taken with relatively easy operations. Figure 3.13 (b) shows the configurations of SEM system. Recently the resolution is improved as ~ 1 nm. Hence, it is widely used for the investigation of nanomaterials [73].

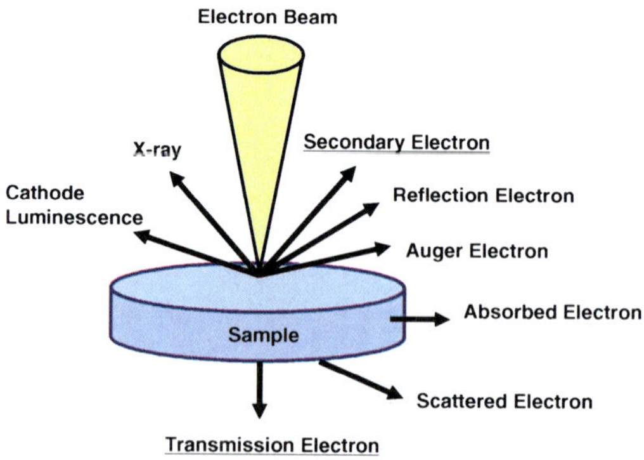

Fig. 3.11 Various information from sample by applied electron beam

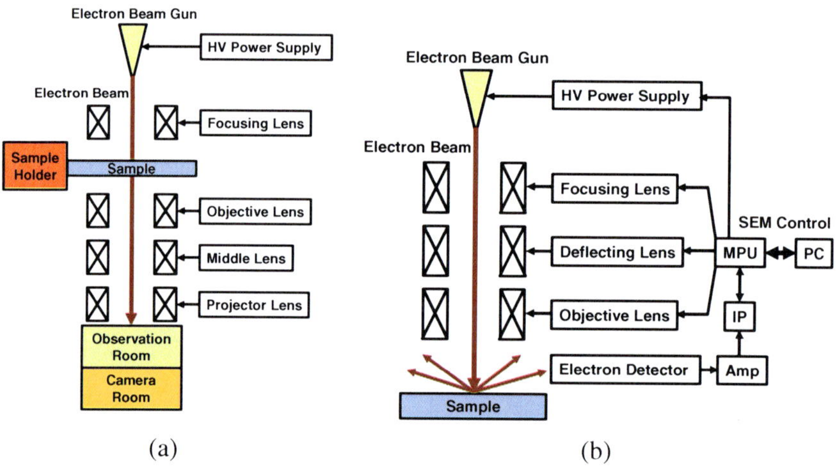

Fig. 3.12 Configurations of electron microscope (a)TEM, (b) SEM

The working principle of E-SEM has two important backbones, i.e. a differential pumping system and a gaseous scanning electron detector (GSED). The overview of E-SEM is shown in Figure 3.13.The differential pumping system enables the electron gun and upper parts of the column to be held at high vacuum (1.33×10^{-5} Pa), whilst the level of vacuum becomes progressively lower further down the column. Pressure limiting apertures allow the electron beam to pass through, but minimize the leakage of gases between zones pumped at different rates. Within the specimen chamber, pressures of up to 2.67 kPa can be maintained at maximum [58].

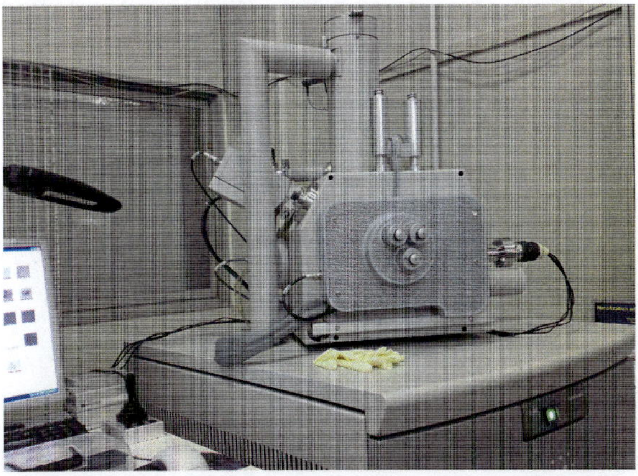

Fig. 3.13 An environmental scanning electron microscope (E-SEM)

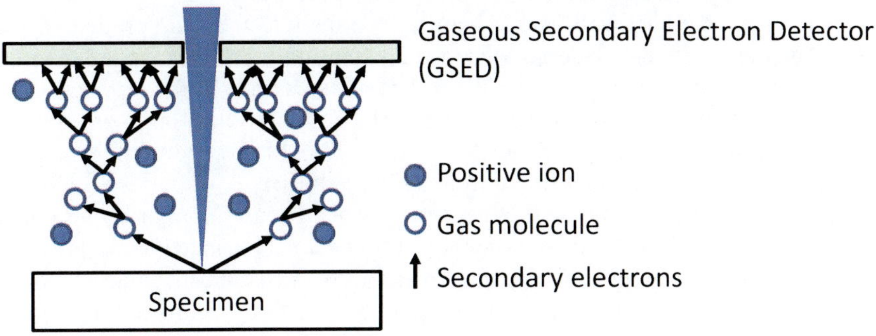

Fig. 3.14 Working principle of GSED under E-SEM condition

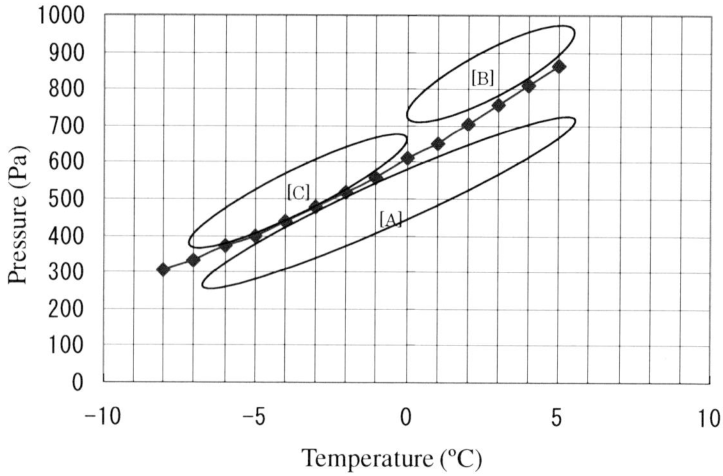

Fig. 3.15 Saturation Water Vapor Pressure Curve [A]: Wet Condition, [B]: Super-saturation state (Water), [C]: Super-saturation state (Ice)

An equally important feature of the E-SEM is the gaseous secondary electron detector (GSED), which replace the customary Everhart-Thornley and other solid state backscattered electron detectors used in conventional SEM as shown in Figure 3.14. The GSED actually relies on the presence of the gas between itself and the sample, as the gas acts as a means of amplifying the signal generated by secondary electrons. The amplification of secondary electrons arises via a gas cascade [59], initiated as the emitted secondary electrons collide with and ionize gas molecules on their way to the positively biased detector. Further electrons are produced as a result, and these will also undergo ionizing collisions, thus propagating the cascade and amplifying the signal. A by-product of the gas cascade is the production of positive ions that fall towards the sample surface, thereby helping to neutralize negative charge build-up. Furthermore, the gas is acting like a conditioning medium that preventing evaporation of liquid from a sample [60].

Hydrated specimens can be stabilized by means of careful control of sample temperature and chamber pressure, in order to maintain a saturated vapor pressure of water above the sample. The temperature control of specimens is achieved by using the Peltier cooling stage. In general, the Peltier cooling stage uses a thermoelectric module to alter temperature. The thermoelectric module is a small wafer composed of PN semiconductor elements. When current passes through these elements, one side of the wafer heats and the opposite side cools. When current polarity reverses, the sides of the wafer reverse heating and cooling. This is referred to as the Peltier effect [58]. The maximum temperature range of the Peltier cooling stage is from -25 °C to +55 °C.

The saturated vapor pressure curve of water in Figure 3.15 indicates the criteria needed for stabilization of the sample. These conditions are particularly suited to the observation of organic/biological specimens in their natural state, especially as many such samples contain water. Specimens may also be deliberately hydrated or dehydrated in situ, by shifting conditions in favour of condensation or evaporation of water. This can be achieved by adjusting temperature and/or pressure.

3.3 Fabrication Techniques in Micro-Nano Scale

3.3.1 Photolithography

Photo lithography is a process of transferring patterns an objects in a micro fabrication. This process is simplified to show as Figure 3.16. First, spin coater is used to coat a resist (photosensitive organic membrane) in constant thickness on the well washed workpiece (base), and s applied to it. While this carries out the spin (rotation) of the base, application material is extended using centrifugal force and film pressure is decided by the concentration of resist, viscosity, and the drying rate of number of rotations and a solvent. The application thickness of typical resist and the relation of number of rotations are shown in a figure.

After a resist coating, pre-baking is operated in order to make a film precise. An oven or a hot plate is used for baking powder. Since the way of a hot plate causes a crack by a film with rise descent of temperature but thick early simple, heat contact changes with the states of the surface of a plate or it is easy to be subject to the influence of the temperature unevenness of a plate, cautions are required.

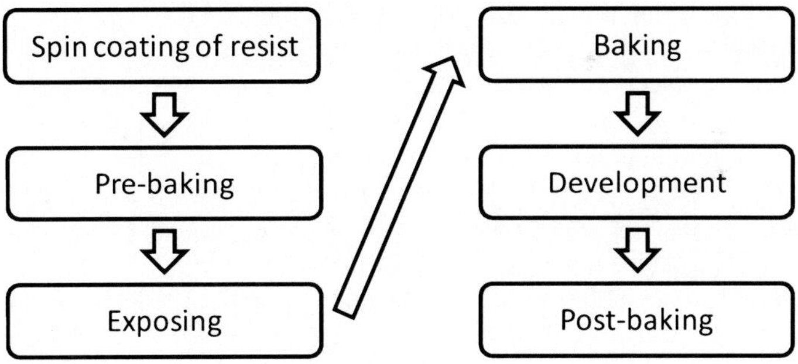

Fig. 3.16 Process of Photolithography

Next, a pattern is transferred by exposure. There is projection exposure which generally carries out as the method of optical exposure using engineering systems, such as proximity exposure performed by 10~30 μm Separating a substrate from a mask, and the adhesion exposure and the mask on which a substrate is pasted up, and a lens, by carrying out reduction projection of the image of a mask. The resolution of projection exposure is proportional to an exposure wavelength. Therefore, short wavelength-ization is following the exposure wavelength to KrF excimer laser light (248 nm) or ArF excimer laser light (193 nm) [31]. Moreover, the light and the electron beam with which improvement in the resolution by the electron beam lithography using not light but an electron beam is also performed bridge the macromolecule in resist, and have the action make it hard to melt into a developing solution, and the action make a link easy to cut and solve conversely. The former used negative resist and the latter is called positive resist. The resist of the portion exposed in the positive resist falls out, a ground is exposed, and it is reverse in negative resist.

The PDMS is siloxane system polymer (rubber) which has a methyl group in a side chain, and it polymerizes by adding the catalyst solution which generally contains platinum (bridge construction). For example, since mobility is high compared with a thermoplastic plastic which is used for injection moulding, it is possible to trespass also upon detailed crevices, such as a seal for waterproofing and insulation of a circuit, and to stiffen them. If the structure of a mold is transferred by applying this character, manufacturing the mold which reversed unevenness of a required micro structure beforehand, and pouring in unpolymerized PDMS there, micro structure can be formed on PDMS.

The concrete process of manufacturing PDMS micro structure by such molding is shown in Figure 3.17. Although what patterned photoresist on the silicon wafer is used as a mold here, It is also possible to use what gave micromachining to the silicon wafer itself by the dry etching (deep reactive ion etching, DRIE). First, on a silicon wafer, the spin coat of the photoresist is carried out and this is patterned. Thereby on a silicon wafer, a convex type structure which reversed the micro channel to actually build is formed. Since it is necessary to thicken the layer of photoresist when manufacturing deep micro channels, such as 50 μm or 100 μm, SU-8 of super-thick film resist, etc. is used well. Processing for making easy to release the mold, such as plasma treatment, is performed to this mold.

The mixed un-polymerized PDMS is solidified by hating. The polymerization initiator is poured in to the mold after processing. Hardened PDMS is torn off from a mold and it sticks on the flat surfaces, such as a glass base. Since PDMS has self-adhesiveness to the flat surface, in the case of the purpose of use which does not need so big internal pressure, it can be used now enough.

In this case, since adhesion on PDMS and a glass base can be desorbed, it can tear off, wash or re-use a chip according to the purpose. Moreover, after processing the adhesion side of silicone rubber with O_2 plasma beforehand to adhere a glass base and PDMS strongly, semipermanent junction is also possible by sticking on a base. In addition, although the method established about processing of the access port for introducing solution etc. on a chip does not exist, how to puncture by punch etc. is common to PDMS which carries out puncturing processing to the glass base etc. beforehand, or was manufactured.

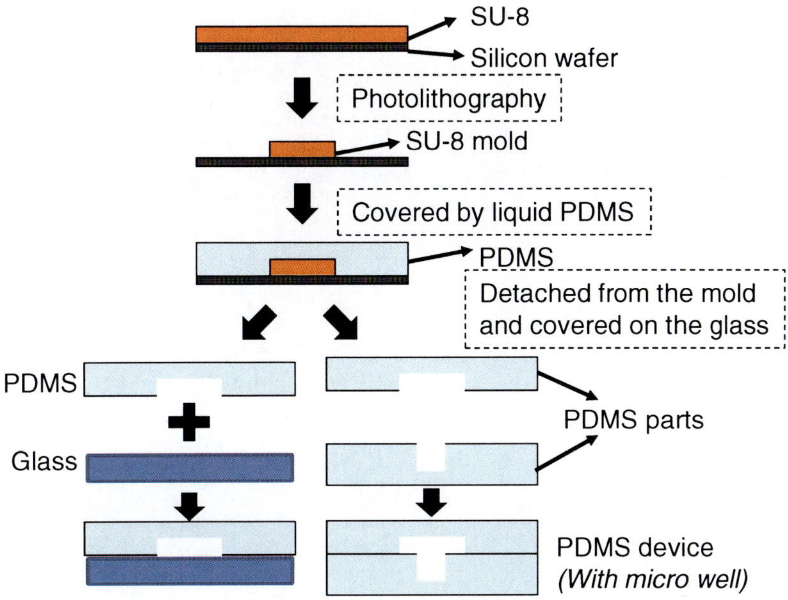

Fig. 3.17 Process of PDMS mold by Photolithography process

3.3.2 Electron-Beam-Induced Fabrication System

Electron-beam-induced fabrication includes deposition and etching based on a FE-SEM. Both of the system has different gas supply system. It is caused by the dissociation of molecules adsorbed to a surface by high energy electrons. The dissociation mechanism is complex and not well understood because of the huge number of excitation channels available even for small molecules. Further complications are induced by the close neighborhood of the substrate. These facts explain why there is no analytical solution for the problem without gross approximations. Therefore, only some special aspects of the decomposition process have been investigated. Dissociative attachment by low energy electrons is, for instance, a very efficient mechanism for the fragmentation of adsorbed molecules.[74] The sequence of destruction steps has been shown in experiments for carbonyl materials.[79]

Direct electron-beam-induced deposition has been demonstrated as a new technique of lithography for many applications. It has developed into "additive lithography" with nanometer resolution employed in scanning electron microscopes, in dedicated lithography systems, in reducing image projection systems, and in scanning tunneling microscopes. The technique allows deposition of nanometer- to micrometer-size structures with nanometer precision in three dimensions without supplementary process steps such as lift-off or etching procedures. Early applications were for

Table 3.3 Materials, characteristics and applications of EBID

Precursor	State of Precursor	Vapor Pressure at RT	Deposit	Application, Achievement	Reference
Pump-oil	Liquid	$<10^{-6}$ Torr	Contamination (Amorphous Carbon)	Masks for ion milling	[68]
WF_6	Gas		W	Tip, letters, nm gates	[69]
$W(CO)_6$	Solid	5-20 mTorr	Metal carbides	STM lithography, 10 nm resolution	[71]
$W(CO)_6$	Solid	17 mTorr	Carbide	Tips, etch masks, mask repair	[70]
$Cr(C_6H_6)_2$	Liquid	10 Torr	Cr carbide		
$Me_2Au(acac)$,	Solid, Liquid	8 mTorr	Au compounds		
$Me_2Au(acac)$,	Solid, Liquid	8 mTorr	Au compounds	3D nanostructures	[72]
Metal acac's	Solid, Liquid		Metal-carbides	Tip probing tips on spotter chip	[73]
$Me_2Au(acac)$,	Solid, Liquid	8 mTorr	Au compounds	3D-AFM probe tips	[74, 77]
Metal acac's	Solid, Liquid		Metal-carbides	STM and AFM tips	[75]
HMDS, HMCTS	Liquids	>10 Torr	Carbon	AFM-tips, insulating, smooth	[76]
Metal acac's	Solid, Liquid		Metal-carbides	3-dimensional deposits	[79]
$Fe(CO)_5$	Liquid		Carbide/oxide	Seeding + thermal deposition	[80]
MIBK, IPA	Liquids	> 10 Torr	Carbonaceous	Dots for stitching marks	[81]
$Mo(CO)_6$	Solid	78 mTorr	Mo, MoC		[82]
$Cu(hfac)_2$	Solid		Cu oxide		[83]
$C_5H_5Pt(CH_3)_3$	Solid	54 mTorr	Pt compounds		[82]

metrology of probe astigmatism and image storage. Recent applications include etch masks for ion milling [75], and the first metal nanostructures [76]. Other experiments have shown the ability to grow material across a whole perpendicular to the beam. Furthermore the superior quality of deposited lines as etch masks for reactive ion etching (RIE) has been demonstrated. [77] Employing additive lithography in a scanning tunneling microscope (STM) rendered lines and dots of a few nanometers in size.[78] Employing the technique to construct high-aspect-ratio probes with tips in x-, y-, and z-directions enabled 3-dimensional metrology with nm precision using atomic force microscopes (AFM) to characterize deep trenches. [79] In a similar manner, tips for imaging AFM- and STM-probes were produced, placed onto a spotter chip, where the tips could be located in micrometer-sized fields addressed with a barcode. [80] Probe tips for scanning force microscope (SFM) for differing metrology tasks and with custom-designed tip radius are now routinely applied. [81-84]. Table 3.3 summarizes the materials used for electron induced deposition (EBID), their characteristics and applications.

In our study, electron-beam-induced deposition is investigate with two kinds of organometallic compounds including tungsten hexacarbonyle ($W(CO)_6$) and dimethyl-gold ($(CH_3)_2(C_5H_7O_2)Au$) as precursors. $W(CO)_6$ is white crystalline powder at room temperature and can sublime at this temperature. The properties of it are listed in Table 3.4 and that of dimethyl-gold are listed in Table 3.5.

Table 3.4 Physical and chemical characteristics of $W(CO)_6$

Items	Data	Source of Data
Chemical Name	Tungsten Hexacarbonyl	Aldrich Inc. Material Safety Data Sheet (MSDS)
CAS Registry Number	14040-11-0	Gelest Inc. MSDS
Appearance and Odor	White Crystalline Powder	Gelest Inc. MSDS
Molecular Weight	351.92	Gelest Inc. MSDS
Specific Gravity	2.650 g/cm^3	Aldrich Inc. MSDS
Melting Point	150°C	Aldrich Inc. MSDS
Boiling Point	451.3 °C @ 760.00mmHg	Acros Organics MSDS
Flash Point, COC	200°C	Gelest Inc. MSDS
Autoignition Temp.	300°C	Gelest Inc. MSDS
Flammability Limits- LEL	5.7 g/m^3	Gelest Inc. MSDS
Vapor Pressure	1.2 mm Hg@67°C	Aldrich Inc. MSDS
Vapor Density (Air=1)	12.1	Aldrich Inc. MSDS
Solubility in water	Insoluble	Gelest Inc. MSDS

Table 3.5 Physical and chemical characteristics of $(CH_3)_2(C_5H_7O_2)Au$

Items	Data	Source of Data
Chemical Name	Dimethyl(acetylacetonate)gold(III)	Strem Chemicals, Inc. Material Safety Data Sheet (MSDS)
CAS Registry Number	14951-50-9	Strem Chemicals, Inc. MSDS
Appearance and Odor	White to off-white xtl.	Strem Chemicals, Inc. MSDS
Molecular Weight	326.6	Strem Chemicals, Inc. MSDS
Boiling Point	subl. ~25°C/0.01mm	Strem Chemicals, Inc. MSDS
Decomposition Products	Carbon monoxide, carbon dioxide, and gold	Strem Chemicals, Inc. MSDS
Solubility in water	Reacts with water	Strem Chemicals, Inc. MSDS

3.3.3 *Focused Ion Beam System*

The focused ion beam system uses a gallium ion beam to raster over the surface of a sample in a similar way as the electron beam in a scanning electron microscope. The generated secondary electrons are collected to form an image of the surface of the sample. The ion beam allows the milling of small holes in the sample at well localized sites, so that cross-sectional images of the structure can be obtained or that modifications in the structures can be made as shown in Figure 3.18. The FIB system has a 5-axis motorized eucentric tilting stage and can work from 5 to 30 kV. It achieves the smallest beam diameter of 4 nm at 30 kV. Expect the scanning ion microscope function, sputter etching function the system also has the deposition function by squirting source gas and ion beam irradiation. The deposits can be carbon or tungsten.

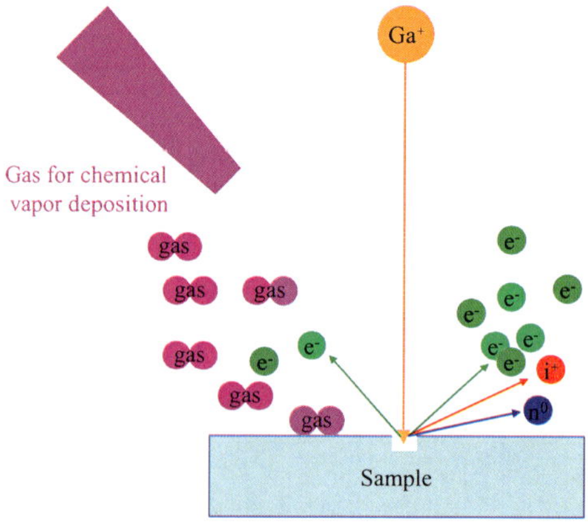

Fig. 3.18 Principle of the focused ion beam (FIB) operation

All of the nanoprobes used in the single cell analysis experiments reported in this thesis were fabricated using focused ion beam (FIB) techniques. FIB techniques was used to modified the tip of standard AFM cantilever to arrive at some specific shapes, such as flat tip, nanoneedle, buckling nanoneedle, nanofork and dual nanoprobe.

In general, the FIB instrument is similar to a scanning electron microscope (SEM), except that the beam that is rastered over the sample is an ion beam rather than an electron beam. Secondary electrons are generated by the interaction of the ion beam with the sample surface and can be used to obtain high-spatial-resolution images [70]. The most popular ion source being used is the Liquid-metal ion sources, especially gallium (*Ga*) ion sources.

Basically, there are three main operations of the FIB, i.e. imaging, etching and deposition. As can be seen in Figure 3 18, the gallium (Ga^+) primary ion beam, which accelerated with an energy of 30 keV, hits the sample surface and sputters a small amount of material, which leaves the surface as either secondary ions (i^+ or i^-) or neutral atoms (n^0). The primary beam also produces secondary electrons (e^-). As the primary beam rasters on the sample surface, the signal from the sputtered ions or secondary electrons is collected to form an image. When the high-energy Ga^+ strikes the sample, they will sputter atoms from the surface, the so called etching operating. FIB can also be used to deposit material via ion beam induced deposition. FIB-assisted chemical vapor deposition (CVD) occurs when a gas, such as tungsten hexacarbonyl ($W(CO)_6$) is introduced to the vacuum chamber and allowed to chemisorb onto the sample. By scanning an area with the beam, the precursor gas will be decomposed into volatile and non-volatile components; the non-volatile component, such as tungsten, remains on the surface as a deposition. The nanoprobes were fabricated using FIB's etching and/or deposition. The nanoprobes can be categorized into two, i.e. soft nanoneedle (buckling nanoneedle) and hard nanoneedle.

3.3.4 Nano-imprinting System

The thickness and height of the line structure are about 1 μm each. The line structure on the substrate was fabricated using nanoimprinting lithography technology. The fabrication steps of the line-patterned substrate are shown in Figure 3.19. In general, there are two steps in nanoimprinting process. The first step is setup and heating of the resist to above glass transition temperature (T_g). In our experiment, the resist is based on a polymer material. The mold is then imprinting and holds on the resist. The second step is cooling of the resist temperature below the T_g. Figure 3.20 (a) shows optical microscope and SEM images of the line-patterned substrate which were successfully fabricated.

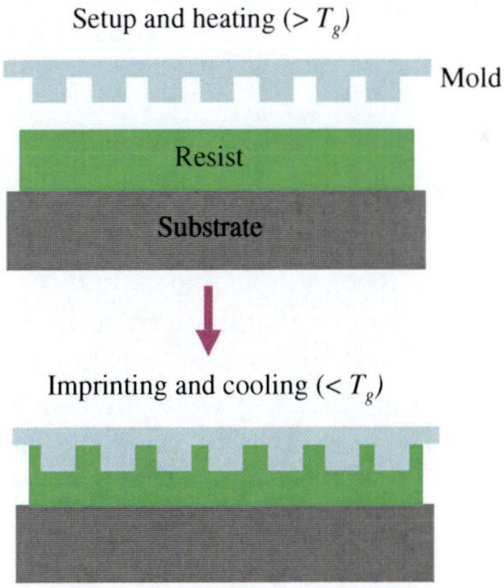

Fig. 3.19 Fabrication of the line-patterned substrate using nanoimprinting technique

In order to successfully observe the line-patterned substrate under SEM, the substrate needs to be coated with a conductive material. The conductive coating is needed to minimize, if not eliminate, the charging effect which prevent clear observations under SEM. Figure 3.20 (b) shows a SEM image of the line-patterned substrate before coating with Au material. It can be noticed that the charging effect reduces the clarity of the SEM image of a non-conductive line-patterned substrate. In order to make the line-patterned substrate conductive, gold (Au) material was coated on the line-patterned substrate. The Au coating on the line-patterned substrate was performed using sputtering technique. The thickness of the Au coating on the line-patterned substrate is about 10 nm which determined by controlling the sputtering time, i.e. 10 nm/minute. The Au coating on the line-patterned substrate improves the clarity of the SEM image as shown in Figure 3.20 (c).

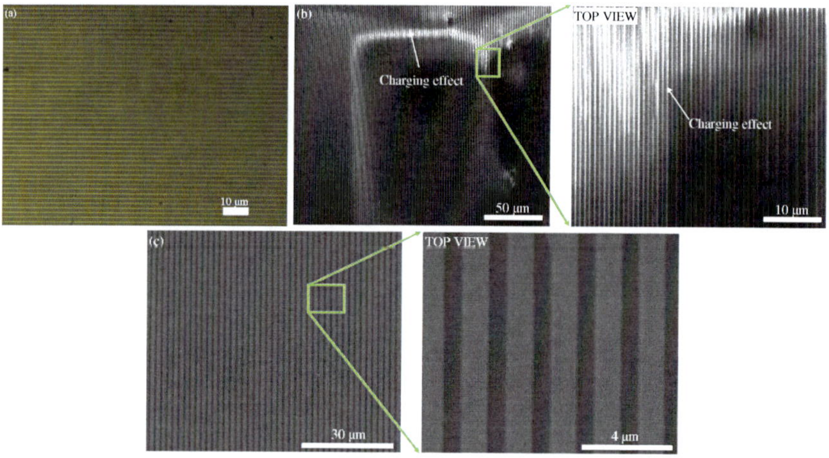

Fig. 3.20 Images of the line-patterned substrate observed under (a) optical microscope and (b)-(c) SEM. (b) shows charging effect on SEM images of non-conducive line-patterned substrate. (c) shows the improved quality of the line-patterned substrate after coating with Au material.

3.3.5 Self-Assembly Techniques

Self-assembly is a technique to assemble the structures with multi-components organized in an order under a specific condition by itself. This type of phenomenon is widely observed in nature such as patterns of clouds, skin pattern of animal, crystal of ice and so on. From engineering view point, this technique is effectively used to assemble the designed structures for example, as shown in Figure 3.23 [90]. A fluidics self-assembly method was proposed with a unique shape of perforated brocks [91]. The simulation and preliminary

3.4 Sensing and Actuation in Micro-Nano Scale

3.4.1 Sensing in Micro-Nano Scale

The sensing mechanisms are summarized in Figure 3.21 and Table 3.6 as examples. The local and high-accuracy sensing mechanisms are important for the micro-nano mechatronics applications. To select the sensing mechanism, the various limitations need to be considered such as 1) environments (wet/dry/vacuum…, sterile or non-sterile environments…), 2) space, and 3) energy supply limitations. To construct the control system, the sensing mechanism needs to be combinable with actuation mechanism as micro-nano mechatronics system.

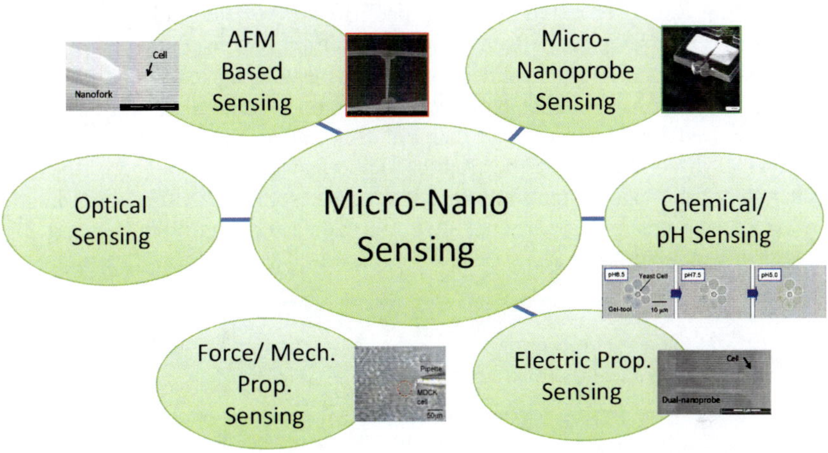

Fig. 3.21 Examples of micro-nano sensing technology

Table 3.6 Examples of sensing mechanisms in micro-nanometer scale

Micro-Nano Sensing Categories	Examples
Positioning Sensing	Laser meter, Ultrasonic sound meter, Piezo sensing, Microscopic imaging, …
Force Sensing	Strain gauge, Electric capacitance, Electro-static gauge, Piezo cantilever…
Electromagnetic Sensing	Flow-meter, Particles charges, Photo-detectors …
Chemical Sensing	Gas sensing, Ionic sensing, Biosensing, Bioindicator,…
pH Sensing	Indicator, Potential difference mechanism, Field effect transistor, …
Temperature Sensing	Indicator, Infra-red intensity, Resistive change, …
Cell Activity Sensing	Indicator, Fluorescent dye, Fluorescent protein, Physical/Chemical interactions, …

3.4.2 Actuation in Micro-Nano Scale

The actuation mechanisms are summarized in Figure 3.22 and Table 3.7 as examples. From the view point of physical interactions, the actuation technology is divided into contact or non-contact ways mainly. For micro-nanomanipulation system, the piezo driven mechanism are widely used because of their good performance in the positioning accuracy. For the biological cell manipulation, the manipulation system is required to approach to the targeted cells under fluidics which is needed to keep the cells in a viable condition. By the contact manipulation, the fluidics surrounding the targeted cell is affected during manipulation to make some turbulence. On the other hand, the non-contact manipulation mechanism is effectively used for biological cell manipulation under wet environments without the physical direct contact to the fluidics. The electrical and magnetically driven mechanisms have lager actuation forces by comparison with optical manipulation. However, the optical manipulation is superior on the local auction of targeted object by its narrow actuation range.

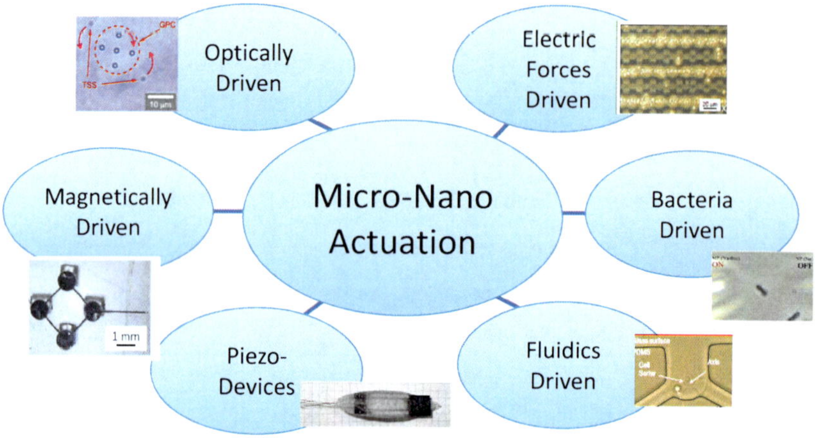

Fig. 3.22 Examples of micro-nano actuation technology

Bio Actuators

Bio actuators are natural actuators in micro-nanometer -meter scale. Their energy sources are chemical reactions. So, there is the possibility to downsize the actuator and power supply system for the micro-nanorobots.

The one of the most important feature of bio-actuator is the efficiency of actuation. The sizes of biomaterial is nano-meter scale fine structures for example actin filaments: ~6-7 nm (dia.), myosin: ~30 nm (dia.), kinesin: ~80 nm (dia.), flagellar motor: ~40 nm (Ring dia.) and so on. The energy efficiency of bioactuator is almost 100% in the flagellar motor case, higher than 70 % in the muscle system (Actin-Myosin). On the other hand, the internal-combustion engine is less than 40 %. This difference is generally caused by generation though the steps Chemical Energy ⇒ Thermal Energy ⇒ Mechanical Energy in the Macro world, but in Nano world (Biological Actuator) the steps is only Chemical Energy ⇒ Mechanical Energy.

Table 3.7 Examples of actuation mechanisms in micro-nanometer scale

Micro-Nano Actuation Mechanism	Physical interaction	Accuracy	Actuation range	Durability	Environmental Condition
Electrical Motor Driven	Contact	++	+++	+++	Dry
Hydraulic pressure Driven	Contact	++	+++	+	Dry
Piezo-Driven	Contact	+++	+	++	Dry
Biological Cell Driven	Contact	+	+	+	Wet
Bacteria Driven	Contact	+	+	+	Wet
Optically Driven	Non-contact	+++	+	++	Dry, Wet
Electric Driven	Non-contact	++	++	++	Dry, Wet
Magnetically Driven	Non-contact	++	++	++	Dry, Wet

The micro-nanometer scale actuation systems using the bio actuators have been already proposed and constructed. For examples, uni-directional movement of DNA activated by enzymatic ligation and specific scission processes (called "DNA walker") [92, 93], alignment motion of actin filament by myosin-rich/myosin-poor surfaces' boundary [94], one dimensional myosin-driven motion of actin filaments with patterned surface of positive polyelectrolytes [95], selective movements of chemically cross-linked polymer-actin complex gel on a myosin coated surface [96] or a chemically cross-linked polymer-myosin gel [97] directional selective movement of kinesin on microtubule by chemical switch [98] and the biased movements of kinesin-driven microtubule along the microlithographic tracks with arrowhead shapes as rectifiers [99] have been proposed and demonstrated. In these researches, the bio actuators successfully worked and the moving direction was controlled. However, nothing (but the transporter itself) was transported by the movements of the biomolecular motors.

Bacteria as Bio Actuators

The intact bacteria have advantages to be used as actuators for the micro-nanorobots. In the first place, the bacteria self-propagate unlike the molecular motors which need the reconstitution. Many bacteria, such as *Escherichia coli (E. coli) and Vibrio alginolyticus (V. alginolyticus)*, can multiply in a few dozen minutes to a few hours time scale depending on the environmental condition, such as temperature and nutrients. So, the mass production of the actuators is quite easy and low time-cost. And the bacteria have the metabolic function. The micro-nano electro mechanical actuators are easily impaired in a short period with the friction of the rotational or the sliding motion. And, the biomolecular motors are also easily impaired in the short period [100]. So, the self-repairing function of the actuators is necessary for the actual utilization of the micro-nanorobots. The bacteria can produce the proteins and can replace the new proteins, i.e. self-repairing, by their metabolic function.

Second, as mentioned before, the bacteria have the biomolecular motors, bacterial flagellar motors, which convert chemical reactions into mechanical energy, i.e. propulsion, near 100 % energy conversion efficiency at maximum. The internal-combustion engines convert the fuels (thermal energy) into the driving force in low efficiency around 30~40 %. The electric motors convert the electric power into the driving force ~90 % at maximum. However, there is a lot of energy loss in the electric power generation when the thermal energy is converted into the electric energy. Therefore, the biomolecular motors, especially the bacterial flagellar motors, excel in the energy efficiency.

Third, most of the bacteria naturally have sensors and systems to regulate the propulsion in response to the environment. Bacteria do not move randomly but have a taxis system to move towards or away from environmental stimuli [101], such as chemical substances (chemotaxis), temperature (thermotaxis), light (phototaxis), magnetic field (magnetotaxis) and so on. For all of these reasons, the intact bacteria are one of the promising candidates to be used as the actuators for the micro-nano objects.

In addition, some of the flagellated bacteria such as *E. coli* and *V. alginolyticus* were genetically and behaviourally studied well. These bacteria can be modified genetically to add or delete some functions depending on the requirements for the micro-nanorobots. The taxis system is well characterized. So, it is expected that the taxis system can be utilized to control the bacterial propulsion artificially. Therefore, I focused on the flagellated bacteria as the actuators for the micro-nanorobots.

So, the intact bacteria have attracted much attention as easy-to-implement bio actuators for the micro-nanorobots. The actuations of the micro-nano scale objects using the intact bacteria were already reported, for example, the microrotary motor was driven by the gliding bacterium *Mycoplasma mobile (M. mobile)* [102]

3.5 Control Techniques in Micro-Nano Scale

With the advancement of nanotechnology, the objects to be manipulated become smaller and smaller, and recently entered nanometer scale from micrometer scale. Nanomanipulation, or position control at the nanometer scale, is one of the most promising ways towards molecular nanotechnology. Although AFM is capable of suitably small motions (tenths of an angstrom), but cannot repeatedly position to, and hold, sub-angstrom accuracy due to hysteresis and drift. It is also currently limited to three degrees of freedom, with no orientation control. As a scanning probe, it works quite well; as a construction device for nanotechnology, it is practically limited to 2D plane.

It is much significant to manipulate nanometer scale objects in 3D space for constructing nano structures and devices. In order to realize such manipulations, robotic manipulators with nanometer resolutions will be useful tools. The basic requirements for a nanorobotic manipulator for 3D manipulations include nano-scale positional resolution, a relative large working space, enough DOFs for 3D positioning of the end effectors, and usually multi end-effectors for complex operations and necessary measurements. One of the key techniques will be how to design the end-effectors because of the specialty of the physical phenomena in micro-nano order world. It has been shown that it is extremely difficult to apply a micro gripper to realize the pick-and-place operations for nano-order objects because the electromagnetic interactions between the gripper and the objects become larger than gravity that cause the departure of the object from the gripper to be difficult or even impossible. Hence, it is a much more promising strategy to realize nano manipulation by controlling the interactions between tools and the objects rather than by using a gripper. Several strategies for controlling the interactions have been presented [103,104].

3.5.1 Master-Slave Control System for Micro-Nano Manipulation

For control of micro-nano manipulator, a "master-slave control method" is widely applied. Because, in many cases, 1) the space for manipulator is limited caused by imaging system for objects, 2) the environment of sample are set in certain temperature, pressure, and so on. The master-slave control method is basically the human operates a master system as input device and the micro-nano manipulator is worked as a slave system. In this situation, the micro-nano manipulator is working in a small scale, on the other hand the human is operating in a macro scale. Hence, one of the fundamental problems is the scale ratio of position and force is different from manipulation and operation spaces. In here, we did not discuss on the general problem such as time-delay problems to focus on the fundamental problems.

Generally, the master-slave control method is divided into two types, one is unilateral control and the other is bilateral control. The bilateral control has some merits especially the operator can find a reaction force during manipulation.

When a 6 axis force sensor is built into the 3 axis master arm, the applying forces by operator is possible to measure. Based on this configuration, the bilateral control system was designed based on the parallel type bilateral control system under the condition which the phase lag is small by single manipulator as shown in Figure 3.23.This system is designed to compute the target displacement magnitude by a master scale and multiplying the difference of the value which multiplied the power which the operator applied to the master, and the reaction force committed at the tip of a tool by the power feedback ratio by a force gain.

About the master side, the torque of DC motor is computed by multiplying the target displacement magnitude by a force gain. At the slave side, a target position is computed from target displacement magnitude, and the position command voltage for multiplying by a position gain, making it a slave scale with movement transmissibility, and making piezo-drive is impressed.

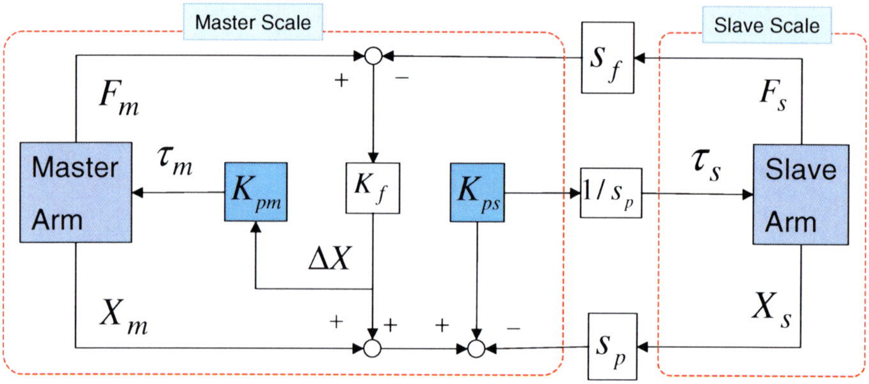

Fig. 3.23 Block diagram of bilateral control system

F_m, F_s: Sensor output of master/slave
X_m, X_s: Position of master/slave
ΔX: Target movement
τ_m, τ_s: Control input for master/slave
S_f, S_p: Transmissibility of force/position

The driving forces of a master and a slave are given by the following equations.

$$\tau_m = K_{pm}\Delta X \tag{3.2}$$

$$\tau_s = K_{ps}(X_m + \Delta X - s_p X_s)/s_p + X_s \tag{3.3}$$

The target displacement magnitude is given by the following equation,

$$\Delta X = K_f(F_m - s_f F_s) \tag{3.4}$$

Since the master of a flexible system and the dynamics of a slave are given by the following formula,

$$\tau_m + F_m = M_m \ddot{X}_m + B_m \dot{X}_m \tag{3.5}$$

$$\tau_s - F_s = M_s \ddot{X}_s + B_s \dot{X}_s \tag{3.6}$$

M_m, M_s: mass of master/slave
B_m, B_s: friction forces of master/slave

From a formula, the flattery nature of power and a position is given by the following formula from (3.2) (3.3).

$$F_m = \frac{K_{pm}K_f}{1 + K_{pm}K_f} s_f F_s + \frac{1}{1 + K_{pm}K_f}(M_m \ddot{X}_m + B_m \dot{X}_m) \tag{3.7}$$

$$X_m - s_p X_s = \frac{1}{K_{ps}} s_p \{(M_s \ddot{X}_s + B_s \dot{X}_s) + F_s\} - K_f (F_m - s_f F_s) \quad (3.8)$$

From (3.7), the slave is tracked the master forces by $F_m - s_f F_s \to 0$ at $K_{pm} K_f \to \infty$. It becomes if it carries out and a slave follows the power of a master. Furthermore from (3.8), the slave is tracked the master positions by $X_m - s_p X_s \to 0$ at $K_{ps} \to \infty$ and $k_f \to 0$.

3.5.2 Control of Master-Slave Control System for Laser Manipulation

The laser trapping force needs to calculate for the bilateral control for laser manipulation. To measure a laser trapping constant, some techniques are proposed for example, 1) to measure the releasing force by known external force, 2) to calculate by Stork's viscosity resistance force, 3) to calculate by Brownian motion. After determining the laser trapping constant, it is needed to measure the displacement of center position between focusing potion and laser beam position by photo-diode or CCD camera image.

It is possible to construct control system of laser trap by the parallel type bilateral control, however the noise in measurement force is not easy to reduce, because the laser trapping force is normally pN order. The block diagram of force accommodation bilateral control is shown in Figure 3.24. The parameters are as follows.

X_s: Focusing position of laser beam
Z: Center position of particle trapped by laser beam
X: Displacement of center position between focusing potion and laser beam position
X_m, X_s: Displacement of master/slave
F_s: Applying force to object by slave
τ_m, τ_s: Control input for master/slave
S_f, S_p: Transmissibility of force/position

There is no typical problem for operability, when we design the master system with better backdrivability.

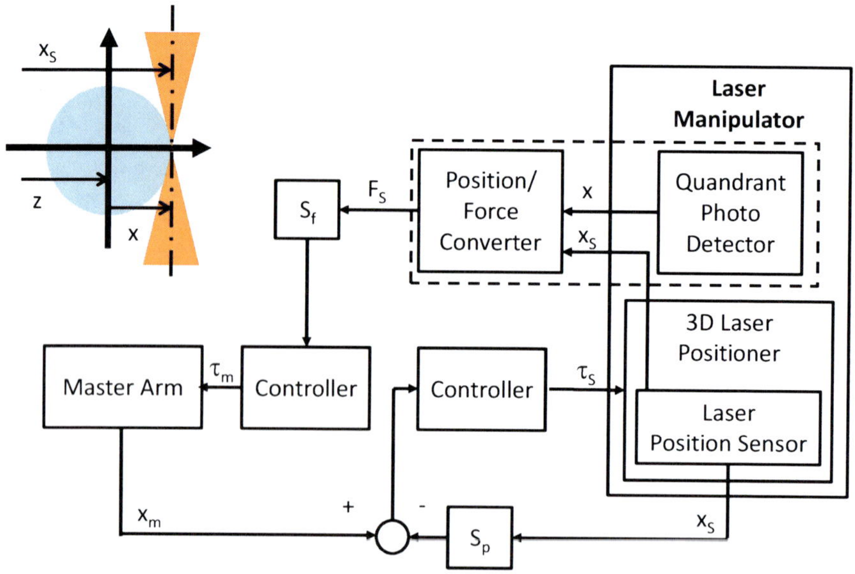

Fig. 3.24 Block diagram of bilateral control system for laser manipulation system

3.6 Assembly Techniques in Micro-Nano Scale

3.6.1 2D Assembly Technique in Micro-Nano Scale

2D manipulations of nanotubes on a surface are first performed with AFMs by contactive pushing on a substrate. Figure 3.25 shows the typical methods for 2D pushing. Although similar to that shown in Figure 3.25 (b), the same manipulation caused various results because nanotubes cannot be regarded as a 0D point. The first demonstration was given by Lieber and co-workers as measuring the mechanical properties of a nanotube [105]. The method was used i.e., to bend a nanotube by pushing one end of it and fixing the other end. The same strategy was used for the investigation of the behaviours of nanotubes under large strain [106]. Dekker and co-workers applied the strategies shown in Figure 3.25 (c) and (d) to get a kinked and crossed junction of nanotubes [107,108]. Avouris and co-workers combined this technique with an inverse process, namely straightening, by pushing along a bent tube, and realized the translation of the tube to another place [109]. This technique was also used to place a tube on another tube to form a SET with cross-junction of nanotubes [110]. Pushing induced breaking (Figure 3.25 (d)) has also been demonstrated for an adsorbed nanotube. The simple assembly of two bent tubes and a straight one formed a Greek letter "θ". To investigate the dynamics of rolling at the atomic level, rolling and sliding of a nanotube (as shown in Figure 3.25 (e) and (f)) are performed on graphite surfaces using an AFM [111, 112].

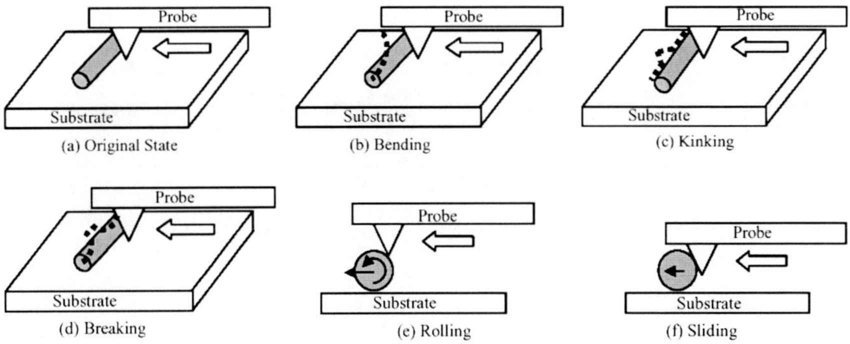

Fig. 3.25 Schematic diagram and example of 2D nanomanipulation of carbon nanotube

3.6.2 3D Assembly Technique in Micro-Nano Scale

Manipulations of CNTs in 3D space are very important techniques for assembling them into structures and devices. The basic step for this is to pick up a single tube from nanotube soot (Figure 3.26 (a)). This has been shown first by using dielectrophoresis through nanorobotic manipulations [113] (Figure 3.26 (b)). The interaction between a tube and the atomic flat surface of AFM cantilever tip has been shown to be strong enough for picking up a tube onto the tip [114] (Figure 3.26 (c)). By using electron-beam-induced deposition (EBID), it is possible to pick up a nanotube onto a probe [115,116, 117] (Figure 3.26 (d)). For handling a tube, weak connection between the tube and the probe is desired. Bending and buckling a CNT as shown in Figure 3.26 (e) and (f) are important for the in situ property characterizations of a nanotube [118], which is a simple way to get the Young's modulus of a nanotube without damaging the tube if performed in elastic range and hence can be used for the selection of a tube with desired properties. Plastic bending or buckling can generate intramolecular kink junctions of CNTs [119, 120]. Combined bending and buckling with shape fixing with EBID can be used for the shape modifications of a nanotube. Stretching a nanotube between two probes or a probe and a substrate has brought out several interesting results. The first demonstration of 3D nanomanipulations of nanotubes took this one as an example to show the breaking mechanism of a MWNT [109]. By breaking a MWNT in a controlled manner, some interesting nanodevices have been fabricated. This technique—destructive fabrication—has been presented to get sharpened and layered structures of nanotubes, and to improve the length control of a nantube [111]. Bearing motion has also been observed in an incompletely broken MWNT. The reverse process, namely the connection of broken tubes, has been demonstrated, and the mechanism is revealed as rebonding of unclosed dangling bonds at the ends of a broken tube [121]. Based on this interesting phenonmenon, mechanochemical nanorobotic manipulations have been presented [122].

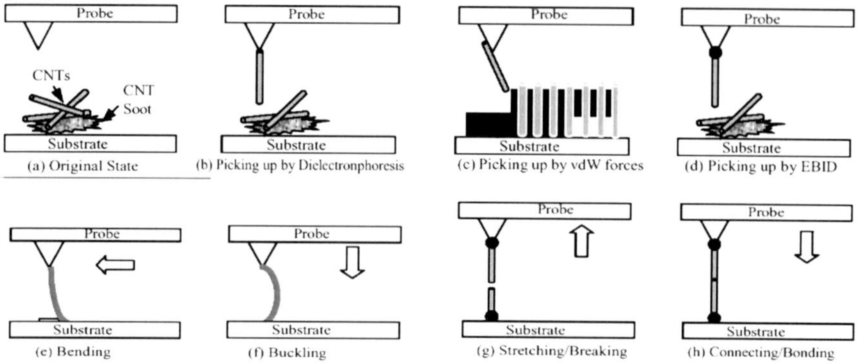

Fig. 3.26 Schematic diagram and example of 3D nanomanipulation of carbon nanotube

References

[1] Iijima, S.: Helical microtubules of graphitic carbon. Nature 354, 56–58 (1991)

[2] Geim, A.K., Novoselov, K.S.: Nature Materials 6, 183 (2007)

[3] Iijima, S., Ichihashi, T.: Single-Shell Carbon Nanotubes of 1-nm Diameter. Nature 363, 603–605 (1993)

[4] Bethune, D.S., Kiang, C.H., De Vries, M.S., Gorman, G., Savoy, R., Vazquez, J., Beyers, R.: Cobalt-Catalysed Growth of Carbon Nanotubes with Single-Atomic-Layer Walls. Nature 363, 605–607 (1993)

[5] Wang, N., Tang, Z.K., Li, G.D., Chen, J.S.: Single-Walled 4Å Carbon Nanotube Arrays. Nature 408, 50–51 (2000)

[6] Pan, Z.W., Xie, S.S., Chang, B.H., Wang, C.Y., Lu, L., Liu, W., Zhou, W.Y., Li, W.Z., Qian, L.X.: Very Long Carbon Nanotubes. Nature 394, 631–632 (1998)

[7] Zhu, H.W., Xu, C.L., Wu, D.H., Wei, B.Q., Vajtai, R., Ajayan, P.M.: Direct Synthesis of Long Single-Walled Carbon Nanotube Strands. Science 296, 884–886 (2002)

[8] Saito, R., Dresselhaus, G., Dresselhaus, M.S.: Physical Properties of Carbon Nanotubes. ImperialCollege Press, London (1998)

[9] Baughman, R.H., Zakhidov, A.A., de Heer, W.A.: Carbon Nanotubes—the Route Toward Applications. Science 297, 787–792 (2002)

[10] Treacy, M.J., Ebbesen, T.W., Gibson, J.M.: Exceptionally High Young's Modulus Observed for Individual Carbon Nanotubes. Nature 381, 678–680 (1996)

[11] Wong, E.W., Sheehan, P.E., Lieber, C.M.: Nanobeam Mechanics: Elasticity, Strength, and Toughness of Nanorods and Nanotubes. Science 277, 1971–1975 (1997)

[12] Poncharal, P., Wang, Z.L., Ugarte, D., de Heer, W.A.: Electrostatic Deflections and Electro- mechanical Resonances of Carbon Nanotubes. Science 283, 1513–1516 (1999)

[13] Yu, M.F., Lourie, O., Dyer, M.J., Moloni, K., Kelley, T.F., Ruoff, R.S.: Strength and Breaking Mechanism of Multiwalled Carbon Nanotubes under Tensile Load. Science 287, 637–640 (2000)

[14] Krishnan, A., Dujardin, E., Ebbesen, T.W., Yianilos, P.N., Treacy, M.M.J.: Young's Modulus of Single-Walled Nanotubes. Phys. Rev. B 58, 14013–14019 (1998)

[15] Salvetat, J.-P., Briggs, G.A.D., Bonard, J.-M., Bacsa, R.R., Kulik, A.J., Stockli, T., Burnham, N.A., Forro, L.: Elastic and Shear Moduli of Single-Walled Carbon Nanotube Ropes. Phys. Rev. Lett. 82, 944–947 (1999)

[16] Yu, M.F., Files, B.S., Arepalli, S., Ruoff, R.S.: Tensile Loading of Ropes of Single Wall Carbon Nanotubes and Their Mechanical Properties. Phys. Rev. Lett. 84, 5552–5555 (2000)

[17] Walters, D.A., Ericson, L.M., Casavant, M.J., Liu, J., Colbert, D.T., Smith, K.A., Smalley, R.E.: Elastic Strain of Freely Suspended Single-Wall Carbon Nanotube Ropes. Appl. Phys. Lett. 74, 3803–3805 (1999)

[18] Saito, R., Fujita, M., Dresselhaus, G., Dresselhaus, M.S.: Electronic Structure of Graphene Tubules Based on C60. Phys. Rev. B 46, 1804–1811 (1992)

[19] Ebbesen, T.W., Lezec, H.J., Hiura, H., Bennett, J.W., Ghaemi, H.F., Thio, T.: Electrical Conductivity of Individual Carbon Nanotubes. Nature 382, 54–56 (1996)

[20] Dai, H.J., Wong, E.W., Lieber, C.M.: Probing Electrical Transport in Nanomaterials: Conductivity of Individual Carbon Nanotubes. Science 272, 523–526 (1996)

[21] Liang, W.J., Bockrath, M., Bozovic, D., Hafner, J.H., Tinkham, M., Park, H.: Fabry-Perot Interference in a Nanotube Electron Waveguide. Nature 411, 665–669 (2001)

[22] Frank, S., Poncharal, P., Wang, Z.L., de Heer, W.A.: Carbon Nanotube Quantum Resistors. Science 280, 1744–1746 (1998)

[23] Kim, P., Shi, L., Majumdar, A., Mceuen, P.L.: Thermal Transport Measurements of Individual Multiwalled Nanotubes. Phys. Rev. Lett. 87, 215502 (2001)

[24] Dai, H.J., Hafner, J.H., Rinzler, A.G., Colbert, D.T., Smalley, R.E.: Nanotubes as Nanoprobes in Scanning Probe Microscopy. Nature 384, 147–150 (1996)

[25] Hafner, J.H., Cheung, C.L., Lieber, C.M.: Growth of Nanotubes for Probe Microscopy Tips. Nature 398, 761–762 (1999)

[26] Nishijima, H., Kamo, S., Akita, S., Nakayama, Y., Hohmura, K.I., Yoshimura, S.H., Takeyasu, K.: Carbon-Nanotube Tips for Scanning Probe Microscopy: Preparation by a Controlled Process and Observation of Deoxyribonucleic Acid. Appl. Phys. Lett. 74, 4061–4063 (1999)

[27] Hafner, J.H., Cheung, C.-L., Oosterkamp, T.H., Lieber, C.M.: High-Yield Assembly of Individual Single-Walled Carbon Nanotube Tips for Scanning Probe Microscopies. J. Phys. Chem. B 105, 743–746 (2001)

[28] Kim, P., Lieber, C.M.: Nanotube Nanotweezers. Science 286, 2148–2150 (1999)

[29] Cumings, J., Zettl, A.: Low-Friction Nanoscale Linear Bearing Realized from Multiwall Carbon Nanotubes. Science 289, 602–604 (2000)

[30] Cumings, J., Collins, P.G., Zettl, A.: Peeling and Sharpening Multiwall Nanotubes. Nature 406, 586 (2000)

[31] Kong, J., Franklin, N.R., Zhou, C.W., Chapline, M.G., Peng, S., Cho, K.J., Dai, H.J.: Nanotube Molecular Wires as Chemical Sensors. Science 287, 622–625 (2000)

[32] Tans, S.J., Verchueren, A.R.M., Dekker, C.: Room-Temperature Transistor Based ona Single Carbon Nanotube. Nature 393, 49–52 (1998)

[33] Bachtold, A., Hadley, P., Nakanishi, T., Dekker, C.: Logic Circuits with Carbon Nanotube Transistors. Science 294, 1317–1320 (2001)

[34] Saito, R., Dresselhaus, G., Dresselhaus, M.S.: Tunneling Conductance of Connected Carbon Nanotubes. Phys. Rev. B 53, 2044–2050 (1996)

[35] Chico, L., Crespi, V.H., Benedict, L.X., Louie, S.G., Cohen, M.L.: Pure Carbon Nanoscale Devices: Nanotube Heterojunctions. Phys. Rev. Lett. 76, 971–974 (1996)

[36] Menon, M., Srivastava, D.: Carbon Nanotube 'T Junctions': Nanoscale Metal-Semiconductor-Metal Contact Devices. Phys. Rev. Lett. 79, 4453–4456 (1997)

[37] Yao, Z., Postma, H.W.C., Balents, L., Dekker, C.: Carbon Nanotube Intramolecular Junctions. Nature 402, 273–276 (1999)

[38] Postma, H.W.C., Teepen, T., Yao, Z., Grifoni, M., Dekker, C.: Carbon NanotubeSingle-Electron Transistors atRoom Temperature. Science 293, 76–79 (2001)

[39] Fuhrer, M.S., Nygård, J., Shih, L., Forero, M., Yoon, Y.-G., Mazzoni, M.S.C., Choi, H.J., Ihm, J., Louie, S.G., Zettl, A., Mceuen, P.L.: Crossed Nanotube Junctions. Science 288, 494–497 (2000)

[40] Rueckes, T., Kim, K., Joselevich, E., Treng, G.Y., Cheung, C.L., Lieber, C.M.: Carbon Nanotube-Based Nonvolatile Random Access Memory for Molecular Computing Science. Science 289, 94–97 (2000)

[41] Franklin, N.R., Li, Y.M., Chen, R.J., Javey, A., Dai, H.J.: Patterned Growth of Single-Walled Carbon Nanotubes on Full 4-Inch Wafers. Appl. Phys. Lett. 79, 4571–4573 (2001)

[42] Kuttel, O.M., Groening, O., Emmenegger, C., Schlapbach, L.: Electron field emission from phase pure nanotube films grown in a methane/hydrogen plasma. Appl. Phys. Lett. 73, 2113–2115 (1998)

[43] Rinzler, A.G., Hafner, J.H., Nikolaev, P., Lou, L., Kim, S.G., Tomanek, D., Nordlander, P., Colbert, D.T., Smalley, R.E.: Unraveling nanotubes: field emission from an atomic wire. Science 269, 1550–1553 (1995)

[44] de Pablo, P.J., Howell, S., Crittenden, S., Walsh, B., Graugnard, E., Reifenberger, R.: Correlating the location of structural defects with the electrical failure of multiwalled carbon nanotubes. Appl. Phys. Lett. 75, 3941–3943 (1999)

[45] Saito, Y., Hamaguchi, K., Hata, K., Tohji, K., Kasuya, A., Nishina, Y., Uchida, K., Tasaka, Y., Ikazaki, F., Yumura, M.: Field emission from carbon nanotubes: purified single-walled and multi-walled tubes. Ultramicroscopy 73, 1–6 (1998)

[46] Wang, Q.H., Corrigan, T.D., Dai, J.Y., Chang, R.P.H., Krauss, A.R.: Field emission from nanotube bundle emitters at low fields. Appl. Phys. Lett. 70, 3308–3310 (1997)

[47] Kuttel, O.M., Groning, O., Emmenegger, C., Nilsson, L., Maillard, E., Diederich, L., Schlapbach, L.: Field emission from diamond, diamond-like and nanostructured carbon films. Carbon 37, 745–752 (1999)

[48] Bonard, J.M., Stockli, T., Maier, F., De Heer, W.A., Chatelain, A., Salvetat, J.P., Forro, L.: Field-emission-induced luminescence from carbon nanotubes. Phys. Rev. Lett. 81, 1441–1444 (1998)

[49] Tamura, R., Tsukada, M.: Electronic states of the cap structure in the carbon nanotube. Phys. Rev. B 52, 6015–6026 (1995)

[50] Carroll, D.L., Redlich, P., Ajayan, P.M., Charlier, J.C., Blase, X., De Vita, A., Car, R.: Electronic structure and localized states at carbon nanotube tips. Phys. Rev. Lett. 78, 2811–2814 (1997)

[51] Cassell, A.M., Franklin, N.R., Tombler, T.W., Chan, E.M., Han, J., Dai, H.J.: Directed growth of free-standing single-walled carbon nanotubes. J. Am. Chem. Soc. 121, 7875–7976 (1999)

[52] Dean, K.A., Groening, O., Kuttel, O.M., Schlapbach, L.: Nanotube electronic states observed with thermal field emission electron spectroscopy. Appl. Phys. Lett. 75, 2773–2775 (1999)

[53] Dean, K.A., Chalamala, B.R.: Current saturation mechanisms in carbon nanotube field emitters. Appl. Phys. Lett. 76, 375–377 (2000)

[54] Dean, K.A., von Allmen, P., Chalamala, B.R.: Three behavioral states observed in field emission from single-walled carbon nanotube. J. Vac. Sci. Technol. B 17, 1959–1969 (1999)

[55] Obraztsov, A.N., Volkov, A.P., et al.: Role of the curvature of atomic layers in electron field emission from graphitic nanostructured carbon. JETP Lett. 69, 411–417 (1999)

[56] Bonard, J.M., Stockli, T., Maier, F., De Heer, W.A., Chatelain, A., Salvetat, J.P., Forro, L.: Field-emission-induced luminescence from carbon nanotubes. Phys. Rev. Lett. 81, 1441–1444 (1998)

[57] Tamura, R., Tsukada, M.: Electronic states of the cap structure in the carbon nanotube. Phys. Rev. B 52, 6015–6026 (1995)

[58] Carroll, D.L., Redlich, P., Ajayan, P.M., Charlier, J.C., Blase, X., De Vita, A., Car, R.: Electronic structure and localized states at carbon nanotube tips. Phys. Rev. Lett. 78, 2811–2814 (1997)

[59] Lovall, D., Buss, M., Graugnard, E., Anders, R.P., Reifenberger, R.: Electron emission and structural characterization of a rope of single walled carbon nanotubes. Phys. Rev. B 61, 5683–5691 (2000)

[60] Gao, R., Pan, Z.W., Wang, Z.L.: Work function at the tips of multiwalled carbon nanotubes. Appl. Phys. Lett. 78, 1757–1759 (2001)

[61] Ootsu, M.: Nano ・Photonics. Yoneda Inc. (1999) (in Japanese)

[62] Kawata, S.: Photonics and Nanotechnology. Kubapro (2002) (in Japanese)

[63] Igarashi, K., Uchihashi, T., Koivula, A., Wada, M., Kimura, S., Okamoto, T., Penttila, M., Ando, T., Samejima, M.: Traffic Jams Reduce Hydrolytic Efficiency of Cellulaseon Cellulose Surface. Science 333, 1279–1282 (2011)

[64] Bogner, A., Jouneau, P.-H., Thollet, G., Basset, D., Gauthier, C.: A history of scanning electron microscopy developments: Towards "wet-stem" imaging. Micron 38, 390–401 (2007)

[65] Breton, P.J.: From microns to nanometers: Early landmarks in the science of scanning electron microscope imaging. Scanning Microsc. 13, 1–6 (1999)

[66] Danilatos, G.D.: Foundations of environmental scanning electronmicroscopy. Advances in Electronics and Electron Physics 71, 109–250 (1988)

[67] Donald, A.M.: The use of environmental scanning electron microscopy for imaging wet and insulating materials. Nature Materials 2, 511–516 (2003)

[68] Stokes, D.J.: Characterisation of soft condensed matter and delicate materials using environmental scanning electronmicroscopy (esem). Advanced Engineering Materials 3, 126–130 (2001)

[69] Stokes, D.J., Donald, A.M.: In situ mechanical testing of dry and hydrated breadcrumb in the environmental scanning electron microscope (esem). Journal of Materials Science 35, 599–607 (2000)

[70] Horiuchi, S., Hirotsu, S., Asakura, K.: Electron Microscope Q&A, Agune-shouhu-sha (1996) (in Japanese)

[71] Surface Science Society of Japan, Transmission Electron Microscope, Maruzen (1999) (in Japanese)

[72] Kondo, D., Oikawa, T.: Analytical Electron Microscopes for Material Evaluation, Kyoritsu shuppan (1999) (in Japanese)
[73] Surface Science Society of Japan, Scanning Electron Microscope for Nanotechnology, Maruzen (2004) (in Japanese)
[74] Sanche, L.: Dissociative attachment and surface reactions induced by low-energy electrons. J. Vac. Sci. & Technol. B 10, 196–200 (1992)
[75] Broers, A.N., Molzen, W.W., Cuomo, J.J., Wittles, N.D.: Electron-beam fabrication of 80-Å metal structures. Appl. Phys. Lett. 29, 596–598 (1976)
[76] Matsui, S., Mori, K.: New selective deposition technology by electron beam induced surface reaction. J. Vac. Sci. Technol. B 4, 299–304 (1986)
[77] Koops, H.W.P., Weiel, R., Kern, D.P., Baum, T.H.: High-resolution electron-beam induced deposition. J. Vac. Sci. Technol. B 6, 477–481 (1988)
[78] Mccord, M.A., Kern, D.P., Chang, T.H.P.: Direct deposition of 10-nm metallic features with the scanning tunneling microscope. J. Vac. Sci. Technol. B 6, 1877–1880 (1988)
[79] Lee, K.L., Hatzakis, M.: Direct electron-beam patterning for nanolithography. J. Vac. Sci. Technol. B 7, 1941–1946 (1989)
[80] Weber, M., Koops, H.W.P., Gortz, W.: Scanning probe microscopy of deposits employed to image the current density distribution of electron beams. J. Vac. Sci. Technol. B 10, 3116–3119 (1992)
[81] Lee, K.L., Abraham, D.W., Secord, F., Landstein, L.: Submicron Si trench profiling with an electron-beam fabricated atomic force microscope tip. J. Vac. Sci. Technol. B 9, 3562–3568 (1991)
[82] Keller, D.J., Chi-Chung, C.: Imaging steep, high structures by scanning force microscopy with electron beam deposited tips. Surf. Sci. 268, 333–339 (1992)
[83] Gortz, W., Kempf, B., Kretz, J.: Resolution enhanced scanning force microscopy measurements for characterizing dry etching methods applied to titanium masked InP. J. Vac. Sci. Technol. B 13, 34–39 (1995)
[84] Griesinger, U.A., Kaden, C., Lichtenstein, N., Hommel, J., Lehr, G., Bergmann, R., Menschig, A., Schweitzer, H., Hillmer, H., Koops, H.W.P., Kretz, J., Rudolph, M.: Investigations of artificial nanostructures and lithography techniques with a scanning probe microscope. J. Vac. Sci. Technol. B 11, 2441–2445 (1993)
[85] Koops, H.W.P., Kretz, J., Rudolph, M., Weber, M.: Constructive three-dimensional lithography with electron-beam induced deposition for quantum effect devices. J. Vac. Sci. Technol. B 11, 2386–2389 (1993)
[86] Kunz, R.R., Mayer, T.M.: Electron beam induced surface nucleation and low-temperature decomposition of metal carbonyls. J. Vac. Sci. Technol. B 6, 1557–1564 (1988)
[87] Koops, H.W.P., Kretz, J., Weber, M.: Combined lithographies for the reduction of stitching errors in lithography. J. Vac. Sci. & Technol. B 12, 3265–3269 (1994)
[88] Cho, C.C., Bernasek, F.L.: Molybdenum deposition from the decomposition of molybdenum hexacarbonyl. J. Appl. Phys. 65, 3035–3043 (1989)
[89] Weber, M., Koops, H.W.P., Rudolph, M., Kretz, J., Schmidt, G.: New compound quantum dot materials produced by electron-beam induced deposition. J. Vac. Sci. Technol. B 13, 1364–1368 (1995)
[90] Whitesides, G.M., Grzybowski, B.: Self-Assembly at all Scales. Science 295, 2418–2421 (2002)

[91] Tolley, M.T., Kalontarov, M., Neubert, J., Erickson, D., Lipson, H.: Stochastic Modular Robotic Systems: A Study of Fluidic Assembly Strategies. IEEE Trans. Robotics 26, 518–530 (2010)

[92] Beissenhirtz, M.K., Willner, I.: DNA-based machines. Org. Biomol. Chem. 4, 3392–3401 (2006)

[93] Kelly, T.R.: Molecular Motors: Synthetic DNA-Based Walkers Inspired by Kinesin. Angew. Chem. Int. Ed. 44, 4124–4127 (2005)

[94] Nicolau, D.V., Suzuki, H., Mashiko, S., Taguchi, T., Yoshikawa, S.: Actin Motion on Microlithographically Functionalized Myosin Surfaces and Tracks. Biophys. J. 77, 1126–1134 (1999)

[95] Jaber, J.A., Chase, P.B., Schlenoff, J.B.: Actomyosin-Driven Motility on Patterned Polyelectrolyte Mono- and Multilayers. Nano Lett. 3(11), 1505–1509 (2003)

[96] Kakugo, A., Shikinaka, K., Gong, J.P.: Integration of Motor Proteins - Towards an ATP Fueled Soft Actuator. Int. J. Mol. Sci. 9, 1685–1703 (2008)

[97] Osada, Y., Gong, J.P.: Nano-Biomachine from Actin and Myosin Gels. Polymer Sci., Ser. A 51(6), 689–700 (2009)

[98] Konishi, K., Uyeda, T.Q.P., Kubo, T.: Genetic engineering of a Ca2+ dependent chemical switch into the linear biomotor kinesin. FEBS Letters 580, 3589–3594 (2006)

[99] Hiratsuka, Y., Tada, T., Oiwa, K., Kanayama, T., Uyeda, T.Q.P.: Controlling the Direction of Kinesin-Driven Microtubule Movements along Microlithographic Tracks. Biophys. J. 81, 1555–1561 (2001)

[100] Soong, R.K., Bachand, G.D., Neves, H.P., Olkhovets, A.G., Craighead, H.G., Montemagno, C.D.: Powering an Inorganic Nanodevice with a Biomolecular Motor. Science 290, 1555–1558 (2000)

[101] Homma, M., Oota, H., Kojima, S., Kawagishi, I., Imae, Y.: Chemotactic Responses to an Attractant and a Repellent by the Polar and Lateral Flagellar Systems of Vibrio Alginolyticus. Microbiol. 142, 2777–2783 (1996)

[102] Hiratsuka, Y., Miyata, M., Tada, T., Uyeda, T.Q.P.: A Microrotary Motor Powered by Bacteria. Proc. Natl. Acad. Sci. 103(37), 13618–13623 (2006)

[103] Arai, F., Andou, D., Nonoda, Y., Fukuda, T., Iwata, H., Itoigawa, K.: Integrated Microendeffector for Micro-Manipulation. IEEE/ASME Trans. Mechatronics 3(1), 17–23 (1998)

[104] Saito, S., Miyazaki, H., Sato, T.: Pick and Place Operation of Micro Object with High Reliability and Precision Based on Micro Physics under SEM. In: Proc. ICRA 1999, pp. 2736–2743 (1999)

[105] Wong, E.W., Sheehan, P.E., Lieber, C.M.: Nanobeam Mechanics: Elasticity, Strength, and Toughness of Nanorods and Nanotubes. Science 277, 1971–1975 (1997)

[106] Falvo, M.R., Clary, G.J., Taylor, R.M., Chi, V., Brooks, F.P., Washburn, S., Superfine, R.: Bending and Buckling of Carbon Nanotubes Under Large Strain. Nature 389, 582–584 (1997)

[107] Postma, H.W.C., Sellmeijer, A., Dekker, C.: Manipulation and Imaging of Individual Single-Walled Carbon Nanotubes with an Atomic force Microscope. Advanced Materials 12, 1299–1302 (2000)

[108] Postma, H.W.C., De Jonge, M., Yao, Z., Dekker, C.: Electrical Transport Through Carbon Nanotube Junctions Created by Mechanical Manipulation. Phys. Rev. B 62, R10653–R10656 (2000)

[109] Hertel, T., Martel, R., Avouris, P.: Manipulation of Individual Carbon Nanotubes and Their Interaction with Surfaces. J. of Phys. Chem. B 102, 910–915 (1998)

[110] Ahlskog, M., Tarkiainen, R., Roschier, L., Hakonen, P.: Single-Electron Transistor Made of Two Crossing Multiwalled Carbon Nanotubes and Its Noise Properties. Appl. Phys. Lett. 77, 4037–4039 (2000)

[111] Falvo, M.R., Taylor II, R.M., Helser, A., Chi, V., Brooks Jr., F.P., Washburn, S., Superfine, R.: Nanometre-Scale Rolling and Sliding of Carbon Nanotubes. Nature 397, 236–238 (1999)

[112] Falvo, M.R., Steele, J., Taylor II, R.M., Superfine, R.: Gearlike Rolling Motion Mediated by Commensurate Contact: Carbon Nanotubes on HOPG. Phys. Rev. B 62, R10665–R10667 (2000)

[113] Dong, L.X., Arai, F., Fukuda, T.: 3D Nanorobotic Manipulation of Nano-Order Objects Inside SEM. In: Proc. of the 2000 International Symposium on Micromechatronics and Human Science, Nagoya, Japan, pp. 151–156 (2000)

[114] Hafner, J.H., Cheung, C.-L., Oosterkamp, T.H., Lieber, C.M.: High-Yield Assembly of Individual Single-Walled Carbon Nanotube Tips for Scanning Probe Microscopies. J. Phys. Chem. B 105, 743–746 (2001)

[115] Yu, M.F., Dyer, M.J., Skidmore, G.D., Rohrs, H.W., Lu, X.K., Ausman, K.D., Von Ehr, J.R., Ruoff, R.S.: Three-Dimensional Manipulation of Carbon Nanotubes Under a Scanning Electron Microscope. Nanotechnology 10, 244–252 (1999)

[116] Dong, L.X., Arai, F., Fukuda, T.: Electron-Beam-Induced Deposition with Carbon Nanotube Emitters. Appl. Phys. Lett. 81, 1919–1921 (2002)

[117] Dong, L.X., Arai, F., Fukuda, T.: Three-Dimensional Nanoassembly of Multi-Walled Carbon Nanotubes Through Nanorobotic Manipulations by Using Electron-Beam-Induced Deposition. In: Proc. of the 1st IEEE Conf. of Nanotechnology (IEEE NANO 2001), Maui, Hawaii, pp. 93–98 (2001)

[118] Dong, L.X., Arai, F., Fukuda, T.: 3D Nanorobotic Manipulations of Multi-Walled Carbon Nanotubes. In: Proc. of the 2001 IEEE International Conf. on Robotics and Automation (ICRA 2001), Seoul, Korea, pp. 632–637 (2001)

[119] Dong, L.X., Arai, F., Fukuda, T.: Inter-Process Measurement of MWNT Rigidity and Fabrication of MWNT Junctions Through Nanorobotic Manipulations. In: American Institute of Physics Conference Proceedings 590: Nanonetwork Materials: Fullerenes, Nanotubes, and Related Materials, pp. 71–74 (2001)

[120] Fukuda, T., Arai, F., Dong, L.X.: Fabrication and Property Analysis of MWNT Junctions Through Nanorobotic Manipulations. Int'l J. of Nonlinear Sciences and Numerical Simulation 3, 753–758 (2002)

[121] Dong, L.X., Arai, F., Fukuda, T.: 3D Nanoassembly of Carbon Nanotubes Through Nanorobotic Manipulations. In: Proc. of the 2002 IEEE Int'l Conf. on Robotics & Automation (ICRA 2002), Washington, U.S.A., May 11-15, pp. 1477–1482 (2002)

[122] Dong, L.X., Arai, F., Fukuda, T.: Mechanochemical Nanorobotic Manipulations of Carbon Nanotubes. Jap. J. Appl. Phys. 42(1 Pt. 1), 295–298 (2003)

Chapter 4
Micromanipulation System under Optical Microscope

4.1 Biomicromanipulation Methods for On-Chip Cell Experiments

History of Biomicromanipulation

Technologies required for the biomicromanipulation can be classified as follows: Transportation, separation, controls of position and attitude, immobilization, detection, and measurement. Figure 4. 1 shows the history of the development of biomicromanipulation for biotechnology and bioengineering.

Electrical force is famous for bio-micromanipulation. Electrophoresis is a useful separation technique which has been used for years especially in biochemistry and biology. This method is suitable for the separation of large charged molecules such as proteins and DNA. This method is based on the differential migration of charged species under the influence of an electrical field. Although this concept can be implemented in various ways, it has been mainly used as gel electrophoresis in which separation is performed in slabs or tubes of gel.

In 1981, Jorgenson developed new electrophoresis in glass capillaries with an internal diameter of 75 μm and a length of 100 cm. This method extended the concept of conventional electrophoresis and achieved rapid protein separation [1]. A special fluorescence detector was used to detect fluorescent zones while they migrated inside the capillary. With the application of 30 kV potentials to this system, rapid and efficient separations of amino acids, peptides, and urinary amines were demonstrated.

In 1978, H. A. Pohl developed dielectrophoresis (DEP), which is electrical force based on different principle from capillary electrophoresis by D. C. electrical field [2]. This phenomenon was defined as the lateral motion imparted on uncharged particles as a result of polarization induced by non-uniform electrical fields. Many theoretical and experimental studies have been investigated such as microelectrode structures, a capability of DEP force, electrorotational force and traveling field force [3, 4, 5]. These studies were applied to perform a wide range of applications to bio-processing functions such as manipulation, trap, transportation, separation, cultivation of biological molecules [6, 7, 8].

In 1970, A. Ashkin developed new manipulation by optical force and this manipulation method was called as "Optical tweezers" [9]. While electrical force works in filed, optical tweezers works at laser focal point. Therefore, the manipulation of selected target is possible. This technique was applied to bio-microsamples such as a virus, microorganism and DNA molecule. Multiple objects can be trapped simultaneously by scanning the laser focal point. The laser scanning approach is suitable for the attitude control of target and the trajectory control of the object since it is easy to create and change multiple laser trap potentials [10]. The time-sharing approach where the beam is repetitively scanned in the sample by rapidly moving a piezoelectric mirror was proposed [11]. However, not much attention has been paid the independent trajectory control of each trapped object by using a single laser spot. In 2002, F. Arai developed time shared scanning laser manipulation (TSS) for an independent trajectory control of multi-objects by using single laser with a scanning system of its focal point [12], [13].

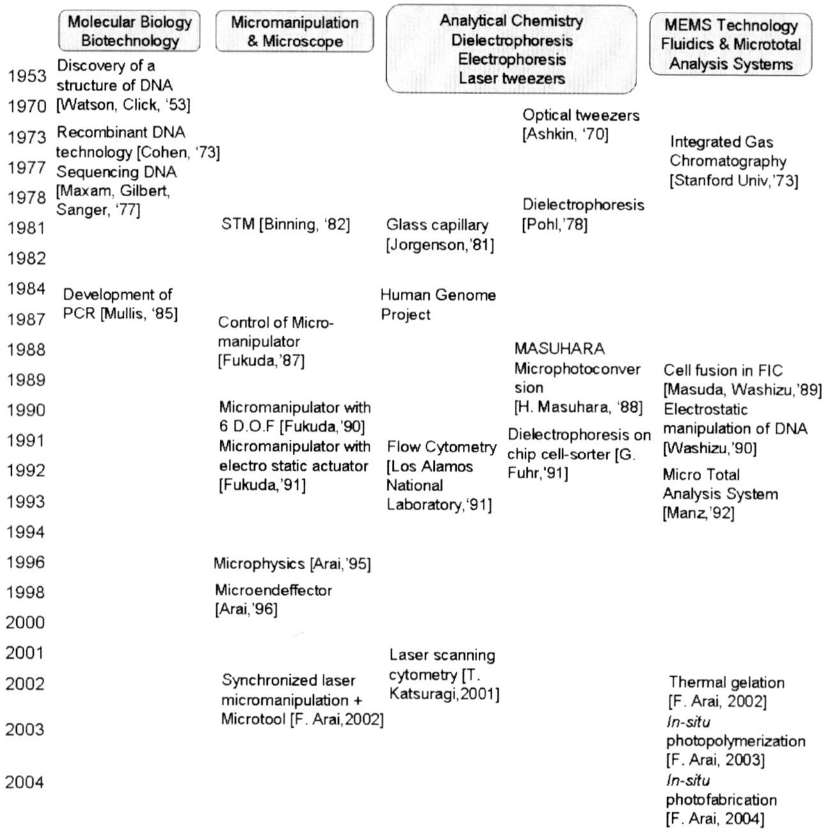

Fig. 4.1 History of Bio-Micromanipulation in MEMS

In 1992, A. Manz proposed "Micro Total Analysis System (μTAS)" [14, 15, 16, 17]. Remarkable progress of micromachining such as photolithography and etching enabled integration of microchannles and other microdevices for biological and chemical analysis in microchip. Microchips which have microchannels and reservoirs enabled to reduce the volume of the sample solution which was used for analysis [18, 19, 20, 21]. Microfabrication and microfluidic techniques have been progressed toward small volume handling and integrated microdevices for chemical and biochemical analysis [22, 23, 24, 25].

The trend in the microchip technology [26] has appeared to be toward smaller devices. The novel application of these microtechnologies and miniaturization techniques to the pharmaceutical industry provided powerful tools and techniques for nano-biotechnology such as DNA separation, molecular surgery, the drug discovery system, clinical diagnostics [27, 28, 29, 30]. Research works with highly integrated microchips [31, 32, 33, 34] will make a great contributions in the future.

Previous Researches on Cell Manipulation in Microchip for On-Chip Cell Experiment

On-Chip Screening

In 1991, G. Fuhr et. al. developed on chip flow cytometer [35]. In recent years, the micromachining technology was improved and the research works of microbe/cell manipulation in on-chip flow cytometer were spread widely.

The 3-D microelectrode system was proposed by consisting of two layers of electrode structures separated by a 40 μm thick polymer spacer forming a flow channel [36]. The electrode elements such as a funnel, aligner, cage and switch were driven by AC voltage and rotating electrical fields. Electrodes were designed to focus, trap and separate cells or latex particles with a diameter of 10–30 μm using negative DEP. The electrode width is ~10 μm and active electrode surfaces have been minimized to reduce heating of the solution. The electrodes of the funnel, aligner or the switch were operated with 5–11 V at 5–15 MHz, efficient handling of particles could be achieved with flow rates up to 3500 μm/s. Cells could be aligned effectively at flow rates up to 300 μm/s in PBS. A latex particle could be retained within the dielectric field cages (DFC) or aligner against a laminar flow of 40–200 μm/s. Cells could be held at flow rates up to 50 μm/s. This technique could sort a large number of cells sequentially at high speed. However, this technique is not suitable for single cell analysis because the position information on the sorted cell is not known.

On the other hand, the disposable microfabricated fluorescence-activated cell sorter (μFACS) for sorting various biological samples [37]. Compared with conventional FACS machines, this μFACS device achieved higher sensitivity, no cross-contamination, and lower cost. The μFACS microchips have been used to obtain substantial enrichment of micro-sized fluorescent bead populations of differing colors. In this paper, E. coli cells expressing green fluorescent protein were separated from a background of non-fluorescent E. coli cells and shown that the bacteria were viable after extraction from the sorting device by using this μFACS chip. This technique could sort a large number of cells with detecting cell properties by fluorescent observation. However, this technique is also sequential process, so that some ideas are needed for applying single cell analysis.

A novel system for random separation of a single cell was proposed by the microfluidic device under the microscope by integrating the optical tweezers and DEP force [38]. An arbitrarily selected single cell could be isolated in a microchannel from a large number of cells in a solution. After the target cell was trapped by optical tweezers, the trapped cell was excluded from the sample chamber. In this process, unnecessary cells are environmental disturbance. These cells are moved by both edges of microchannel by DEP. The transported cell was extracted from microchip by liquid flow. The unique character of this separation system is that flow speed in the microchannel is adjusted and balanced to realize high-speed and high-purity extraction of the target. In this system, total separation time was less than 20 s. This system could realize high-throughput random cell screening. However, photo damage caused by direct laser irradiation to cell may be serious problem to cell activity.

We proposed the *in-situ* formation of gel microbeads made of a thermoreversible hydrogel for indirect cell manipulation as shown in Figure 4. 2 [39]. Irradiation of a 1064 nm laser to a solution mixed with poly-N-isopropylacrylamide generated a gel microbead at the laser focus point by heating of the solution. The gel microbead was manipulated by the laser, and the gel microbead was used for cellmicromanipulation. However, the laser power for formation of the gel bead was generally too strong to perform manipulation by the conventional method. In this paper, Reduction of the laser power for formation of the gel bead using the poly-N-isopropylacrylamide aqueous solution was achieved by addition of additives was developed. YPD Broth (yeast extract, peptone, and dextrose) were selected as an additive and measured the relationship between the laser power, irradiation time, and diameter of the gel microbead. Manipulations of a yeast cell and DNA with laser trapped gel microbead were succeeded. This method can achieve indirect manipulation of cell. However, additives are needed for generating the gel mcirobead with low laser power.

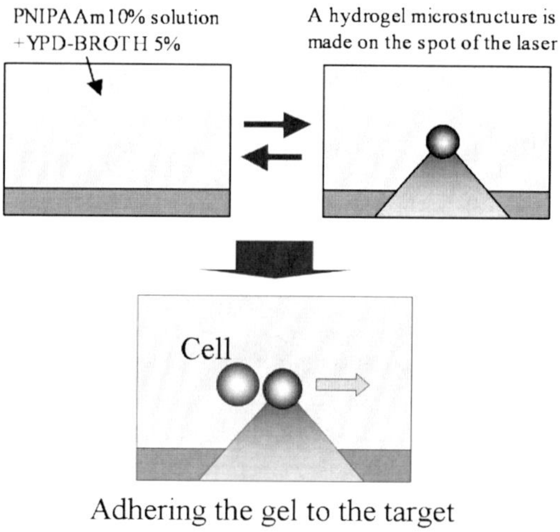

Fig. 4.2 Indirect cell manipulations with gel microbead

A micro dynamic array cytometer (μDAC) device was proposed for parallel luminescent single cell assays that can sort populations upon the basis of dynamic functional responses to stimuli [40]. This device is composed of a regular array of non-contact single cell traps. These traps use DEP to confine cells and hold them against disrupting fluid flow. Using quantitative modeling, designed traps with novel asymmetric extruded-quadrupole geometries were achieved. This new trap can be physically arrayed and electrically addressed. Situating an array of these traps in a microchannel, cells were introduced into the array and observation of fluorescent dynamic responses. Such a device has potential for investigating functional processes such as temporal behavior in large numbers of single cells. This technique can achieve single cell analysis. However, the arrays of this chip have to be reconstructed by changing the target size.

Cell Capture and Monitoring Methods

An on-chip microculture system was proposed for single cell analysis [74]. The advantages of the system are continuous cultivation of a series of isolated single cells or a group of cells under contamination free conditions, continuous observation and comparison of those cells with 0.2 μm spatial resolution by a phase-contrast or fluorescent microscopy system with digital image processing. The core of the system was an array of chambers. Holes were 20–70 μm in diameter and 5–30 μm in deep. These holes are etched into a biotin-coated 0.17 mm thick glass slide. The biotin-coated glass slide was covered with the streptavidin coated cellulose semipermeable membrane, which is fixed on the surface of the glass slide by streptavidin–biotin attachment. A single cell or group of cells could be isolated from environment, and the medium in each chamber could be changed within the diffusion time (< 1/30 s). By using this system, increase in the length of isolated *E. coli* between cell divisions regardless of the chamber volume was observed. The system was useful for single cell level analysis because the direct descendants of single cells can be cultured and compared in the isolated microchambers, and the physical properties of the cells in each microchamber can be continuously observed and compared. However, the observation of the single arbitrary cell is difficult because the chambers are closed.

We proposed a random screening and capture of target microorganisms from a large heterogeneous population by using a local viscosity control as shown in Figure 4. 3 [41]. A thermal sol-gel transformation material is mixed with the sample liquid and the state from sol to gel and gel to sol was controlled reversibly based on the temperature control by heating of the microelectrode with the electrical current and focused laser irradiation. The selected cells are immobilized on the bottom plate by gel since the viscosity around the target is temporally increased by the local heating by the focused laser. The other cells are easily extracted by the liquid flow. Processes of immobilization, cleaning, isolation and extraction of the target cell were possible in very short time. This cell immobilization method can capture arbitrary cells easily in microchip. However, energy supply is needed for keeping the cell immobilization.

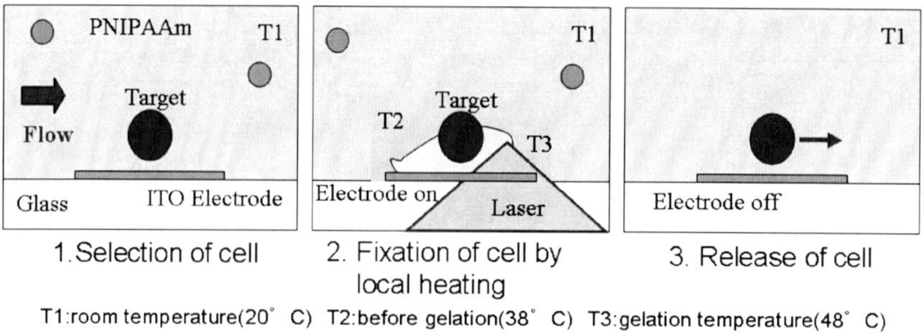

Fig. 4.3 Cell immobilization system by thermal sol-gel transformation

A cell cultivation system was proposed based on photo-thermal etching has been developed for the on-chip cultivation of living cells using an agars microchamber array [42]. This microchip has flexibility of the chamber structure by photo-thermal etching, even during the cultivation of cells, depending upon the progress in cell growth. An infrared (1064 nm) focused laser is used as a heat source to etch agar gel at the heated spot on a thin chromium layer. The size of the photo-thermally etched area depended almost linearly on the power of the irradiated laser beam from 2 μm to 50 μm. Narrow tunnel-shaped channels between the microchambers were easily fabricated by using photo-thermal etching with laser power even during cell cultivation.

An on-chip microculture system as proposed by employing optical tweezers [43]. This system is composed of four parts: microchamber array, optical tweezers system, buffer exchange unit, and phase contrast microscope. The microchamber array is microfabricated on a glass slide from thick SU-8. To prevent the cells escaping from the microchamber the array was sealed with a semipermeable membrane decorated with streptavidin, and the glass slide was decorated with biotin to ensure tight sealing. Key features of the microchamber structure were four observation areas at the center of the microstructure, and two discarding areas at both ends of the microstructure. Optical tweezers were used for non-contact cell manipulation. By using optical tweezers for cell manipulation, cell property analysis which is impossible by using only cell monitoring was achieved. This microculture system employing optical tweezers can achieve single cell analysis with control of the cell position and cell number in chamber. However, photo damage caused by direct laser irradiation to cell may be serious problem to cell activity.

On-Chip Environment Measurement Methods

A microchip was integrated ion selectivity electrode (ISE) [44]. The solvent polymeric membranes were prepared with a mixture of ionophore 3 wt% TD19C6, 67 wt% plasticizers BBPA or 67 wt% TOTM, 30 wt% PVC and anionic additives K-TCPB or 10 mol% TFPB. PVC was gradually dissolved in 2ml THF with a sonic shaker for two hours, followed by the addition of anionic additives, ionophore (dissolved in 1ml of THF) and finally plasticizer. The cocktail of the PVC-based sensor membranes was fabricated using an auto dispenser or manual dispenser. This type sensor enables on-chip measurement. However, measurement is limited within sensor exist area.

Luminescence and fluorescence are also used for environment measurement [45]. The ruthenium (II) diimine complexes such as ruthenium (II) tris(bipyridyl), $Ru(bpy)_3^{2+}$, possess highly luminescent excited states that are not only readily quenched by oxygen but also by an increase in temperature. The former effect can be rendered insignificant by encapsulating the complex in some oxygen impermeable polymer although encapsulation often leads also to a loss of temperature sensitivity. The luminescence properties of $Ru\ (bpy)_3^{2+}$ encapsulated in PVA were studied as a function of oxygen concentration and temperature and found to be independent of the former, but still very sensitive towards the latter. Luminescence and fluorescence are useful methods for on-chip environment measurement. However, degradation is a serious problem for long time measurement by fluorescence.

A wireless micro-sensor was developed for the simultaneous measurement of pH, temperature, and pressure [46]. In response to a magnetic field impulse, magnetostrictive magnetoelastic sensors mechanically vibrate. These vibrations are detected in several ways: Optically from the amplitude modulation of a reflected laser, acoustically using a microphone or hydrophone, and by using a pickup coil to detect the magnetic flux emitted from the sensor. Earlier work has shown that the resonant frequency of magnetoelastic sensor shifts in response to different environmental parameters, including temperature, pressure, fluid flow velocity and mass loading, with each parameter determined in an otherwise constant environment. To extend the utility of the sensor technology in this work, the fabrication and application of a miniaturized array of four magnetoelastic sensors that enable the simultaneous remote measurement of pH, temperature, and pressure were developed. This type sensor is useful for measuring various environments. However, integration of such microfabricated sensors into microchip is high cost and making the integrated microchip disposable is difficult.

4.2 Multiple Trapping by Optical Tweezers

4.2.1 Time Shared Scanning (TSS) Laser Trapping System

For advancedmicromanipulation with micro-tool, we proposed the Synchronized Lasermicromanipulation for the trajectory control of the multiple targets independently by the single laser. TSS shown in Figure 4. 4 (b) is different from existing laser scanning micromanipulation shown in Figure 4. 4 (a) in that we can manipulate multiple targets independently by the single laser by changing the discrete laser scanning pattern [13]. Accordingly we can manipulate multiple micro-tools along with each designed trajectory by the proposed method as shown in Figure 4. 5.

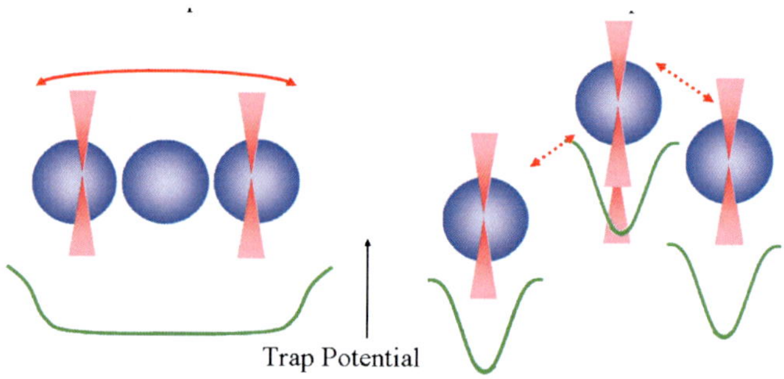

Fig. 4.4 Model of TSS (a) normal laser scanning manipulation, (b) time shared scanning laser (TSS) manipulation

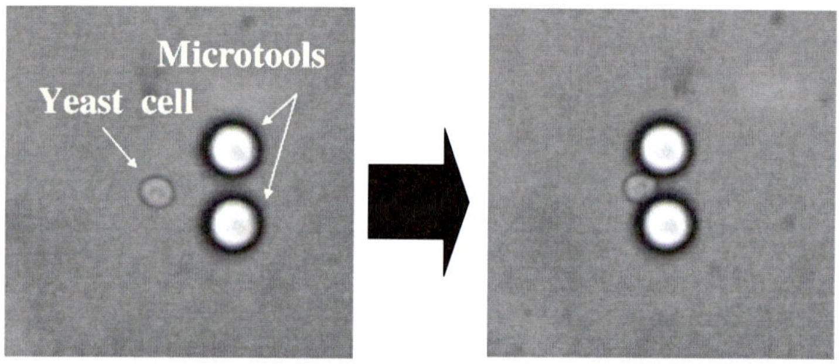

Fig. 4.5 Example of manipulation of a cell using TSS

TSS can be extended to 3D TSS. 3D TSS enables 3D control with 6DOF (position and orientation) of a micro-tool or microorganism. 3D orientation control is effective to improve various cell surgeries such as gene injection.

Figure 4.21 shows a schematic diagram of the optical system used in 3D TSS [47, 48]. A inverted microscope is used for a platform. Nd: YVO_4 laser emission (1064 nm, continuous wave, max: 4.98W) for optical tweezers is introduced into the microscope. The X-Y stage is actuated by a stepping motor. The laser is scanned in 2D observation plane by controlling the angle of a pair of galvanometer mirrors with DC motors. A piezo stage was used to drive an objective lens for scanning the laser in the Z axis. These actuation units are controlled by a computer. By using the galvanometers and the piezo stage simultaneously, 3D laser scanning is achieved.

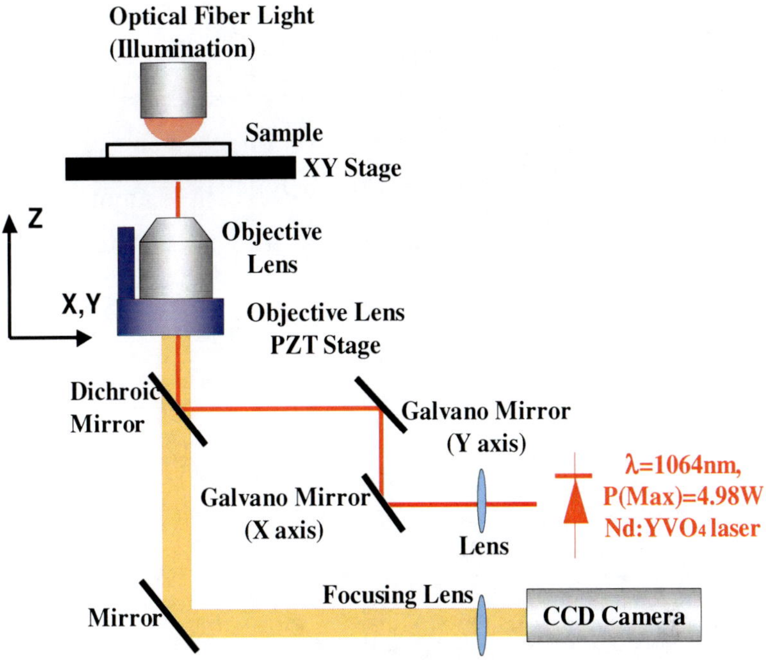

Fig. 4.6 Schematic diagram of optics for 3D TSS

As a demonstration of 3D TSS, Figures 4.22 and 4.23 show 3D orientation control of microbead. The target microbead is rotated indirectly using two micro-tools in contact with the target. It can be clearly seen that the nanobead put inside the target is defocused while it is in focus, indicating that its position had changed. 3D TSS with micro-tool will make a great contribution in cell biology and bioengineering.

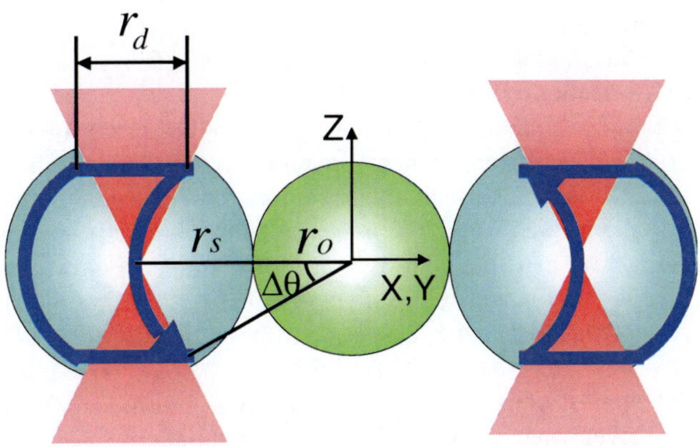

Fig. 4.7 Laser scanning trajectory

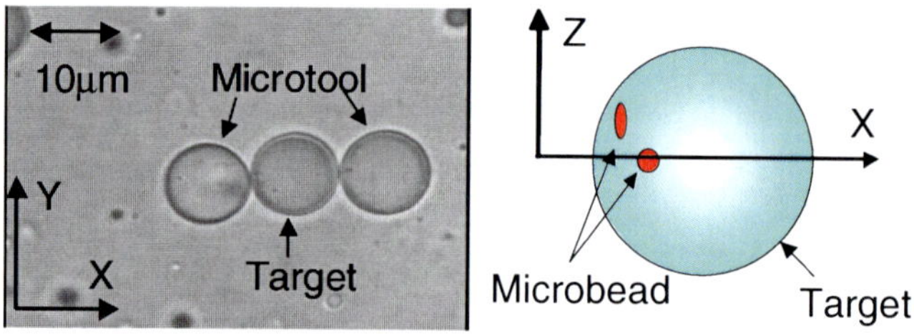

(a) Initial state: 0 sec, Top view (left) and side view (right)

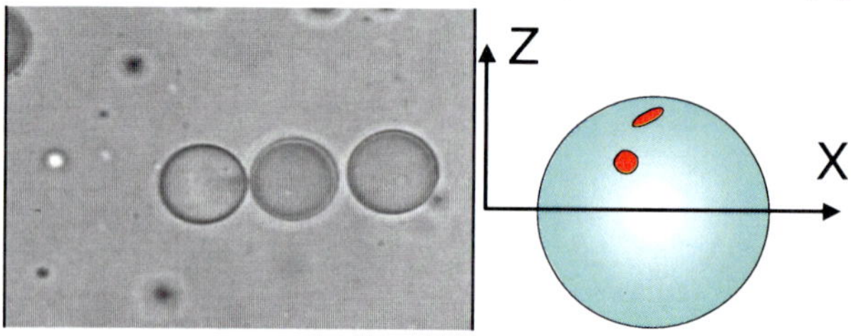

(b) Intermediate state: 60 sec, Top view (left) and side view (right)

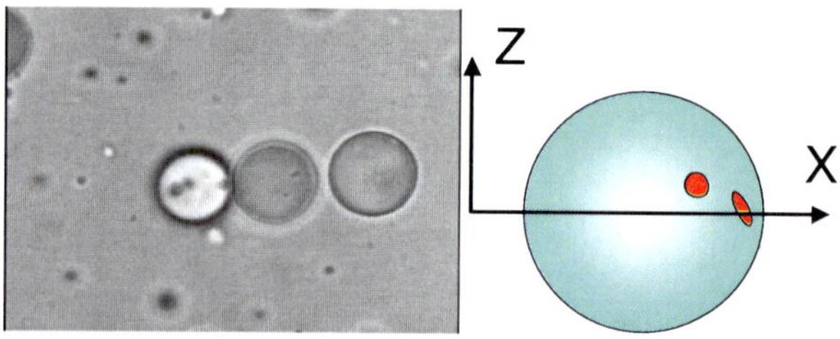

(c) Final state: 180 sec, Top view (left) and side view (right)

Fig. 4.8 Experiment of 3D orientation control

Research which has been described so far controlled the micorotool with optical tweezers. The micro-tool can be controlled by the other non-contact forces such as magnetic force and electrical force besides optical tweezers. For manipulating a big object (> tens of m), optical tweezers is not suitable. However, other forces such as magnetic force are suitable for this purpose. In this reason, we should choose driving force for manipulating micro-tool.

As an example, cell manipulation with microobject controlled by magnetic force is described [49]. The experimental manipulation of PS microbead is shown in Figure 4.24. The object is pushed and transported from its initial position to the reference y_C by the magnetically driven manipulator. The transport strategy has two stages: coarse control and fine control. During the coarse control, the manipulator follows the magnet by fits and starts (stick-slip effect). It pushes the object near the reference y_C, and it uses the fine control. During the fine control, point I is fixed. The displacement of the magnet causes only a modification of the manipulator orientation. The polystyrene (PS) microbead which is 50 μmϕ is manipulated by the magnetically driven manipulator. The micromanipulator used is a piece of electroplated nickel (200 μm × 100 μm × 20 μm). The PS microbead is in the initial position. The microbead is pushed with a coarse control to near the reference. The contact point between the glass pane and the manipulator is fixed and the microbead is pushed with the fine control to the final position. By using this technique, the manipulation of big microobject such as a human oocyte which is 150 μmϕ is achieved.

4.2.2 Computer Generated Hologram (CGH) method

Multi-beam optical tweezers is important technique to manipulate multiple small objects. Computer Generated Hologram (CGH) is one of the techniques and it can trap more than 200 objects in three dimension. For dexterousmicromanipulation, it is useful to apply robotics into optical tweezers.

We designed the optical system and control system of Holographic Optical Tweezers (HOT) [50]. For optical design, we considered working space and positioning resolution. For control design, we applied teleoperation. For online generation of hologram, we developed GCH generator applied General-purpose GPU (GPGPU) acceleration. Figure 4. 9 shows the schematic diagram of teleoperation system using holographic optical tweezers. This system achieved high-speed generation of hologram (250 Hz). For full automatic control, we applied stability control theory to the multi-beam control system (Figure 4. 10). With the built system, the positioning precision and response were revealed with trajectory control and teleoperation experiments. As the result, we checked the effectiveness of HOT enhanced by robotics. The pattering using micro-beads was achieved in a high speed as shown in Figure 4. 11.

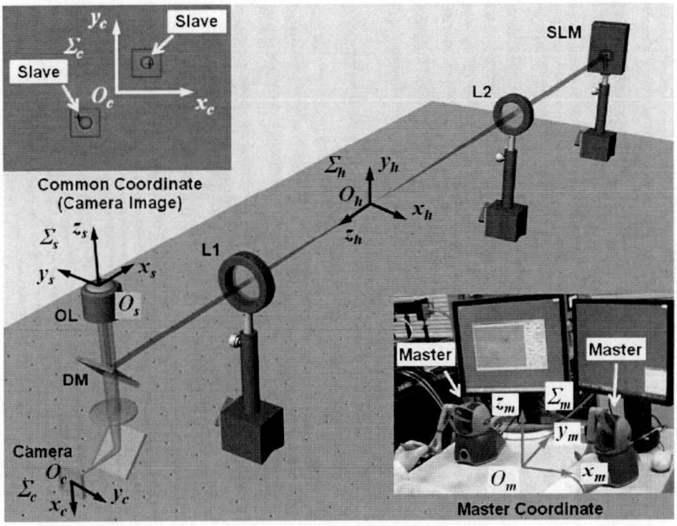

Fig. 4.9 Schematic diagram of teleoperation system using holographic optical tweezers

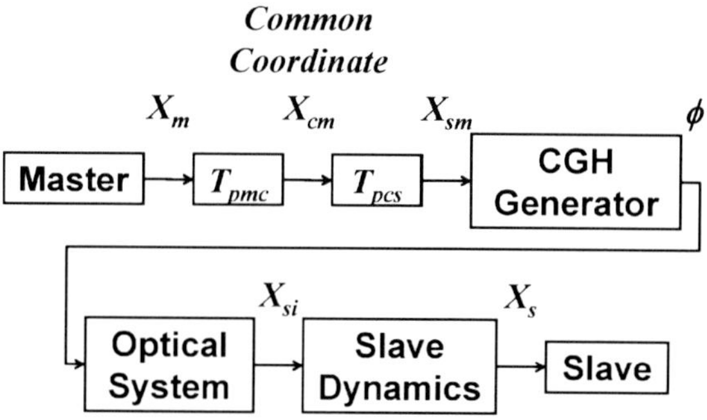

Fig. 4.10 Block diagram of holographic optical tweezers

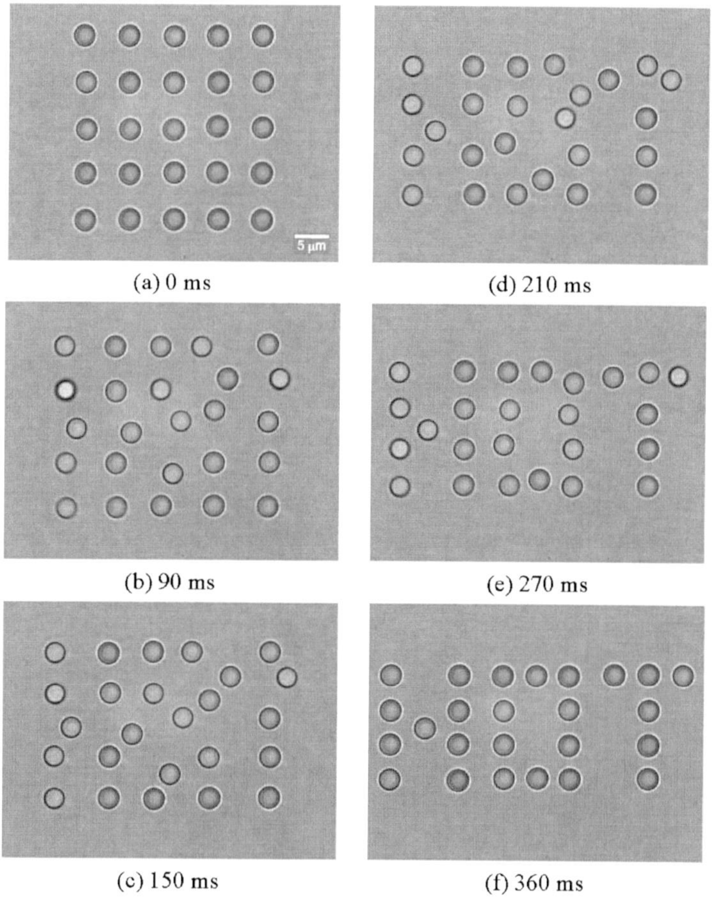

Fig. 4.11 Patterned result of micro-beads by holographic optical tweezers

The block diagram of a bilateral control system is shown in Figure 4. 12. Master position vector X_m is transformed to position X_{cm} in the camera coordinates (common coordinate) system of camera images with T_{pmc}. The positions of masters and slaves are managed in this common coordinate system. Then, X_{cm} is transformed to Xsm in the slave coordinate system with Tpcs. The CGH generator calculates holograms using a GPU in real time.

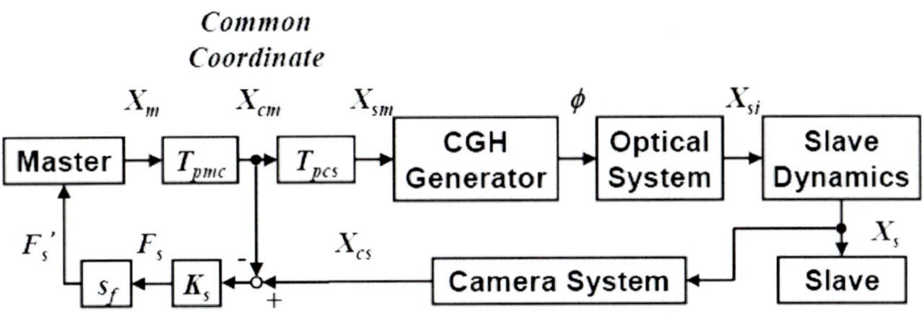

Fig. 4.12 Block diagram of bilateral control system of holographic optical tweezers

For evaluating the effectiveness of the bilateral teleoperation for practical application, the demonstration of touching red blood cells of horses was done. The output laser power was set to 100 mW and the optical trap stiffness was 64 pN/μm. This teleoperation was done by squeezing red blood cells with two slaves as shown in Figure 4. 13. Here, two red blood cells which adhere to each other were caught by the slaves.

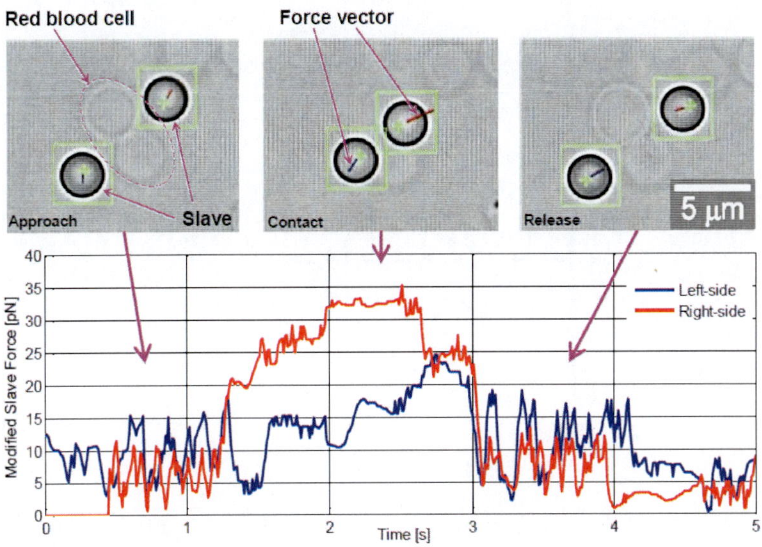

Fig. 4.13 Demonstration of bilateral teleoperation of touching red blood cell

4.3 Configurations of Micro-Fluidics Chips

Because of the great interest in the manipulation of biological cells, many types of micromanipulator were developed. Some micromanipulators were developed for handling small objects of which size was micrometer. Grasping and clamping are required to handle much smaller objects than micromanipulator. Fukuda et. al. have reported a bilateral control method of micromanipulator which is requiring the controls of position and force [51].

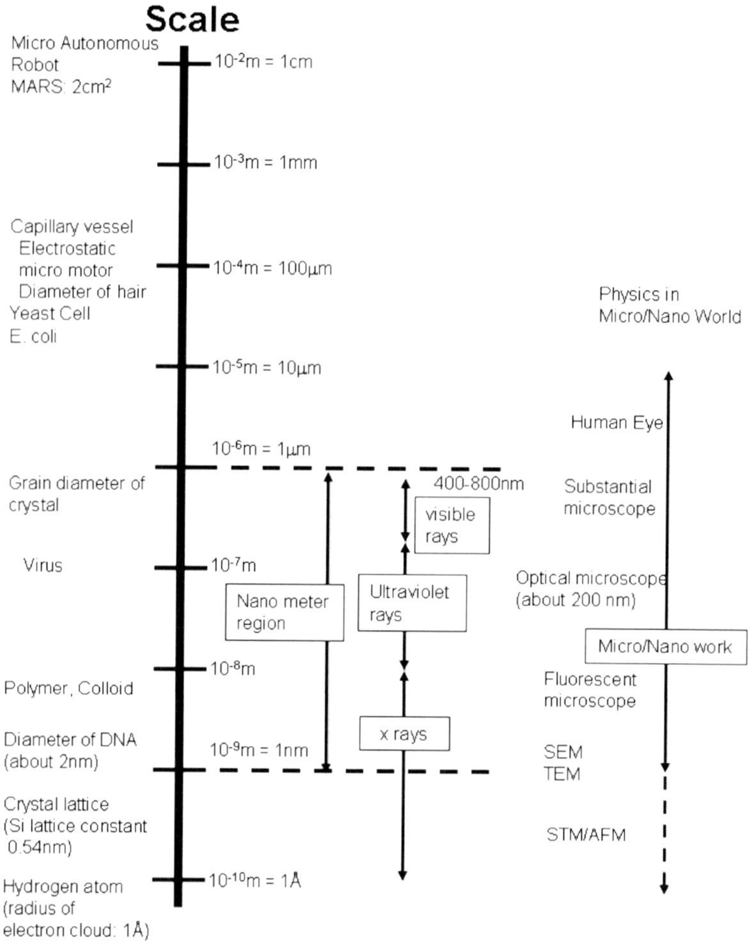

Fig. 4.14 Micro-nano worlds

Most biomanipulation has been operated manually with micromanipulator by skillful operators in petri-dish relying only on visual information from a microscope. However, handling micrometer size objects suspended in a solution with conventional micromanipulator is very difficult even by skillful operators. In micro-nano worlds, we have to use many types of microscope such as SEM and TEM as shown in Figure 4..4. Compared with the general manipulation tasks in the air [52, 53], we have to consider many different kinds of force acting on the cell in a solution such as the gravity, resistance force from the viscosity of the liquid, and Brownian motion. In addition, interactive forces such as Van der Waals' force and electrostatic force depending on the surface charges have to be considered. Moreover, the target is micro-scale object, so that recognition of the correct relative position between the tip of the micromanipulator and target is difficult.

Because of these reasons, operators lost the target easily by the disturbance even by skillful operators in case the manipulation of small bio-microparticles such as microorganisms (from 1-10 μmϕ). For example, the liquid flow due to micropipette motion in the petri-dish acts as a disturbance for cell manipulation as shown in Figure 4. 15. To solve this problem, which is caused by environmental disturbance, manipulation in the microchip can prevent such the disturbance as shown in Figure 4. 16 (a). However, contact manipulation with the micropipette cannot be applied to cell manipulation in a microchip as shown in Figure 4. 16 (b). Therefore, the non-contact manipulation method is required for stable and the effective operation of cells in a microchip.

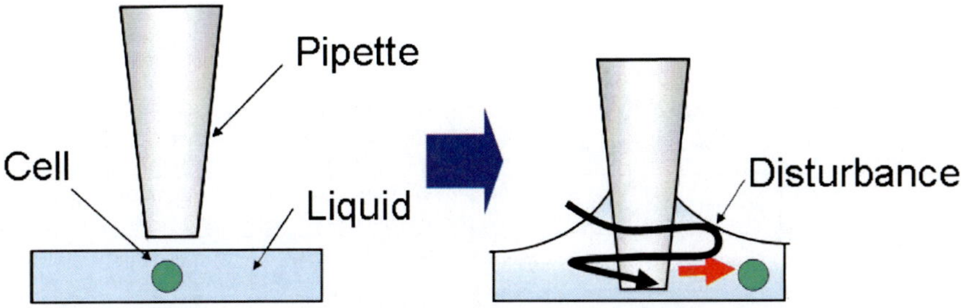

Fig. 4.15 Problem of cell manipulation with contact manipulation

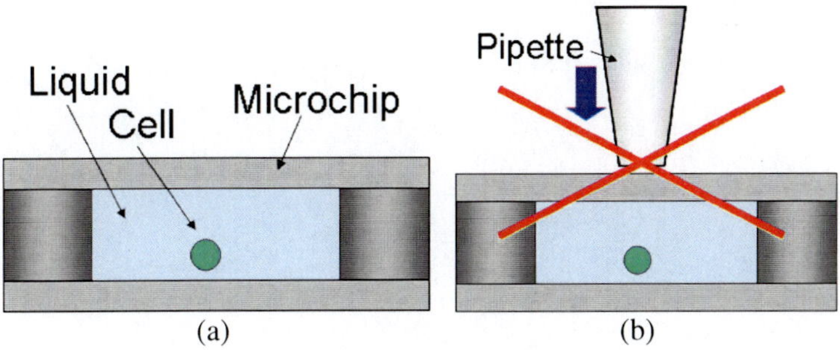

Fig. 4.16 Schematics of cell manipulation in a microchip

4.4 Non-contact Manipulation with Micro-tool

Non-contact manipulation in a microchip is very useful to investigating single cell property [54]. Each non-contact manipulation method has merits and demerits as shown in Tables 4.1 and 4.2. Electrophoresis and DEP are famous for the on-chip manipulation of cell and other bio-microobjects in flow cytometry and gel electrophoresis. These manipulations are operated by field force. Although these forces are suitable for the sequential and continuous separation of cell, the selective manipulation of cell is difficult by employing these forces. Magnetic force has near characteristics to electrical forces. On the other hand, optical tweezers is works at laser focal point. Therefore, this manipulation method is suitable for the random separation of single cell. Ultrasonic force can be disturbed the environment with liquid flow in a microchip. Moreover, 2 or 3 dimensional manipulations of microobject are very difficult by control the liquid flow with an external pump. For these reasons, our research group has been studied for a random and high speed separation of a single cell by employing optical tweezers, DEP and liquid flow in a microchip [55].

Table 4.1 Methods of non-contact manipulation in microchip

	Application	Control Method	Selectivity
Electric force	Transport Separation	High frequency electric field Electrode pattern	Difference of dielectric Properties of particle
Magnetic force	Transport Separation	Magnitude of magnetic field Electromagnet pattern	Difference of magnetism Properties of particle
Optical force	Trapping single object	3-D confocal scanning laser beam pattern	Properties of particle Refractive index
Ultrasonic force	Filteration Fixation	Configuration of resonator	Difference of size Density of particle
Liquid flow	Transport	Pump Microchannel pattern	Density of particle

Table 4.2 Comparisons of manipulation method

	Transport	Selectivity	Controllability	Resolution	3-D operatability	Safty
Electric force	Fair	Fair	Good	Fair	Good	Fair
Magnetic force	Fair	Good	Good	Fair	Good	Good
Optical force	Fair	Good	Good	Good	Fair	Poor
Ultrasonic force	Poor	Fair	Fair	Poor	Poor	Fair
Liquid flow	Good	Poor	Poor	Poor	Poor	Good

Although optical tweezers enable selectivemicromanipulation of an individual cell and other bio-micromolecules, it may cause damage to the cell by direct irradiation of focused laser. In previous works, photo damage to cells in optical tweezers was investigated [56, 57]. The intensity dependence for photo damage is linear. The wavelength of laser relates cell damage. The reduction of photo damage to cells might be possible by considering the above parameters. However, it is difficult to change the laser wavelength according to the cell properties.

For the safety and secure manipulation of the cell independently, we proposed to manipulate the cell with the laser trapped object as shown in Figure 4..24. We call the object as micro-tool and manipulation technique using micro-tool as indirect lasermicromanipulation (ILM). Direct irradiation of focused laser to the cell is avoided by manipulating the cell with the micro-tool and reduction of photo damage attributed the wavelength and intensity of laser is expected.

Although a control of cell attitude is difficult by optical tweezers, micro-tool enables control of a cell position and the cell attitude arbitrary by using single or multiple micrtools. Of course, it is possible to manipulate micro-tool by not only optical tweezers but also electrical force and magnetic force. Most cells do not have magnetism, so that magnetically controlled micro-tool can manipulate the target cell selectively.

We can extend this concept of micro-tool to improve elements of an on-chip cell experiment as shown in the next section.

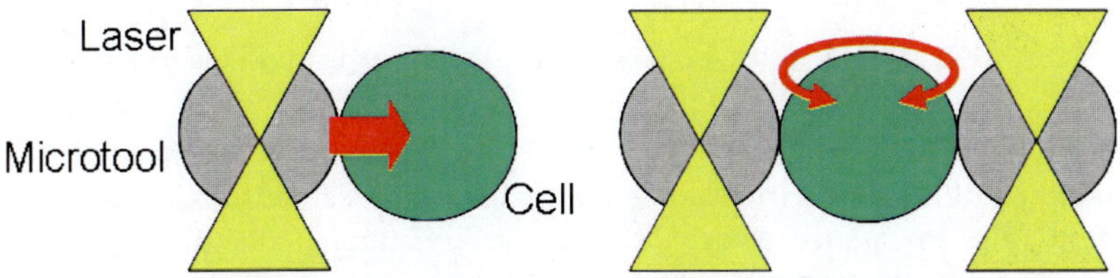

(a) Position control (b) Attitude control

Fig. 4.17 Basic concept of non-contact manipulation with micro-tool

4.4.1 Roles of Micro-tool for On-Chip Cell Experiment System

Figure 4..25 shows the roles of micro-tool for the micro-tool based cell experiment microchip. By using micro-tool, we can achieve elements for the cell experiment such as manipulation of cell, the fabrication of microstructure, and measurement of environment. These elements are very important to operate cell analysis in microchip. Employing micro-tool to these elements enables to add new characteristics to these elements.

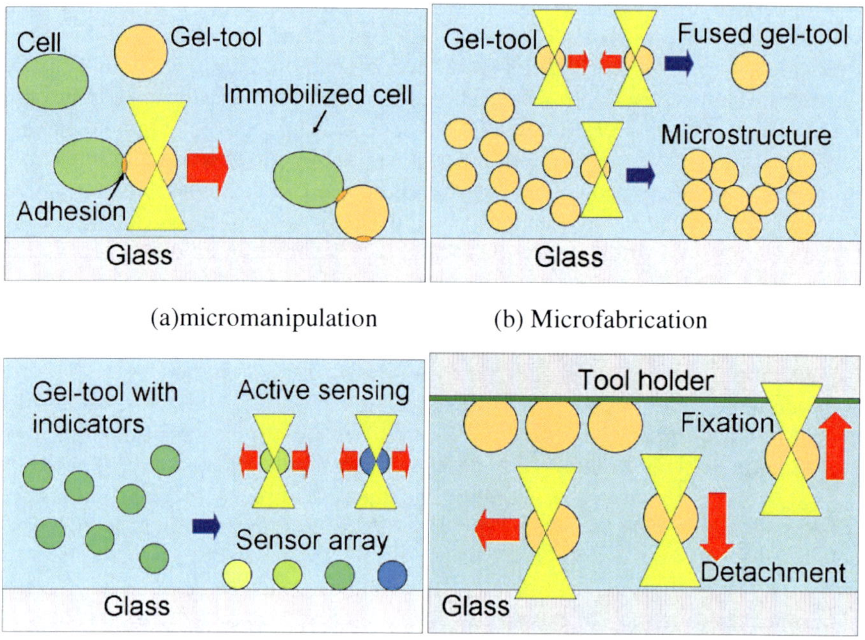

(a)micromanipulation (b) Microfabrication

(c) Local measurement (d) Local injection

Fig. 4.18 Elements for micro-tool based cell experiment microchip

In cell manipulation, Table 4. 3 shows the comparison of manipulation in microchip. Although electrical force and magnetic force are good at rotating the target at a fixed point, these forces cannot rotate the target once the target is away from the fixed point properly. It is difficult to rotate the object by optical tweezers without micro-tool. Micro-tool controlled by optical tweezers enables a control of cell position and cell attitude at the arbitrary point. In microfabrication of microstructure, conventional photofabrication cannot be applied in the microchip in which living cells exist. On the other hand, our proposed method can fabricate microstructures in the microchip in which living cells exist. In local measurement, Table 4. 4 shows the comparison of on-chip environment measurement. In μTAS, fluorescence and luminescence are often employed for on-chip measurement without micro sensor. However, these methods require the continuous supply of reagent. Integrated micro sensors have high precision and resolution. However, these sensors can measure at only fixed area. Microsensors made of micrtool can measure at a desired point of the microchip and arrange an arbitrary pattern of sensors. Local injection of micro-tool is very important to use micro-tool efficiency. Injecting micro-tool into the desired area with a desired number of micro-tool saves a loss of micro-tool and prevents the bad influence to recognition. By developing these techniques of micro-tool, we will be able to achieve a cell experiment system with micro-tool for on-chip cell analysis.

Table 4.3 Comparison of manipulation in microchip

	Position control	Attitude Control	Position control at arbitrary point	Attitude control at arbitrary point
Electric force	Fair	Good	Poor	Poor
Magnetic force	Fair	Good	Poor	Poor
Optical tweezers	Fair	Poor	Fair	Fair
Employing microtool	Fair	Good	Fair	Good

Table 4.4 Comparison of on-chip environment measurement

	In-situ measurement	Precision	Measurement at desired pont	Stability
Fluorecence/ Luminecence	Yes	Good	Yes	Poor
Integrated microsensor	Yes	Good	No	Good
Employing microtool	Yes	Good	Yes	Good

4.4.2 On-Chip Cell Experiment System with Non-contact Manipulation of Micro-tool

Schematics of the micro-tool based cell experiment microchip are shown in Figure 4. 19. There are two type microchips. One is parallel analysis microchip and another is serial cell surgery microchip. In these microchips, each microchannel has a micro-tool chamber. Micro-tool is used for improving the cell experiment in the microchannel. In most on-chip cell experiments, cell is immobilized in the microchannel and treatment of cell is operated by not the manipulation of cell but reagent flow. However, controls of position, attitude, and combination of cells are necessary for the detailed analysis of cell properties. Moreover, these conditions are important in the single cell level experiments. By integrating our proposed technologies of micro-tool into the microchip, effective cell experiment microchip is expected.

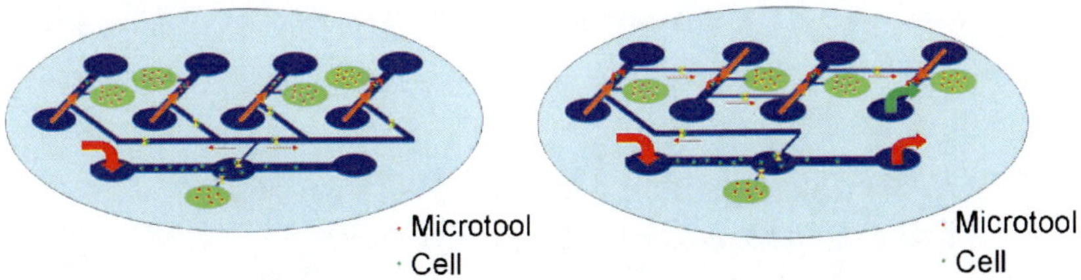

(a) Parallel cell analysis microchip (b) Serial cell surgery microchip

Fig. 4.19 Schematics of micro-tool based cell experiment microchip

To realize the on-chip cell experiment system with micro-tool, the non-contact manipulations of micro-tool such as cell manipulation, cell immobilization, and local environment measurement were performed in a microchip as shown in Figure 4..27. We also performed the pinpoint injection of micorotool into the desired area in a microchip. Micro-tool is manipulated by optical tweezers and electromagnetic force. We can select the control method of micro-tool according to the situation. We proposed two types of the cell immobilization method such as immobilization of cell inside the gel by *in-situ* photopolymerization of photo-crosslinkable resin and immobilization of cell by adhering to the surface of gel micro-tool. The former method is applied when we have to use micro-tool without cell adhesively, the latter method is applied when high-speed transport and immobilization after transport are needed.

In future work, we will realize an on-chip cell experiment system by integrating other devices such as a micropump, microvalve, and the laser scanning cytometer and PCR for detection.

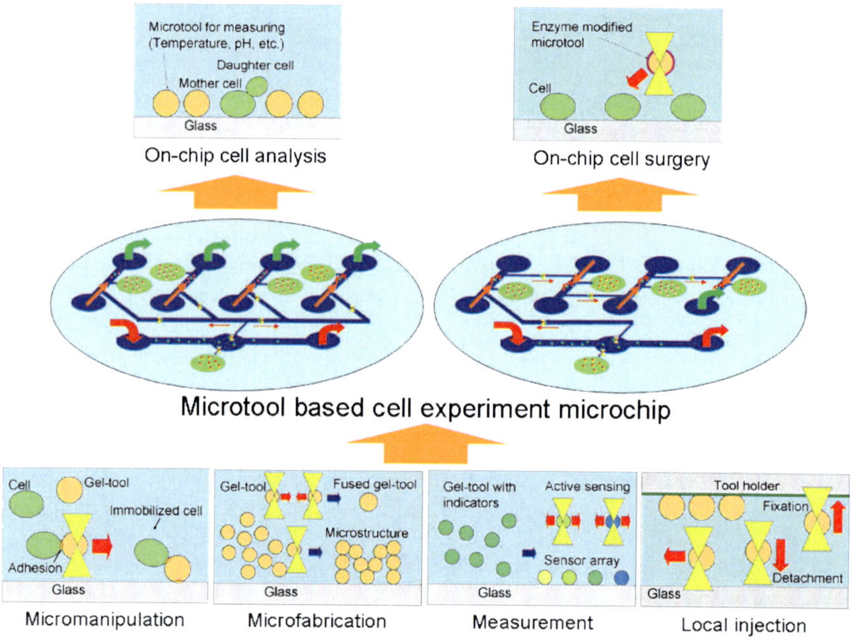

Fig. 4.20 Target of this research

4.4.3 *Reversible Injection Method of Microtool by Dielectrophoretic Floating of Microtool*

We developed the injection method of microtools at desired location in the microchamber [58]. However, the inner installation method (one of the injection methods of microtools) has a problem. In this method, microtools are fixed on the microtool holding chip by the water-soluble layer. Once the microtools are detached after dissolving the detergent, all microtools sink and most of them adhere to the bottom glass surface. After long time passed, they cannot be detached easily by the lasermicromanipulation. Therefore, it is necessary to develop a new method to keep the microtools floating in the microgroove after detaching the microtools from the microtool holding chip.

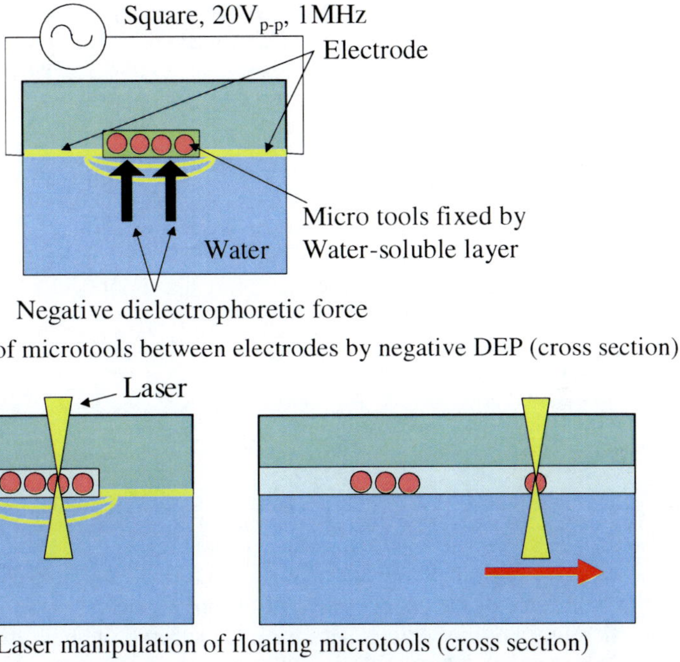

Fig. 4.21 Process of keeping microtools floating and manipulation of microtool by laser

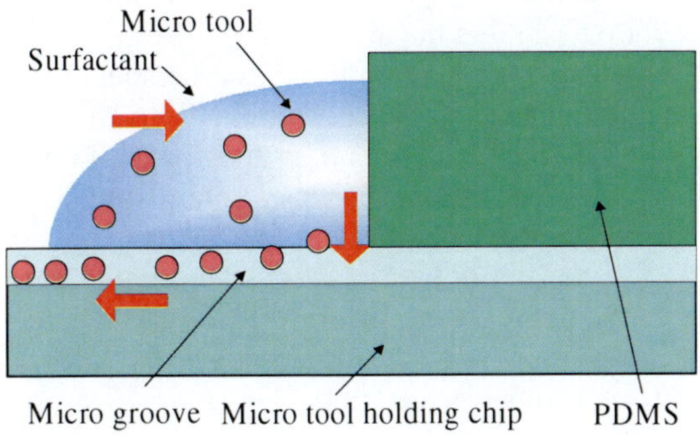

Fig. 4.22 Process of alignment of microtools on the microtool holding chip using surface tension

Figure 4. 21 shows the process of injecting the microtools in the microgroove while preventing them from sinking and adhere to the bottom surface and manipulation of microtools by laser. The microtools are aligned in the microgroove and fixed by the water-soluble layer. In the experiment, we used the PS microbead (nominal diameter: 10μm) as the microtool. The microtools receive the negative DEP force since relative permittivity of microtool (2.5 to 2.7) is smaller than that of water at a room temperature (80). After dissolving the detergent, microtools do not sink but are kept floating in the microgroove by negative DEP force. Floating microtools are manipulated by laser. When microtools exist in the microgroove, microtools are kept floating in the microgroove by negative DEP force since the electrical field becomes weak toward the inside of the microgroove from FEM (Finite Element Method) analysis.

When the DEP force is stronger than the downward force, which is difference in force between the gravity force and buoyancy, working to microtool, microtools are kept in the microgroove. In this chapter, we used 10 m PS microbead as a microtool. The downward force is calculated from equation 3.5. F_{down} represents the downward force working to microtool. The downward force is about 0.26 pN. The DEP force inside the microgroove is at least more than 100 pN.

$$F_{down} = \frac{4}{3}\pi\rho' g\left(\frac{d}{2}\right)^3 - \frac{4}{3}\pi\rho g\left(\frac{d}{2}\right)^3 \tag{4.1}$$

Therefore, PS beads are kept flown inside the microgroove by negative DEP force. When microtools sink from microgroove (10μm below), electrical field around microtools becomes weak up and down. Microtools are moved upward since negative DEP force of the upper part of microtool is larger.

The microtool holding chip is set inside the separation chip after installing the microtools. shows the novel process of the autonomous alignment of microtools in the microgroove of the microtool holding chip using surface tension and gravity. We succeeded in the autonomous alignment of the microtools in the microgroove using surface tension. After fixing the microtools by the water-soluble layer, we set the microtool holding chip bottom up in the microchamber. Then, we applied the square wave voltage between the electrodes from –10V to 10V at 1 MHz. When the water-soluble layer was dissolved in water after soaking the chip, microtools started to sink in the microchamber. However, microtools were kept floating in the microgroove by the negative DEP force. In this way, it is possible to prevent the microtools adhering to the bottom surface. After detaching the microtools, they were kept floating in the microgroove successfully by the negative DEP force. We could manipulate the microtool in the microgroove by laser manipulation and transport it to the microchamber successfully.

4.5 Micro-tools for Lasermicromanipulations

Much research works on cell property analysis have been done for finding and understanding unknown cell properties. Because of the remarkable advancement of biology and bioengineering, single cell analysis has been received great interests. Single cell manipulation and single cell screening are needed for single cell analysis. For this purpose, optical tweezers employed for single cell manipulation because optical tweezers can manipulate an arbitrary single microorganism. Optical tweezers is powerful tool for single cell manipulation. However, this manipulation method has a problem. Direct irradiation of high power laser to the target may cause damage to the target. Moreover, it is difficult to control the attitude of the target by optical tweezers.

For secure manipulation and advanced manipulation of a single cell in a microchip, we proposed a newmicromanipulation of cell with microobject manipulated by non-contact force for the on-chip cell experiment. We call thismicromanipulation with micro-tool as ILM.

In this chapter, we described the necessity of micro-tool and previous research works related on micro-tool. We manipulated polystyrene microbead as a micro-tool. We also manipulated Lactobacillus bulgaricus (LB) as a bio-micro-tool. Microobjects fabricated by microprocessing can be applied as the micro-tool. Micro-tool is controlled by non-contact manipulation methods such as optical tweezers and magnetic force. For control of micro-tool by optical tweezers, TSS was developed. By using this manipulation technique, we can control multiple micro-tools simultaneously in microchip. Magnetic force is suitable for manipulating a big target such as an oocyte (over 100 $\mu m\phi$).

Optical tweezers have been used as a powerful tool in molecular and cell biology. However, their usefulness has been limited, particularly in vivo, by the potential for damage to specimens resulting from the trapping laser.

In previous research, rotating bacterial cell assay were investigated. Single cells of Escherichia coli were tethered to a glass by means of a single flagellum. Such cells rotate at rates proportional to their transmembrane proton potential [56]. Monitoring the rotation rates of cells subjected to laser permits a rapid and quantitative measure of their metabolic state. Employing this assay, photo damage throughout the near infrared region favored for optical trapping (790–1064 nm) was characterized. The trap was turned on at t = 0, after which there was a gradual decrease in rotation rate. The LD_{50} is depicted on the graph as the time at which the rotation rate decreased to 50% of the initial value.

The reduction of photo damage to cells might be possible by considering the above conditions. However, it is difficult to change the laser according to the cell properties of the wavelength dependence for photo damage. For the secure manipulation of the cell independently, we proposed to manipulate the cell with the laser trapped object as shown in Figure 4. 23. We call the object as micro-tool and manipulation technique using micro-tool as ILM. Direct irradiation of focused laser to the cell is avoided by manipulating the cell with micro-tool and reduction of photo damage attributed the wavelength and intensity of laser is expected.

Except for reducing photo damage, micro-tool is expected advanced on-chipmicromanipulation. When manipulating microobject of which refractive index is smaller than that of the solvent, laser scanning is needed for manipulating the object. By using micro-tool, we can manipulate the target regardless of the refractive index of the target as shown in Figure 4. 24. Figure 4. 25 shows advanced manipulations with micro-tool such as multiple points contact by multiple lasers or scanned laser.

Moreover, the concept of micro-tool can be extended not onlymicromanipulation but also other elements of the on-chip cell experiments such as cell immobilization, microfabrication of microstructure, and local environment measurement.

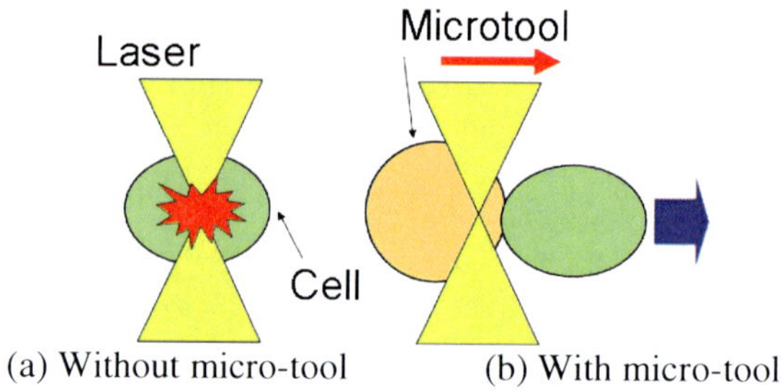

Fig. 4.23 Manipulation regardless of target refractive index

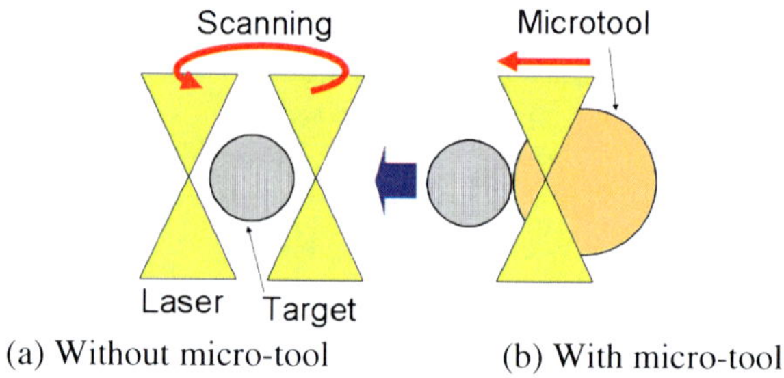

Fig. 4.24 Manipulation regardless of target refractive index

Conventional Research on Micro-tool

Although we use microbead mainly as micro-tool, micro-tool can be fabricated by microprocessing techniques such as photolithography and etching process.

As an example of microstructure made by photolithography, we explain about an optical windmill which is rotated by irradiating laser as shown in Figure 4. 26. Optical tweezers is usually used to trap a relatively symmetrical object. When irradiating laser to the asymmetrical object, laser potential cannot keep the symmetrical property of the potential and the windmill starts to rotate. Nowadays, the fabrication of micromixer based on this phenomenon in μTAS is studied.

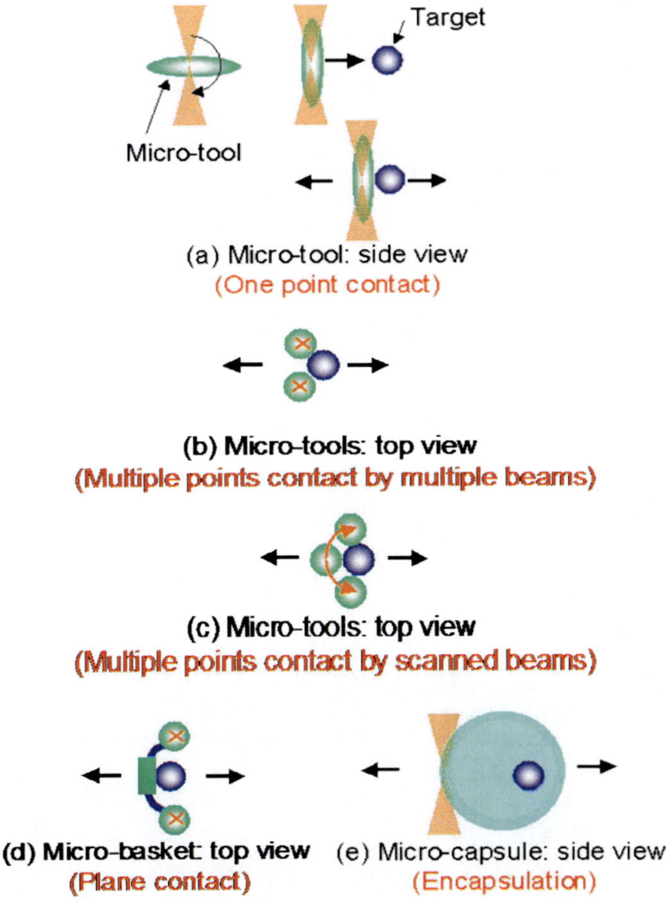

Fig. 4.25 Advanced manipulation method using micro-tools

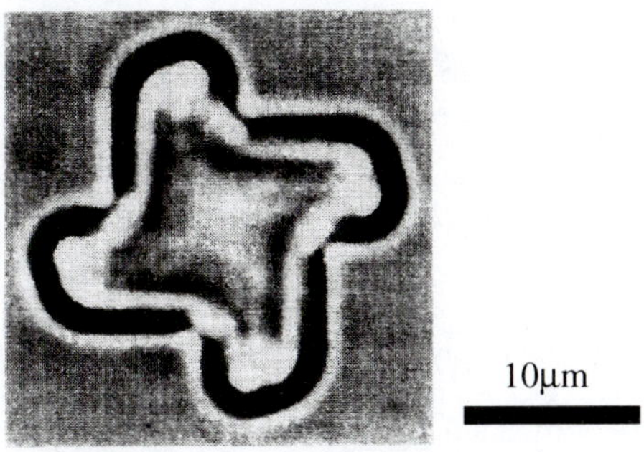

Fig. 4.26 Photograph of optical windmill

A mechanism of rotation is shown in Figure 4. 27. Optical pressure generates from inside to outside when irradiating laser to the windmill from an upper side. This pressure rotates the windmill.

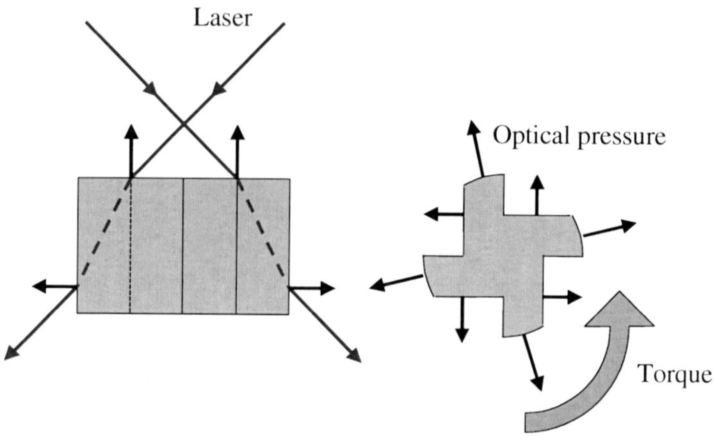

Fig. 4.27 Mechanism of rotating the optical windmill

Figure 4. 28 shows the relation of laser power and rotational frequency. The larger N. A. becomes, the higher rotation speed becomes.

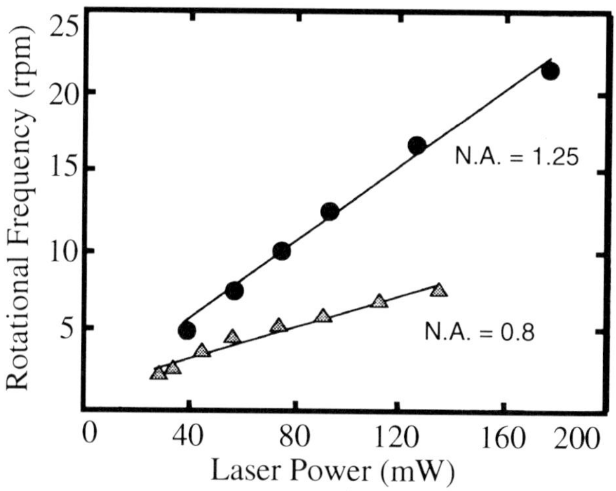

Fig. 4.28 Relation between laser power and rotational frequency

Micro-tool can be fabricated other methods. Next example is a fabrication of micro-tool with three fingers in FESEM as shown in Figure 4. 27. This micro-tool was fabricated by 2 steps: (1) Binding two polystyrene microbead, (2) Depositing on the surface of the microbead by EBID. Purpose of this micro-tool is to manipulate cell with deposited micro fingers.

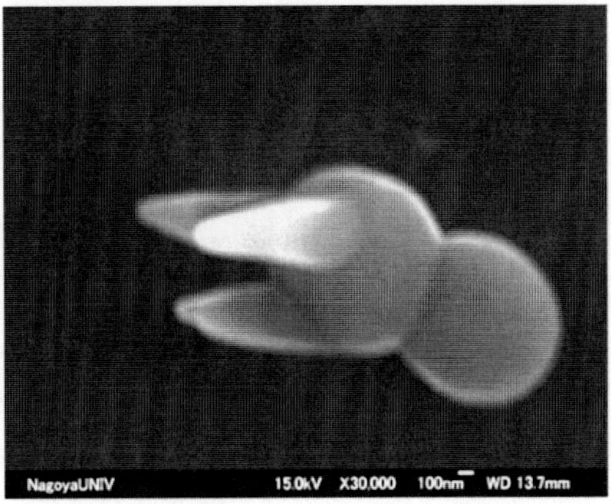

Fig. 4.29 Micro-tool fabricates in FESEM

As this type of unique micro-tool, micro-tool with arbitrary shape can be fabricated by various ways. As a manipulation method of micro-tool, general method is the manipulation of single micro-tool by using optical tweezers. as shown in Figure 4. 30. We also manipulate LB as bio-micro-tool as shown in Figure 4. 31.

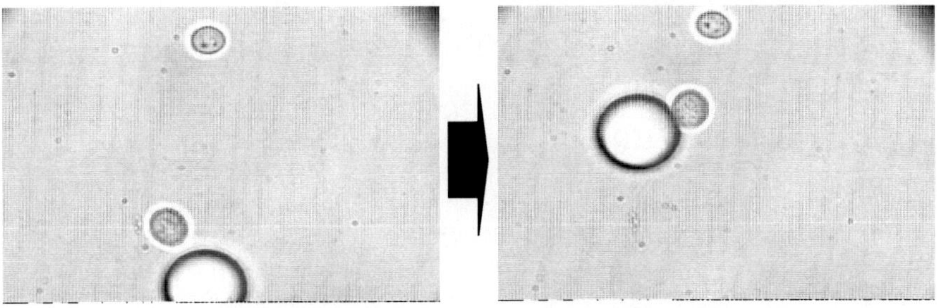

Fig. 4.30 Manipulation using polystyrene bead

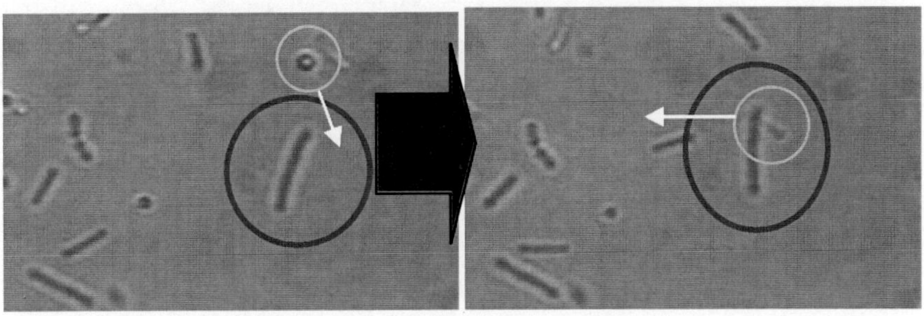

Fig. 4.31 Manipulation using bio – micro-tool

References

[1] Jorgenson, J.W., Lukacs, K.D.: Free-zone electrophoresis in glass capillaries. Clinical Chemistry 27(9), 1551–1553 (1981)

[2] Pohl, H.A.: Dielectrophoresis. Cambridge University Press, Cambridge (1978)

[3] Pethig, R.: Application of A.C. electrical fields to the manipulation and characterization of cells. In: Karube, I. (ed.) Automation in Biotechnology, pp. 159–185. Elsevier (1991)

[4] Schnelle, T., Hagedorn, R., Fuhr, G., Fiedler, S., Müller, T.: Three-dimensional electric field traps for manipulation of cells-calculation and experimental verification. Biochem. Biophys. Acta 1157, 127–140 (1993)

[5] Fuhr, G., Shirley, G.S.: Cell handling and caracterization using micron and submicron electrode arrays: state of the art perspectives of seiconductormicrotools. J. Michromech. Microeng. 5, 77–85 (1995)

[6] Fuhr, G., Glasser, H., Müller, T., Schnelle, T.: Cell manipulation and cultivation under a.c. electric field influence in highly conductive culture media. Biochim. Biophys. Acta 1201, 353–360 (1994)

[7] Müller, T., Gerardio, A., Schnelle, T., Shirley, G.S., Bordoni, F., De-Gasperis, G., Leoni, R., Fuhr, G.: Trapping of micron and submicron particles by high frequency fields and hydrodynamic forces. J. of Physics 29, 340–349 (1996)

[8] Fuhr, G., Hagedorn, R., Göring, H.: Separation of different cell types by rotating electricfields. Plant Cell Physiol. 26(8), 1527–1531 (1985)

[9] Ashkin, A., Dziedzic, J.M., Bjorkholm, J.E., Chu, S.: Observation of a single-beam gradient force optical trap for dielectric particles. Opt. Lett. 11, 288–291 (1986)

[10] Mio, Marr, D.W.M.: Adv. Mater. 12, 917 (2000)

[11] Micro Chemistry, Masuhara Microphotoconversion, Kagakudoujin (1993) (in Japanese)

[12] Arai, F., Sakami, T., Yosikawa, K., Maruyama, H., Fukuda, T.: Proc. of the 2003 IEEE/RSJ Intl. Conference on Intelligent Robots and Systems, 2121 (2003)

[13] Arai, F., Yoshikawa, K., Sakami, T., Fukuda, T.: Synchronized laser micromanipulation of multiple targets along each trajectory by single laser. Applied Physics Letters 85(19), 4301–4303 (2004)

[14] Harrison, D.J., Manz, A., Fan, Z., Lüdi, H., Widmer, H.M.: Capillary Electrophoresis and Sample Injection Systems Integrated on a Planar Glass Chip. Anal. Chem. 64, 1926–1932 (1992)

[15] Saiki, R.K., Scharf, S., Faloona, F., Mullis, K.B., Horn, G.T., Erlich, H.A., Arnhein, N.: Enzymatic Amplification of Beta-Globin Genomic, Sequences and Restriction Site Analysis for Diagnosis of Sickle Cell Anemia. Science 230, 1350–1354 (1988)

[16] Harrison, D.J., Fluri, K., Seiler, K., Fan, Z., Eenhauser, C.S., Manz, A.: Micromachining a Miniaturized Capillary Electrophoresis-Based Chemical Analysis System on a Chip. Science 261 (August 13, 1993)

[17] Seiler, K., Harrison, D.J., Manz, A.: Planar Glass Chips for Capillary Electrophoresis: Repetitive Sample Injection, Quantation, and Separation Efficiency. Anal. Chem. 65, 1481–1488 (1993)

[18] ERenhauser, C.S., Manz, A., Windmer, H.M.: Glass Chips for High-Speed Capillary Electrophoresis Separations with Submicrometer Plate Heights. Anal. Chem. 65, 2637–2642 (1993)

[19] Jacobson, S.C., HergenrAoder, R., Koutny, L.B., Ramsey, J.M.: High-Speed Separation on a Microchip. Anal. Chem. 66, 1114–1118 (1994)

[20] Moore Jr., W., Jacobson, S.C., Ramsey, J.M.: Microchip Separations of Neutral Species via Micellar Electrokinetic Capillary Chromatography. Anal. Chem. 67, 4184–4189 (1995)

[21] Fan, Z.H., Harrison, D.J.: Micromachining of Capillary Electrophoresis Injectors and Separators on Glass Chips and Evaluation of Flow at Capillary Intersections. Anal. Chem. 66, 177–184 (1994)

[22] Effenhauser, C.S., Paulus, A., Manz, A., Widmer, H.M.: High-Speed separation of Antisense Oligonoucleotides on a Micromachined Capillary Electrophoresis Device. Anal. Chem. 66, 2949–2953 (1994)

[23] Jacobson, S.C., Moore, A.W., Ramsey, J.M.: Fused Quartz Substrates for Microchip Electrophoresis. Anal. Chem. 67, 2059–2063 (1995)

[24] Jacobson, S.C., Ramsey, J.M.: Electrokinetic Focusing in Microfabricated Channel Structures. Anal. Chem. 69, 3212–3217 (1997)

[25] Jacobson, S.C., Ramsey, J.M.: Integrated Microdevice for DNA Restriction Fragment Analysis. Anal. Chem. 68, 720–723 (1996)

[26] Fluri, K., Fitzpatrick, G., Chiem, N., Harrison, D.J.: Integrated Capillary Electrophoresis Devices with an Efficient Postcolumn Reactor in Planar Quartz and Glass chips. Anal. Chem. 68, 4285–4290 (1996)

[27] The Incredible Shrinking Laboratory, Research News, Science, vol. 268 (April 7, 1995)

[28] Woolley, T., Mathies, R.A.: Ultra-high-speed DNA fragment separations using microfabricated capillary array electrophresis chips. Proc. Natl. Acad. Sci. USA 91, 11348–11352 (1994)

[29] Woolley, T., Mathies, R.A.: Ultra-high-speed DNA Sequencing Using Capillary Electrophresis chips. Anal. Chem. 67, 3676–3680 (1995)

[30] Wooley, T., Hadley, D., Landre, P., de Mello, A.J., Mathies, R.A., Northrup, M.A.: Functional Integration of PCR Amplification and Capillary Electrophoresis in a Microfabricated DNA Analysis Device. Anal. Chem. 68, 4081–4086 (1996)

[31] Foong, S., Wood, K.L., Busch-Vischniac, I.: Design Assessment of Micro-Electro-Mechanical Systems with Applications to a Microbiology Cell Injector. Micromechanical Systems SME, DSC 46, 49–63 (1993)

[32] Swerdlow, H., Jones, B.J., Wittwer, C.T.: Fully Automated DNA Reaction and Analysis in a Fluid Capillary Instrument. Anal. Chem. 69, 848–855 (1997)

[33] Meldrum, R.: A Biomechatronic Fluid Sample Handling System for DNA Processing. In: Proc. of Advanced Intelligent Mechatronics (1997)

[34] Gavin, P.F., Ewing, A.G.: Continuous Separation with Microfabricated Electrophoresis-Electrochemical Array Detection. J. Am. Chem. Soc. 118, 8932–8936 (1996)

[35] Pohl, H.A.: Dielectrophoresis. Cambridge University Press, Cambridge (1978)

[36] Müller, T., Gradl, G., Howitz, S., Shirley, S., Schnelle, T., Fuhr, G.: A 3-D microelectrode system for handling and caging single cells and particles. Biosensors & Bioelectronics 14, 247–256 (1999)

[37] Fu, A.Y., Spence, C., Scherer, A., Arnold, F.H., Quake, S.R.: A microfabricated fluorescence-activated cell sorter. Nature Biotechnology 17, 1109–1111 (1999)

[38] Arai, F., Ichikawa, A., Ogawa, M., Fukuda, T., Horio, K., Itoigawa, K.: High-speed separation system of randomly suspended single living cells by laser trap and dielectrophoresis. Electrophoresis 22(2), 283–288 (2001)

[39] Ichikawa, A., Arai, F., Yoshikawa, K., Uchida, T., Fukuda, T.: In situ formation of a gel microbead for indirect laser micromanipulation of microorganisms. Applied Phys. Lett. 87, 191108 (2005)

[40] Voldman, J., Gray, M.L., Toner, M., Schmidt, M.A.: A Microfabrication-Based Dynamic Array Cytometer. Anal. Chem. 74, 3984–3990 (2002)
[41] Arai, F., Ichikawa, A., Fukuda, T., Katsuragi, T.: Isolation and extraction of target microbes using thermal sol-gel transformation. Analyst (6), 547–551 (2003)
[42] Moriguchi, H., Wakamoto, Y., Sugio, Y., Takahashi, K., Ippei, I., Yasuda, K.: An agar-microchamber cell-cultivation system: flexible change of microchamber shapes during cultivation by photo-thermal etching. Lab Chip 2, 125–130 (2002)
[43] Wakamoto, Y., Ramsdenb, J., Yasuda, K.: Single-cell growth and division dynamics showing epigenetic correlations. Analyst 130, 311–317 (2005)
[44] Chang, C.-H., Ogawa, H., Oki, A., Takai, M., Nagai, M., Hisamoto, H., Horiike, Y.: Healthcare Chip Based on Integrated Electrochemical Sensors Used for Clinical Diagnostics of Bun. Japanese Journal of Applied Physics 45(5A), 4241–4247 (2006)
[45] Mills, A., Tommons, C., Bailey, R.T., Tedford, M.C., Crilly, P.J.: Luminescence temperature sensing using poly(vinyl alcohol)-encapsulated Ru(bpy)32+ films. Analyst 131, 495–500 (2006)
[46] Jain, M.K., Caiand, Q., Grimes, C.A.: A wireless micro-sensor for simultaneous measurement of pH, temperature, and pressure. Smart Mater. Struct. 10, 347–353 (2001)
[47] Arai, F., Endo, T., Yamauchi, R., Fukuda, T.: 3D 6DOF Manipulation of Microbead by Laser Tweezers. Journal of Robotics and Mechatronics 18(2), 153–159 (2006)
[48] Arai, F., Endo, T., Yamuchi, R., Fukuda, T., Shimizu, T., Kamiya, S.: 3D Manipulation of Lipid Nanotubes with Photo-crosslinkable Gel Micro-beads for Nano-assembly. In: Proc. of mTAS 2006, pp. 233–235 (2006)
[49] Gauthier, M., Piat, E.: An electromagnetic micromanipulation system for single-cell manipulation. Journal of Micromechatronics 2(2), 87–119 (2004)
[50] Onda, K., Arai, F.: Multi-beam bilateral teleoperation of holographic optical tweezers accelerated by general-purpose GPU. Optics Express 20(4), 3642–3653 (2012)
[51] Fukuda, T., Tanie, K., Mitsuoka, T.: A Study on Control of a Micromanipulator (1st Report, The basic feature of Micro Gripper and one method of Bilateral control). In: Proc. Micro Robots and Teleoperators Workshop, Hyannis, MA (November 1987)
[52] Arai, F., Andou, D., Fukuda, T., Noda, Y., Ota, T.: Micro Manipulation Based on Micro Physics, Strategy Based on Attractive Force Reduction and Stress Measurement. In: Proc. Int. Conf. on Intelligent Robotics and Systems (IROS), vol. 2, pp. 236–241 (1995)
[53] Arai, D., Andou, Y., Nonoda, T., Fukuda, H., Iwata, K.: Micro Endeffector with Micro Pyramids and Integrated Piezoresistive Force Sensor. In: Proc. Int. Conf. on Intelligent Robotics and Systems (IROS), vol. 2, pp. 842–849 (1996)
[54] Arai, F., Ichikawa, A., Ogawa, M., Fukuda, T., Horio, K., Itoigawa, K.: High-speed separation system of randomly suspended single living cells by laser trap and dielectrophoresis. Electrophoresis 22(2), 283–288 (2001)
[55] Neuman, K.C., Chadd, E.H., Liou, G.F., Bergman, K., Block, S.M.: Characterization of Photdamage to Escherichia coli in Optical Traps. Biophysical Journal 77, 2856 (1999)
[56] Liang, H., Vu, K.T., Krischnan, P., Tang, T.C., Shin, D., Kimel, S., Bers, M.W.: Wavelength dependence of cell cloning efficiency after optical trapping. Biophysical Journal 70, 1529 (1996)
[57] Sasaki, K., Koshioka, M., Misawa, H., Kitamura, N., Masuhara, H.: Pattern Formation and Flow Control of fine Particles by Laser Scanning Micromanipulation. Optics Letters 16(19), 1463–1465 (1991)
[58] Arai, F., Maruyama, H., Ichikawa, A., Sakami, T., Fukuda, T.: Pinpoint Injection and Dielectrophoretic Floating of Autonomously Aligned Microtools. In: Proc. of μTAS 2002, pp. 548–550 (2002)

Chapter 5
Rotational Speed Control of Single Bacterial Flagellar Motor

5.1 Background of Rotational Speed Control of Single Bacterial Flagellar Motor

5.1.1 Conventional Works

Some researchers proposed the methods to fabricate the bacteria-driven microobjects using various flagellated bacteria. Darnton *et al.* achieved a random delivery of the bacteria-driven microobject using *Serratia marcescens* [1]. Behkam *et al.* improved the directivity of the delivery of the bacteria-driven microobject using *S. marcescens* by limiting the attaching area of the bacteria [2]. Martel *et al.* globally control the direction of the delivery of the bacteria-driven microobjects using *Magnetococcus* by the magnetic field [3,4]. *Chlamydomonas reinhardtii*, which is a beating type flagellated microbe but not a bacterium, was used to control the direction of the delivery of the microobjects globally using the light [5].

These conventional techniques control the directions of the continuously moving bacteria-driven microobjects. If the moving speed can be controlled by precise and direct control of the bacterial propulsion, it is expected to control the positions of the bacteria-driven microobjects more precisely.

5.1.2 Principal of Flagellar Motor

Many bacteria swim actively in liquid by rotating their flagella and migrate toward favorable directions. The bacterial flagellum has a helical filament that acts as a propeller. Each flagellum consists of the helical filament extending from the cell body, the basal body embedded in the cell surface, and the flexible hook that connects them [6,7,8,9]. More than 20 structural proteins are required for this organelle (Figure 5. 1). The bacterial flagellar motor is a molecular machine that converts ion-motive force (IMF) into the mechanical force; the energy source of the rotation is the electrochemical gradient of H^+ or Na^+ ion across the cytoplasmic membrane. The stator of the flagellar motor consists of PomA and PomB in the Na^+-driven motors of *V. alginolyticus* or MotA and MotB in the

H^+-driven motors of *E. coli* [10,11,12]. These stator complexes are thought to function as specific ion channels. Furthermore, it is already revealed that chimeric protein PotB (the C-terminal domain of B subunit of *V. alginolyticus* was exchanged by that of *E. coli*) act as the Na^+-driven type. Recently, detailed single cell analysis revealed the step size of the flagellar motor rotation [13]. However, the mechanism of the flagellar motor rotation is still largely unknown.

To realize the precise and direct control of the bacterial propulsion, firstly it is needed to establish the method to analyze the detailed properties of the bacteria and the bacterial flagellar motors, and the method to manipulate the bacterial propulsion.

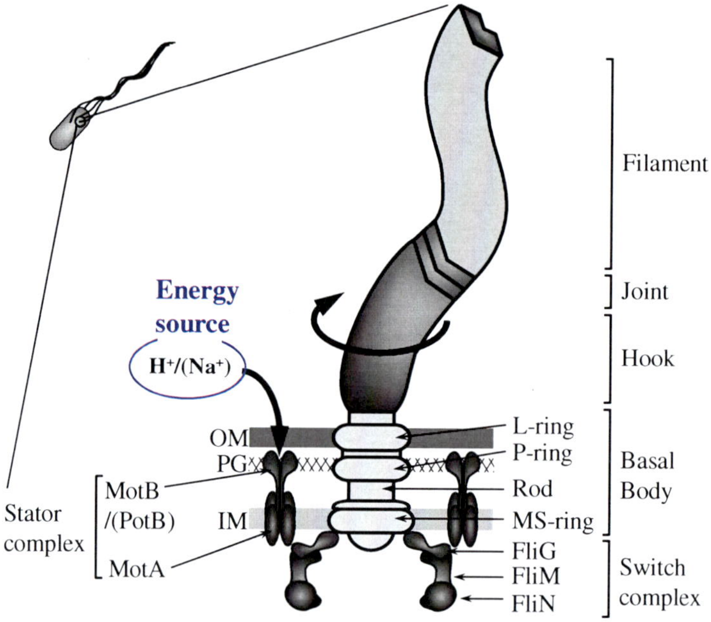

Fig. 5.1 Schematic of the bacterial flagellum. The substructures and components of the flagellum of *E. coli*. The heavily shaded components are essential components for torque generation and switching (MotA, MotB/PotB, FliG, FliM, and FliN). H^+ /Na^+ influx through the stator complex is believed to generate torque at the interface between the stator component of MotA and the rotor component of FliG. OM: Outer membrane; PG: peptidoglycan: IM: inner membrane.

5.1.3 Research Goal of Micro-Nanorobots Using Flagellar Motor

Toward the realization of the micro-nanorobots, recently, the flagellated bacteria are used as the propulsions for the micro-nano objects because of 1) the self-propagation and the metabolic function, 2) the high efficiency energy conversion, 3) the natural sensors and systems to regulate the propulsion, and 4) the possibility to be engineered the functions. To realize the precise control of the bacteria-driven micro-nanorobots, the precise and direct control of the bacterial propulsion is necessary. For the control of the bacterial propulsion, it is needed to develop the methods to analyze the detailed properties of the bacteria, such as the bacterial propulsion, and also to develop the methods to manipulate the bacterial propulsion. So, the local environmental control technique using the nanopipette is considered to be an effective technique to control the local reagents/ions concentrations, and is desired to analyze and manipulate the bacteria and the bacterial flagellar motors.

Therefore, we developed the micro-nanotools, such as nanopipette, to construct novel local environmental control systems. Then I propose and construct the novel local environmental control systems to analyze and manipulate the bacteria and the bacterial propulsion. Finally, toward the realization of the bacteria-driven micro-nanorobots for such as the DDS or the micro-nanosurgery inside the body (Figure 5. 2), I propose and demonstrate the methods to control the bacterial propulsion by the constructed novel local environmental control systems.

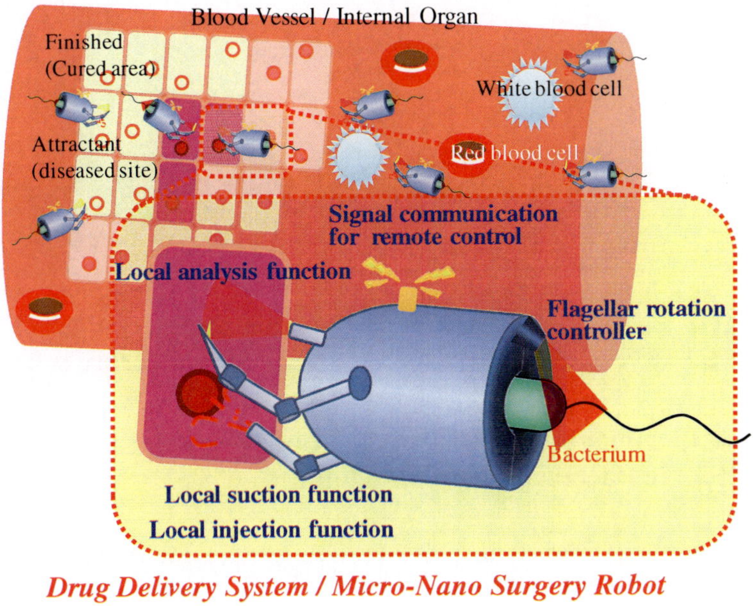

Fig. 5.2 Schematic of the future bacteria-driven medical micro-nanorobots

5.1.4 Driving Force Generated by Flagellum

Under the situation that the bacterial cell is swimming at the constant speed v (Figure 5.19), Magariyama *et al.* defined the driving force generated by the flagellar motor (F_D) as

$$F_D = \alpha_f v + \gamma_f \omega_f \tag{5.1}$$

on the assumption that the flagellar filament is the helical rigid body and rotates about the axis of the helix [92] (where ω_f is the rotational speed of the flagellum, α_f and γ_f are the drag coefficients depending on the shape of the flagellar filament such as the radius (r_f) and the pitch (p_f) of the flagellar helix and the radius (d_f) and the length (L_f) of the flagellar filament).

Under this assumption, the driving force of the flagellar motor (F_D) can be controlled by controlling the swimming speed of the bacterial cell (v) or the rotational speed of the flagellum (ω_f). Generally, the swimming speed of the bacterial cell (v) is roughly proportional to the rotational speed of the flagellum (ω_f) [92]. So, basically, the driving force generated by the flagellum can be controlled by controlling the rotational speed of the flagellar motor. So, in this chapter, I control and measure the rotational speed of Na^+-driven bacterial flagellar motor in single *E. coli* cell to control the flagellar driving force.

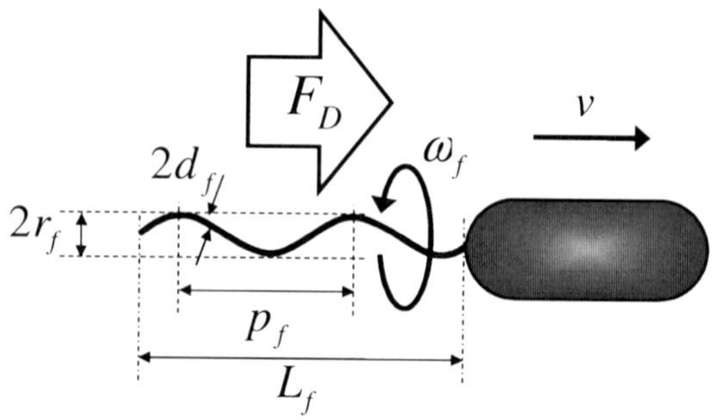

Fig. 5.3 Schematic of the driving force generated by bacterial flagellar motor

5.2 Experimental Set-Up for Rotational Speed Control of Single Bacterial Flagellar Motor

The nanopipette is expected to be an effective tool to analyze more detailed and local information on the single cells [14]. So, in the last two chapters, for the analysis of the bacteria and the bacterial propulsion in single cell level, I developed the nanopipette and theoretically analyzed the discharging amount from the nanopipette.

Piper *et al.* used a glass nanopipette for local and repeatable delivery of water-soluble reagent in ionic solution to make localized controlled changes in reagent concentration on a surface [15]. Then they applied their method to control the Na^+-driven flagellar motor in single *E. coli* cell by dosing Na^+ ions. Observing the dynamic response of the flagellar motor like this will make a great contribution to clarify the mechanism or the control characteristic of the flagellar motor rotation. However, their system was not enough to make an intended dynamic change in the reagents/ions concentrations. When they dose Na^+ ions to accelerate the rotational speed of the Na^+-driven flagellar motor, they can only wait for the diffusion of Na^+ ions to relax the rotational speed. If both drastic increase and decrease in the Na^+ ion concentration can be generated, as shown in Figure 5.4, it will be possible to observe the flagellar dynamic response independent from the influence of the Na^+ ion diffusion. Therefore, the method to control the local reagents/ions concentrations more dynamically and arbitrarily is needed for more detailed analysis of the mechanism or the control characteristic of the flagellar motor rotation.

In addition, the dynamic and arbitrary controlled change in the local reagents/ions concentrations will be useful as a chemical stimulation or a trigger to analyze or control the biological samples, such as DNA, ion channel, protein, neuronal cell and so on [16]. Therefore, in this chapter, I propose the local environmental control system with the micro-nano dual pipettes to produce the dynamic and arbitrary changes in the local reagents/ions concentrations.

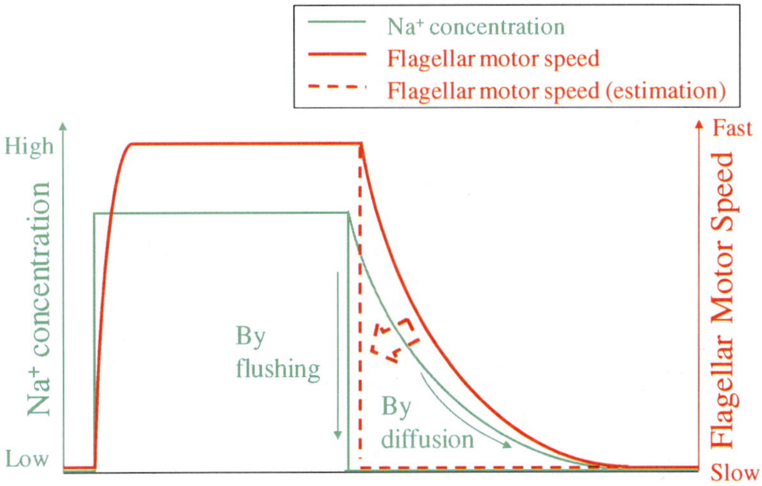

Fig. 5.4 Schematic of the changing aspects of Na^+ ion concentration by diffusion and flushing

5.2.1 Switching Discharge between Micro-Nano Dual Pipettes

When the discharge of certain reagents/ions-containing solution is conducted with only single pipette as Piper *et al.* did, diffusion of the certain reagents/ions is the only way to reduce the certain reagents/ions concentrations (Figure 5.4). In addition, even if the solution is not being discharged, the reagents/ions diffuse through the pipette-bath interface and are unintentionally supplied to the target. In this case, the response of the target is affected by the diffusion rate. In fact, as for the flagellar motor, I confirmed that the rotational speed of the flagellar motor was affected by diffusion of Na^+ ions in the experiment with a single pipette (data not shown). So, for more detailed analysis of the response of the target against the intentional and rapid change (both increase and decrease) of the certain reagents/ions concentrations, such as the response of the Na^+-driven flagellar motor against the intentional and rapid change of the Na^+ ions, I constructed the local environmental control system with micro-nano dual pipettes, as shown in Figure 5.5.

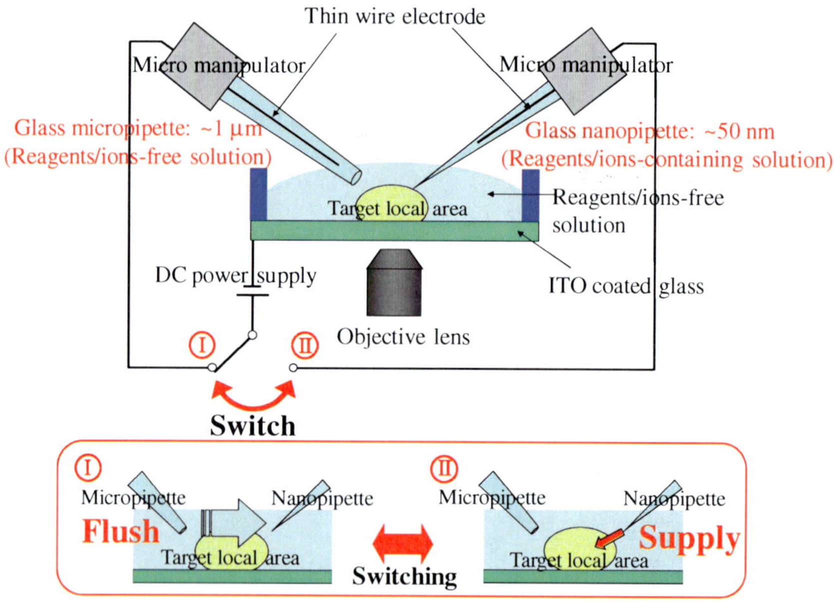

Fig. 5.5 Schematic of the local environmental control system with micro-nano dual pipettes to produce the dynamic and arbitrary changes in local reagent/ion concentration

This system has two pipettes. The glass nanopipette with a ~50 nm inner diameter (Figure 5.6) was used for the discharge of the certain reagents/ions-containing solution. (The glass nanopipettes were fabricated by heating and pulling the borosilicate glass tubes (GD-1, Narishige) with the puller. Generally, the minimum inner diameter of the tip of the glass nanopipette is ~10 nm [17].) In contrast, the glass micropipette with a ~1 μm inner diameter was used for the discharge of the certain reagents/ions-free solution. The bath was filled with the reagents/ions-free solution. The DC voltage was applied between ITO coated glass at the bottom of the bath and the thin wire electrodes inserted into the pipettes. (The appearance of the local environmental control system with micro-nano dual pipettes is shown in Figure 5.7.) I can switch the discharge between the reagents/ions-containing and the reagents/ions- free solutions by switching the applied voltage. Through this dual pipettes system, the certain reagents/ions are supplied to the target in the arbitrary local area by discharge from the nanopipette. Then the reagents/ions diffusing from the nanopipette and the reagents/ions remaining around the target can be flushed to immediately stop the feeding of the reagents/ions to the target by discharge from the micropipette.

These pipettes are fixed on the two micromanipulators, respectively (One is the hydraulic micromanipulator. The other is the piezoelectric micromanipulator. So, it is possible to approach the pipettes to the arbitrary target, and the target-pipette distances can be controlled independently.

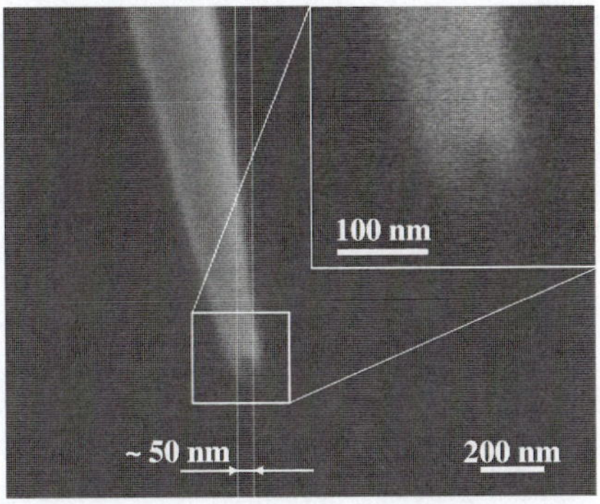

Fig. 5.6 A glass nanopipette (SEM image). The upper right embedded image is the magnified image of the tip.

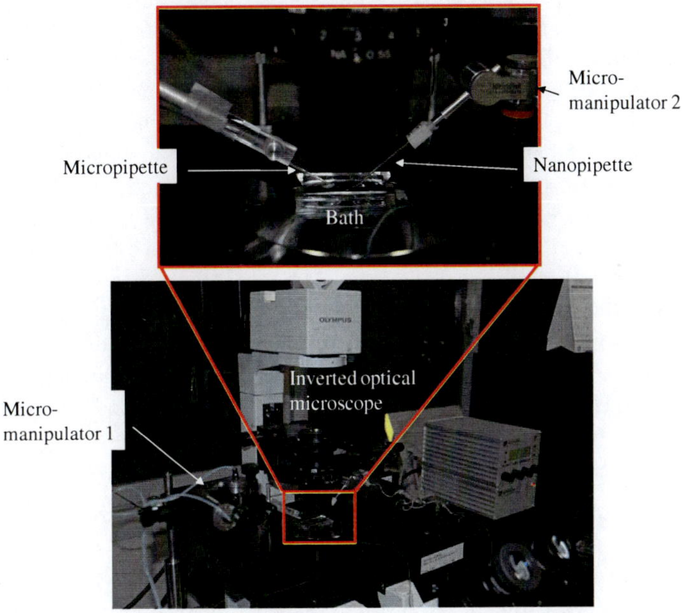

Fig. 5.7 Overview of the appearance of the local environmental control system with micro-nano dual pipettes

5.2.2 Simultaneous Discharge from Micro-Nano Dual Pipettes

The flow velocity (v) of the electro-osmotic flow is estimated by the following Smoluchowski's equation [18],

$$v = \frac{\varepsilon \cdot \zeta}{\eta} E \tag{5.2}$$

where ε : dielectric constant of migrating solution, η : viscosity coefficient of migrating solution, ζ: zeta potential between inner wall and solution, and E : electric field intensity. The product of the flow velocity and the cross-sectional area of the ejection hole gives the flow rate. As the zeta potential (ζ), the value of the zeta potential between glass and 1 mM NaCl aqueous solution 58.4 mV was adopted. The inner diameter of ONT was ~40 nm. And the average of the X-component 1.47×10^4 V/m was used as the electric field intensity (E).

The flow rate was estimated at 0.75 fl/s. This result shows the possibility to control the local environment of the inside of bacteria whose volume is several femto liters or more by the discharge in millisecond order.

With the micro-nano dual pipettes system, when the discharge is continued for a long period, the local concentration of the reagents or ions get into a certain level at any applied DC voltages. It might be caused by the equilibrium between diffusion and supply of the reagents/ions. Therefore, next, I upgrade my local environmental control system with micro-nano dual pipettes to achieve the steady-state control of the local reagents/ions concentrations without changing the concentration of the discharging solution.

As the zeta potential (ζ), the value of the zeta potential between glass and 1 mM NaCl aqueous solution 58.4 mV was assumed. The inner diameter of glass nanopipette was ~50 nm. Then the average of the X-component 2.21×10^3 V/m calculated by electrostatic field analysis was used as the electric field intensity (E). The flow rate of the glass nanopipette was estimated at 0.18 fl/s by equation (3.1). In the same way, the flow rate of the glass micropipette was estimated at 71 fl/s by equation (3.1).

Comparing the flow rates of the glass nanopipette and micropipette, the flow rate of the micropipette is ~400 times as high as the flow rate of the nanopipette. Therefore, it is expected that the reagents/ions diffusing from the nanopipette and the reagents/ions remaining around the target can be flushed immediately by discharge from the micropipette.

5.2.3 Upgrade of Micro-Nano Dual Pipettes System

With the micro-nano dual pipettes system, when the discharge is continued for a long period, the local concentration of the reagents or ions get into a certain level at any applied DC voltages. It might be caused by the equilibrium between diffusion and supply of the reagents/ions. Therefore, next, I upgrade my local environmental control system with micro-nano dual pipettes to achieve the steady-state control of the local reagents/ions concentrations without changing the concentration of the discharging solution.

The upgraded local environmental control system with micro-nano dual pipettes is shown in Figure 5.8. This system also has two pipettes. The glass nanopipette with a ~50 nm inner diameter was used for the discharge of the certain reagents/ions- containing solution. In contrast, the glass micropipette with a ~1 μm inner diameter was used for the discharge of the certain reagents/ions-free solution. The bath was filled with the reagents/ions-free solution. The DC voltage was applied between ITO coated glass at the bottom of the bath and the thin wire electrodes inserted into the pipettes. I can turn on/off the discharge of the reagents/ions-containing and the reagents/ions-free solutions independently, and can change the discharging velocity (volume) of the solutions by changing the applied DC voltages independently.

These pipettes are fixed on the two micro manipulators respectively. So, it is possible to approach the pipettes to the arbitrary target, and the target-pipette distances can be controlled independently.

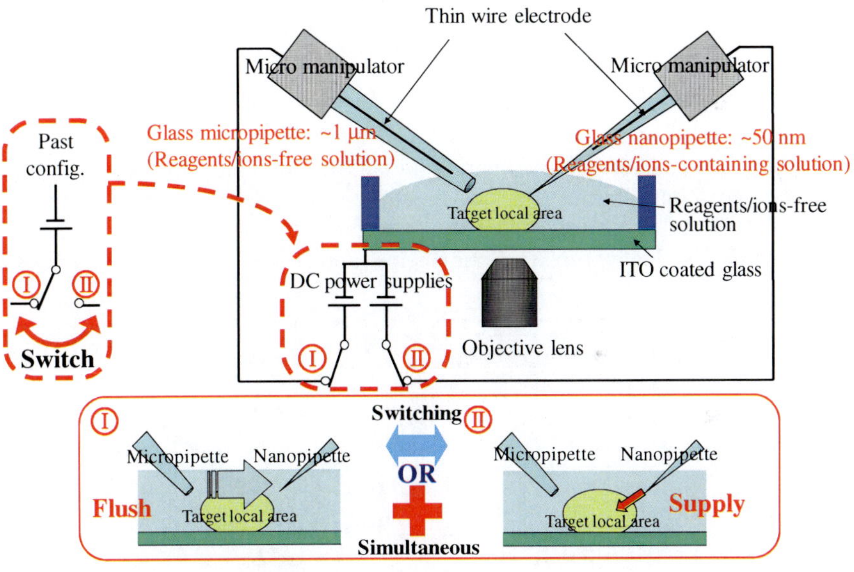

Fig. 5.8 Schematic of the upgraded local environmental control system with micro-nano dual pipettes to achieve the steady-state control of the local reagents/ions concentrations without changing the concentration of the discharging solution

5.2.4 Concept of Simultaneous Discharge

Through this dual pipettes system, certain reagents/ions are supplied to the target in the arbitrary local area by discharge from the nanopipette. And, the reagents/ions diffusing from the nanopipette and the reagents/ions remaining around the target can be flushed by discharge from the micropipette. So, when the both solutions are discharged simultaneously, it is expected that the local reagents/ions concentrations is determined by the difference between the discharging velocities of the reagents/ions-containing and the reagents/ions-free solutions. (And, when the discharge is switched between the reagents/ ions-containing and the reagents/ions-free solutions, the feeding of the reagents/ ions to the target is quickly started/stopped.)

Therefore, the steady-state control of the local reagents/ions concentrations might be realized without changing the concentration of the discharging solution.

5.2.5 *Automation of Voltage Control and Synchronization with Video*

In this section, to eliminate the gap of the voltage control in a few dozen ~ hundred of milli-second caused by the manual control, voltage control of the local environmental control system with micro-nano dual pipettes is automated. And, the voltage control and the recording are synchronized completely to make the analysis accurate.

5.2.6 *Hardware Configuration*

A schematic of the hardware configuration is shown in Figure 5.9. The applied DC voltages were controlled by the analog control signal input to the DC power supplies. The analog control signal was generated through the data acquisition (DAQ) hardware (NI PCI-6230, National Instruments. This device has digital to

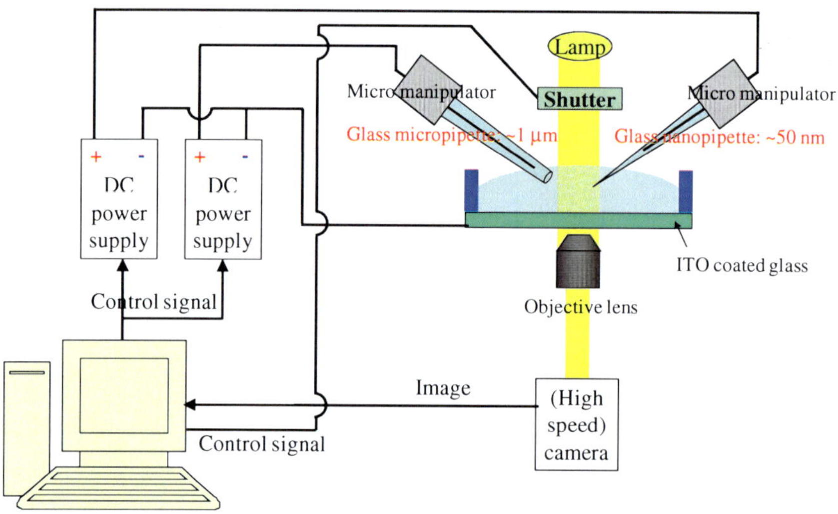

Fig. 5.9 A schematic of the hardware configuration

analog conversion (D/A), analog to digital conversion (A/D), and digital input/output (digital I/O) functions) and regulated. The image was captured continuously after starting to record. To synchronize the voltage control and the video, when the voltage control is started, the shutter which is installed in front of the light for the observation is opened and shut once by the digital control signal from the DAQ hardware.

5.2.7 Rotational Speed Control of Bacterial Flagellar Motor

The micro-nanorobots have been actively studied. If the highly functional micro-nanorobots are realized, it is expected to lead the developments of the novel technologies especially in the medical field such as DDS and micro-nano surgery inside the human body. The size reduction of whole robotic systems, toward the micro-nanometer scale, is hindered by a problem of the size reductions of driving force and its power supply system. To overcome such a problem, recently, the biomolecular motors and the micro living organisms are used as the propulsion for the microobjects [19, 20, 21]. These bio actuators are the natural actuators in micro-nanometer scale. Their energy sources are the chemical reactions. So, there is the possibility to reduce the sizes of the driving force and its power supply system.

The intact flagellated bacteria are one of the promising candidates to be used as the driving forces for the microobjects. The bacteria self-propagate unlike the molecular motors which need the reconstitution and have the metabolic function. The flagella motors convert the IMF (the electrochemical gradient of Na^+ or H^+ ions across the cytoplasmic membrane) into the propulsion in the high efficiency. Bacteria naturally have the sensors and the systems to regulate the rotation of the flagellar motors, the driving force, in response to the environmental stimuli [22]. In addition, some of the flagellated bacteria such as *E. coli* and *V. alginolyticus* can be modified genetically to add or delete some functions depending on the requirements for the micro-nanorobots.

To utilize the driving forces of the flagellar motors more finely, firstly, it is needed to establish the method to analyze the detailed properties and the method to manipulate the driving forces of the flagellar motors. So, in the last chapter, I developed the local environmental control system with the micro-nano dual pipettes to produce the dynamic and arbitrary changes in the local reagents/ions concentrations for the detailed analysis of the flagellar motors.

The rotational speed of the flagellar motor depends on the electrochemical gradient of the Na^+ or H^+ ions across the cytoplasmic membrane (IMF) [23,24]. In most situations, it can be considered that the concentration of the coupling ion inside the membrane is constant, because of the homeostasis of the bacteria. Therefore, the rotational speed of the flagellar motor, the driving force generated by the flagellum, mainly depends on the ion concentration in the external environment. So, the dual pipettes system additionally might be useful to control the propulsion of the flagellar motor by controlling the local reagents/ions concentrations. In this chapter, I apply the dual pipettes system to control the rotational speed of the Na^+-driven bacterial flagellar motor in single *E. coli* cell dynamically and arbitrarily.

5.3 Rotational Speed Measurement of Flagellar Motor

5.3.1 Bacterial Strain

In this study, *E. coli* mutant strain, which has Na^+-driven chimeric flagellar motor, was used for the experiment of flagellar rotational speed control. *E. coli* mutant strain YS34 (*ΔcheY*, *fliC::Tn10*, *ΔpilA*, *ΔmotAmotB*) was transformed with plasmids pYS11 (*fliC* sticky filaments, ampicillin (Amp) resistance) and pYS13 (*pomApotB*, isopropyl-β-Dthiogalactoside (IPTG) inducible, chloramphenicol (Cm) resistance) [25].

Cell samples were cultured and prepared in the following procedure. First, cells were grown overnight at 30 °C with shaking in LB medium (1 %(in weight/volume (w/v)) bacto tryptone, 0.5 %(w/v) yeast extract, and 0.5 %(w/v) NaCl) containing the appropriate antibiotics (50 μg/ml Amp and 25 μg/ml Cm). The overnight culture in LB medium was inoculated into TG medium (1 %(w/v) bacto tryptone, 0.5 %(w/v) NaCl, and 0.5 %(w/v) Grycerol) containing the appropriate antibiotics and the inducer (50 μg/ml Amp, 25 μg/ml Cm, and 100 μM IPTG) at a ten-fold dilution and grown at 30 °C for 3 h. Then, cells were centrifuged by centrifuge (1130, Kubota) in a centrifuging tube at 6000 rpm for 5 min. The sedimented cells were suspended in 1.5 ml Na^+-containing solution (Na^+-motility buffer (10 mM potassium phosphate buffer (pH 7.0), 0.1 mM ethylenediaminetetraacetic acid (EDTA), and 85 mM NaCl) + 100 μM IPTG)

In this study, tethered cells were used for the experiment of the rotational speed control of the Na^+-driven flagellar motor. Generally, it is difficult to directly observe and measure the rotational speed of the flagellum under the optical microscope because of its ultrathin filament and high-speed rotation. "Tethered cells" mean the cells whose flagella are attached onto a substrate [26] as shown in Figure 5. 10. When cells rotate the flagella with attaching onto a substrate, as a result, cell bodies rotate. That is to say, tethered cell is the system with which rotational speed of the flagellum can be easily observed as the rotational speed of the cell body.

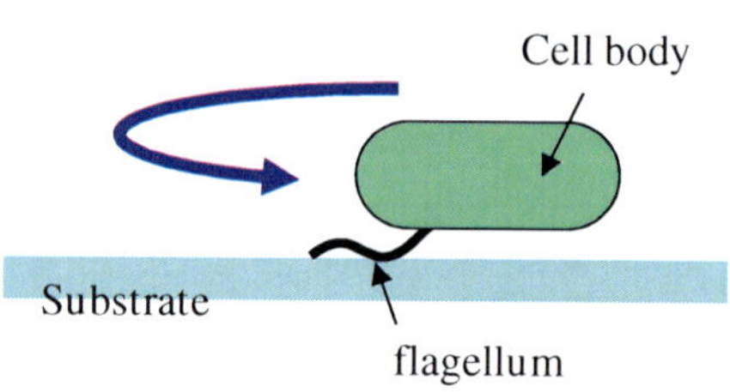

Fig. 5.10 Schematic of the tethered cell

YS34/pYS11/pYS13 used in this study is not only Na^+-driven type, but it is suitable for making tethered cells because of the special flagellar filament called "sticky filament," which easily sticks to a substrate. Its flagella rotate only in the counterclockwise direction by deletion of *cheY* gene, though flagella generally change the rotational direction between clockwise and counterclockwise under CheY protein regulation in response to the surroundings. And the plane of rotation with its tethered cell is approximately parallel to the substrate. Therefore, YS34/pYS11/pYS13 is suitable for the measurement of the rotational speed of the flagellar motor.

The tethered cells were prepared in the following procedure.

(1) Polydimethylsiloxane (PDMS) frame was set onto the substrate. Cell sample was poured into the bath. While it was left for 30 min, cells attached their flagella onto the substrate.
(2) The solution was pumped out to remove the swimming cells, which did not attach their flagella onto the substrate.
(3) The solution was replaced by Na^+-free solution (K^+-motility buffer (10 mM potassium phosphate buffer (pH 7.0), 0.1 mM EDTA, and 85 mM KCl) + 100 μM IPTG) to remove the Na^+ ions in the bath.
(4) After waiting for 1 min to stabilize the cells and the solution, steps 2 and 3 are repeated (total of three times). (Finally, to distinguish the rotatable cells from the cells attached to the substrate, Na^+ ions were added into the bath solution in extremely small amount.)

5.3.2 Experimental Set-Up for Rotational Speed Measurement of Flagellar Motor

In this study, cell bodies of *E. coli* were observed by phase contrast microscopy with the inverted optical microscope. By phase contrast microscopy which applies the phase difference, the living cells can be observed with high contrast, and then cell bodies of *E. coli* look like the small black ellipses.

The rotational speed of the cell body is measured by image processing. Firstly, experimental data was binarized to pick out the cell body as a black ellipse from white background. The centers of the cell body in each frame of the video (29.97 fps) were calculated. The rotational center was determined by circle approximation of the trajectory of the center of the cell body. Then, the angles of the cell body in each frame were determined. Finally, the rotational speeds were calculated frame by frame.

In this study, micro-nano dual pipettes local environmental control system with the first configuration was used (Figure 5.5, Figure 5.11).

This system has two pipettes. The glass nano/micro pipettes with a ~50 nm/~1 μm inner diameter were used for the discharge of the Na^+-containing/Na^+-free solutions, respectively. The bath was filled with the Na^+-free solution. The DC voltage was applied between ITO coated glass at the bottom of the bath and the thin wire electrodes inserted into the pipettes. I can switch the discharge between

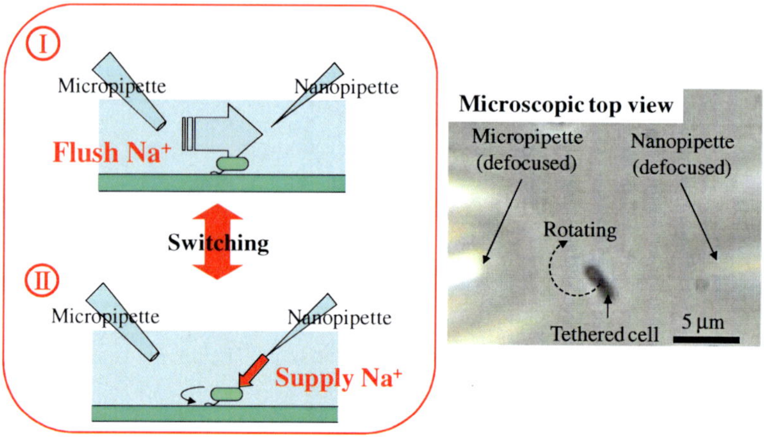

Fig. 5.11 Schematic and experimental appearance of the local environmental control system with micro-nano dual pipettes for the rotational speed manipulation of Na^+-driven flagellar motor

the Na^+-containing and the Na^+-free solutions by switching the applied voltage. Through this dual pipettes system, the Na^+ ions are supplied to the flagellar motor on the arbitrary *E. coli* cell by discharge from the nanopipette. Then the Na^+ ions diffusing from the nanopipette and the Na^+ ions remaining around the cell can be flushed to immediately stop the feeding of the Na^+ ions to the flagellar motor by discharge from the micropipette. These pipettes are fixed on the two micromanipulators, respectively. So, it is possible to approach the pipettes to the arbitrary cell, and the cell-pipette distances can be controlled independently.

5.3.3 Experimental Results of Rotational Speed Measurement of Flagellar Motor

First, we conducted the control, which was done with filling Na^+-free solution into the both pipettes. It was confirmed that there was no change on the rotational speed by applying the DC voltage at 10 V.

Figure 5. 12 shows the experimental result of the rotational speed control of the Na^+-driven flagellar motor in single *E. coli* cell by dual pipettes at 10 V. Figure 5. 12 (a) is the time-series data of the rotational speed of single cell body with switching the discharge between dual pipettes. The shaded area in Figure 5. 12 represents the terms for which the Na^+-containing solution was discharged from the nanopipette. In order to make clear the effect of switching the solutions, the experimental data were processed by the low-pass filter (LPF, threshold frequency:~0.8 Hz) to filter out the fluctuation of the rotational speed caused by the tethered cell and the analysis error (Figure 5. 12 (b)). (With tethered cell, the rotational speed of the flagellum can be easily observed as the rotational speed of the cell body. On the other hand, the rotational speeds of tethered cells often fluctuate because of the high load and the temporal sticking to the substrate.)

From this result, it was shown that the rotational speed of Na^+-driven flagellar motor was iteratively increased/decreased in a near-pulse shape, as expected in Figure 5. 4, by switching the local discharge between Na^+-containing and Na^+-free solutions with micro-nano dual pipettes. That is to say, the transient-state control of the rotational speed of the flagellar motor was achieved.

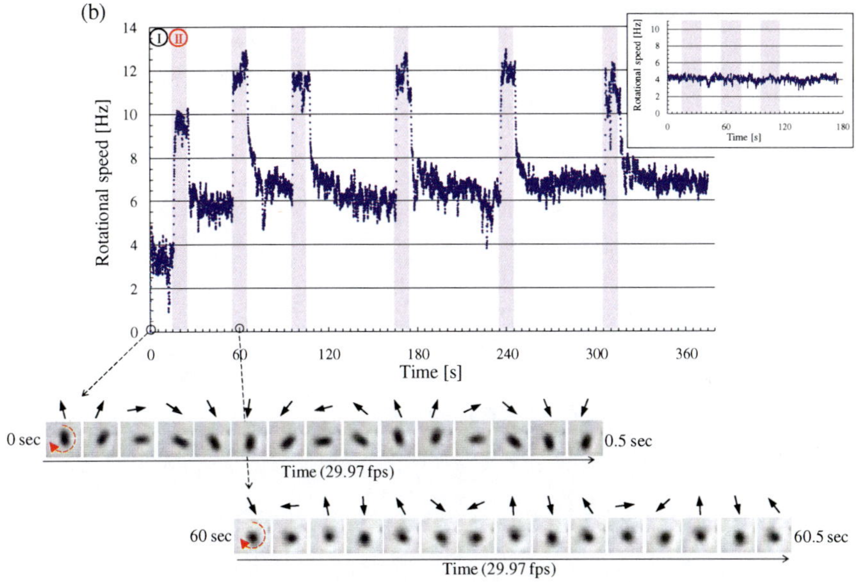

Fig. 5.12 Experimental results of rotational speed control of Na^+-driven flagellar motor. (b) Filtered by low-pass filter (The upper right embedded graph is the result of control filtered by LPF). I: The terms for which the Na^+-free solution was discharged from the micropipette. II: The terms for which the Na^+-containing solution was discharged from the nanopipette.

For evaluation of the flushing effect, as examples, Figure 5. 13 (a) shows the same experimental data in Figure 5. 12 with the magnified time scale from 0 to 55 sec, and Figure 5. 13 (b) shows the time-series data of the rotational speed of another single cell body at 10 V without flushing, (processed by LPF, threshold frequency: ~0.8 Hz). The area marked with diagonal dashed lines in Figure 5. 13 (b) represents the terms for which the applied DC voltages were cut off. By flushing, the rotational speeds of Na^+-driven flagellar motor were relaxed after 2-5 sec of switching. On the other hand, without flushing, the rotational speeds of Na^+-driven flagellar motor were relaxed after 15-20 sec of cutting off the applied DC voltages. This result shows that the local environmental control system with dual pipettes considerably improved the controlling performance in local reagent concentration. This local environmental control system with dual pipettes might be effective to clarify the mechanism of the flagellar motor rotation.

Figure 5.27 shows the experimental results of the rotational speed control of Na^+-driven flagellar motor in single *E. coli* cell with multiplying the DC voltage (the sample was a cell different from the cell in Figure 5. 12 and Figure 5. 13). Figure 5. 14 shows the time-series data of the rotational speed of the single cell body at 10 and 20 V (filtered by LPF). Comparing the rotational speeds in the terms for which the Na^+-containing solution was discharged at each voltage, the rotational speed was increased by multiplying the discharging velocity of Na^+-containing solution. (The averages of the rotational speeds in the terms for which the Na^+-containing solution was discharged are ~7.1 Hz/~7.9 Hz at 10 V/20 V, respectively.) This result shows that the rotational speed of Na^+-driven flagellar motor might be controllable in the short term by controlling the discharging velocity of Na^+-containing solution with changing the applied DC voltage.

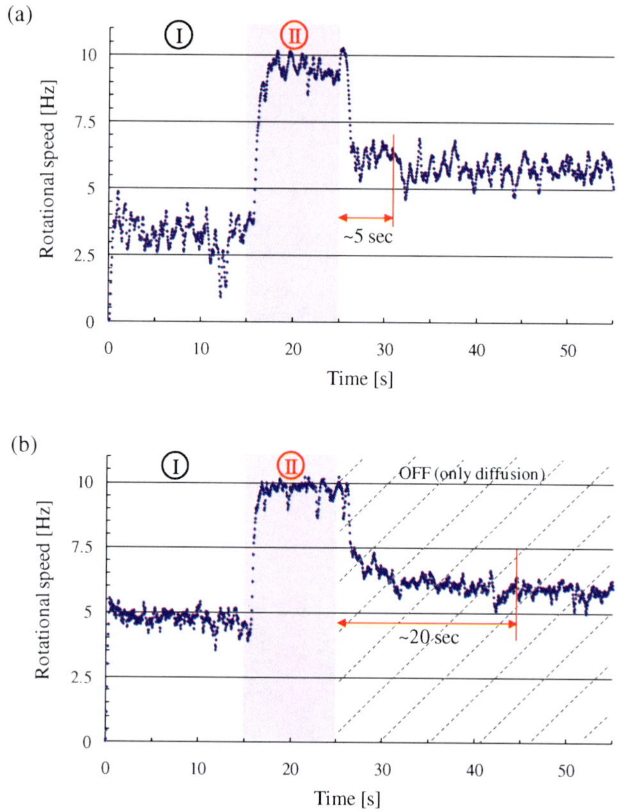

Fig. 5.13 Experimental results of rotational speed control of Na^+-driven flagellar motor: (a) with flushing and (b) without flushing. (Both data were processed by LPF; threshold frequency was ~0.8 Hz.) I: The terms for which the Na^+-free solution was discharged from the micropipette. II: The terms for which the Na^+-containing solution was discharged from the nanopipette. OFF: The term for which the applied DC voltages were cut off.

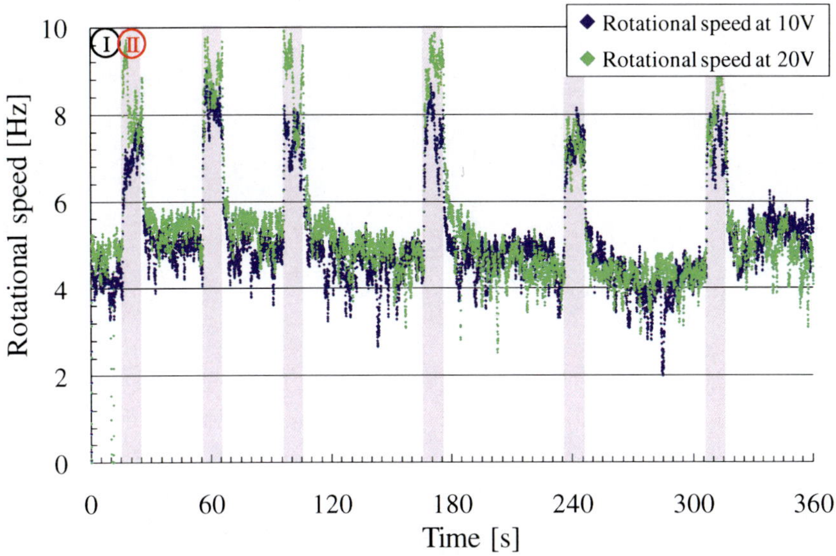

Fig. 5.14 Experimental results of rotational speed control of Na^+-driven flagellar motor with multiplying the applied DC voltages at 10 and 20 V (filtered by LPF). I: The terms for which the Na^+-free solution was discharged from the micropipette. II: The terms for which the Na^+-containing solution was discharged from the nanopipette.

5.3.4 *Discussions of Rotational Speed Measurement of Flagellar Motor*

As aforementioned, the bacterial flagellar motor is a molecular machine that converts IMF into mechanical force. The IMF is defined as

$$IMF = V_m - \frac{k_B T}{q \ln(C_o / C_i)} \tag{5.3}$$

where V_m is the transmembrane voltage (inside minus outside), k_B is Boltzmann's constant, T is the absolute temperature, q is the charge, and C_i and C_o are the inside and outside concentrations of coupling ion, respectively [27]. Torque (N) is defined as

$$N = f\omega \tag{5.4}$$

where f is the rotational drag coefficient and ω is the rotational speed. It can be considered that transmembrane voltage (V_m) and inside concentrations of coupling ion (C_i) are constant because of the homeostasis of bacteria. The rotational torque of flagellar motor is proportional to the IMF. Therefore, the rotational speed of flagellar motor is proportional to the logarithm of the Na^+ ion concentration in the solution. In fact, as the reference data, the relation between the rotational speed of tethered Na^+-driven type *E. coli* cell body and the Na^+ concentration was obtained, as shown in Figure 5. 15.

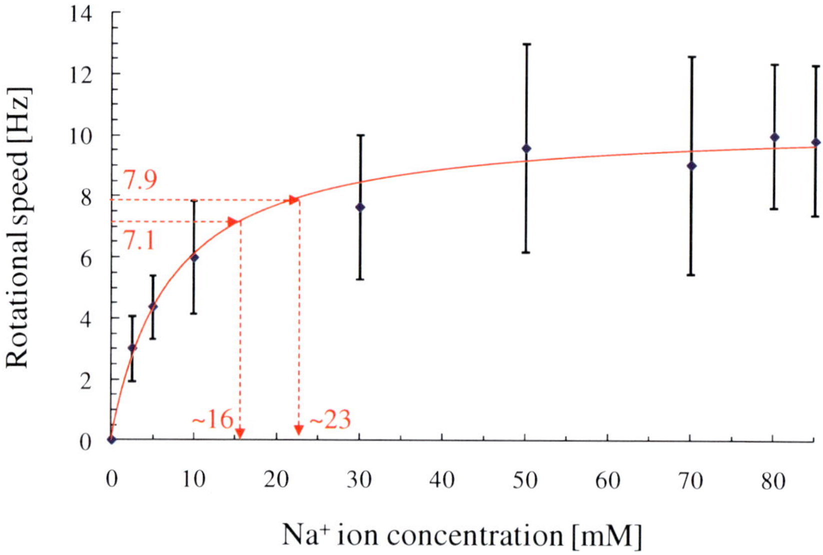

Fig. 5.15 Na^+ ion concentration versus rotational speed of tethered Na^+-driven type *E. coli* cell. The plotted points are mean values of rotational speed with error bars indicating the standard deviation of four cells. The solid curved line is the approximated curve. The dashed arrows indicate the estimation of Na^+ ion concentrations from the experimental results.

Comparing the reference data and the experimental data of Figure 5.14, the local Na^+ ion concentration around the cell was estimated at ~16 mM /~23 mM at 10 V/20 V, respectively. Here, it is assumed that the Na^+-containing solution from the nanopipette flows in y-direction and Na^+ ions diffuse in x-direction (x and y are the axes of the Cartesian coordinate system). The fundamental solution of 1-D diffusion equation is

$$C(x, t) = \frac{1}{\sqrt{4\pi Dt}} \exp\left[-\frac{x^2}{4Dt} \right] \tag{5.5}$$

where C is the concentration, D is the diffusion constant, and t is the time. The time until the solution reaches the cell is given as the division of the distance between the nanopipette and the cell by the flow velocity. The flow velocity is proportional to the electric field intensity (E) or the applied DC voltage (equation (3.1)). Hence, if the time until the solution reaches the cell at 20 V is S, it takes $2S$ to reach the cell at 10 V. Here, the ratio between the Na^+ ion concentrations around the cell at 10 and 20 V is given as the following equation,

$$\frac{C(x, 2S)}{C(x, S)} = \frac{1}{\sqrt{2}} \exp\left[-\frac{x^2}{8DS}\right] \tag{5.6}$$

$$\leq 0.71 \qquad (at\ x = 0)$$

From this result, the difference in the local ion concentrations is certainly caused by the difference of flow velocity with changing the applied DC voltage ($23 \times 0.71 = 16.33$). Therefore, it can be stated that the experimental results largely correspond to the theoretical result. This result supports the earlier conclusion that the rotational speed of Na^+-driven flagellar motor might be controllable in the short term by controlling the discharging velocity of Na^+-containing solution with changing the applied DC voltage.

5.4 Steady-State Control of Rotational Speed of Flagellar Motor

5.4.1 Experimental Set-Up for Steady-State Control of Rotational Speed of Flagellar Motor

In this study, micro-nano dual pipettes local environmental control system with the upgraded second configuration was used (Figure 5. 7). This system has two pipettes. The glass micro-nano pipettes with a ~50 nm/~1 μm inner diameter were used for the discharge of the Na^+-containing/Na^+-free solutions, respectively. The bath was filled with the Na^+-free solution. The DC voltage was applied between ITO coated glass at the bottom of the bath and the thin wire electrodes inserted into the pipettes. We can turn on/off the discharge of the Na^+-containing and the Na^+-free solutions independently, and can change the discharging velocity/volume of the solutions by changing the applied DC voltages independently. Through this dual pipettes system, the Na^+ ions are supplied to the flagellar motor on the arbitrary *E. coli* cell by discharge from the nanopipette. Then the Na^+ ions diffusing from the nanopipette and the Na^+ ions remaining around the cell can be flushed by discharge from the micropipette. So, when the both solutions are discharged simultaneously, it is expected that the local Na^+ ion concentration is determined by the difference between the discharging velocities of the Na^+-containing and the Na^+-free solutions. These pipettes are fixed on the two micromanipulators, respectively. So, it is possible to approach the pipettes to the arbitrary cell, and the cell-pipette distances can be controlled independently.

With the micro-nano dual pipettes system, when the discharge of the Na^+-containing solution is continued for a long period, the rotational speeds of flagellar motor get into a certain level at any applied DC voltages as shown in Figure 5. 16. This is attributed to the fact that the rotational speeds of the tethered cells hit the ceiling at the high ion concentration of the discharging solution. Another supposable cause is the saturation of the local Na^+ ion concentration by

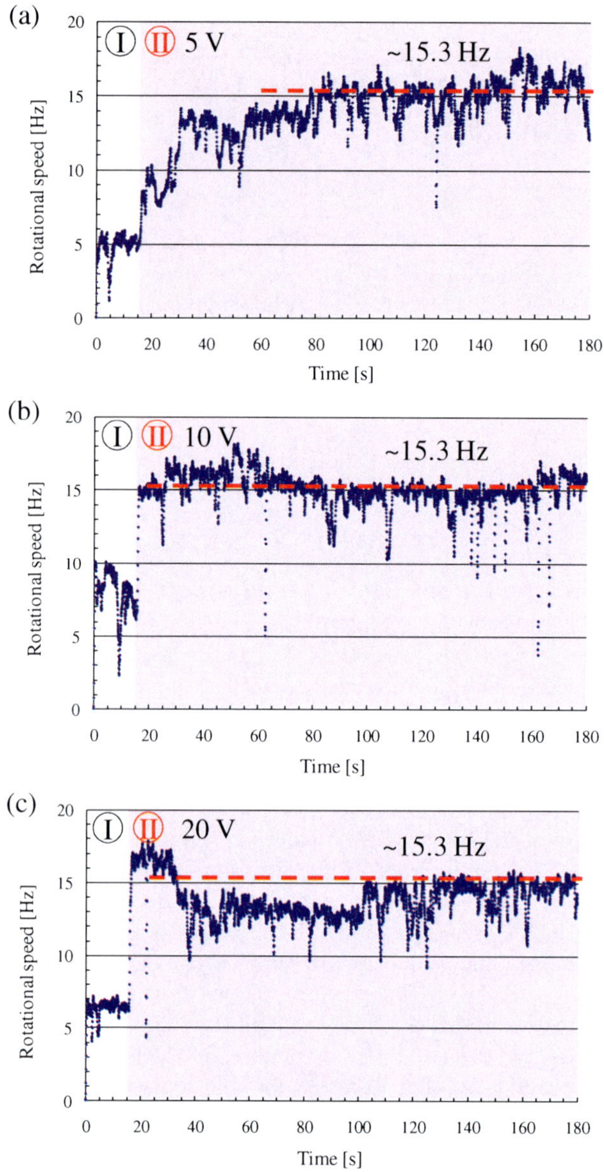

Fig. 5.16 The rotational speeds of flagellar motor after the long periods of time with applied DC voltages of (a) 5 V, (b) 10 V, (c) 20 V. I; The terms for which the Na^{+}-free solution was discharged from the micropipette. II; The terms for which the Na^{+}-containing solution was discharged from the nanopipette.

the equilibrium between diffusion and supply of Na^+ ion as a consequence of the increase of the Na^+ ion concentration around the target. In both cases, the steady-state of the rotational speed of the tethered cells cannot be controlled by the first local environmental control system with micro-nano dual pipettes. So, for the steady-state control of the rotational speed of the Na^+-driven flagellar motor, in this section, I used the upgraded local environmental control system with micro-nano dual pipettes.

5.4.2 Experimental Results of Steady-State Control of Rotational Speed of Flagellar Motor

In the control experiment, the discharge of Na^+ ions was continued for a long period at 20 V (Figure 5. 17 (a)). Figure 5. 17 (a) is the time-series data of the rotational speed of single cell body. (In order to make clear the effect of switching the solutions, the experimental data was processed by LPF (threshold frequency; ~0.8 Hz) to filter out the fluctuation of the rotational speed caused by the tethered cell and the analysis error. With tethered cell, the rotational speed of the flagellum can be easily observed as the rotational speed of the cell body. On the other hand, the rotational speeds of tethered cells often fluctuate because of the high load and the temporal sticking to the substrate.) As a result, the rotational speed got into ~6.9 Hz.

Figure 5. 17 (b) shows the experimental result of the rotational speed control of Na^+-driven flagellar motor in single *E. coli* cell. The rotational speed was controlled by simultaneous local discharges of Na^+-containing and Na^+-free solutions with controlling the discharging velocities independently. Figure 5. 17 (b) is the time-series data of the rotational speed of single cell body. From this result, when the Na^+-containing and Na^+-free solutions were discharged simultaneously, the rotational speed was increased from the base level ~5.0 Hz to ~6.3 Hz. It was slightly slower than the rotational speed in the control (~6.9 Hz, in Figure 5. 17 (a)). After multiplying the applied DC voltage of the micropipette, the rotational speed slightly decreased to ~5.6 Hz. This decreasing of the rotational speed might be caused by the increasing of the discharging velocity of Na^+-free solution. However, because the Na^+-containing solution was still being discharged, it was still faster than the base level (~5.0 Hz).

Therefore, as expected, the steady-state control of the rotational speed of the Na^+-driven flagellar motor was realized by simultaneous local discharges of the Na^+-containing and Na^+-free solutions with controlling the discharging velocities independently. In addition, when the switching and the simultaneous discharging methods are combined, the local environmental control system with nano-micro dual pipettes will become a powerful tool to control the bacterial flagellar motor, bio driving force.

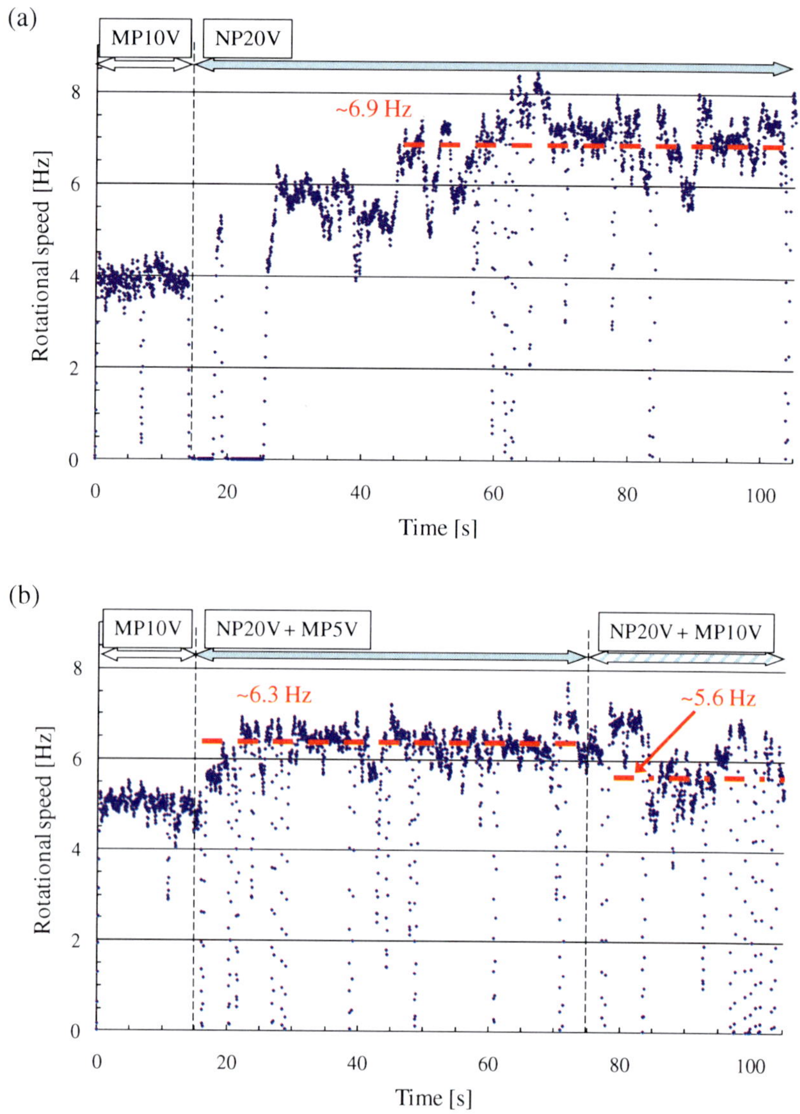

Fig. 5.17 Experimental results of (a) control, (b) steady-state control of the rotational speed of Na^+-driven flagellar motor by simultaneous discharges with changing the applied voltages independently. NP: Nanopipette (Na^+-containing). MP: Micropipette (Na^+-free).

5.4.3 Estimation of Torque Generated by Flagellar Motor

In this study, for the fundamental estimation as the bio driving force, the rotational torque generated by the flagellar motor was estimated. The rotational torque generated by the flagellar motor was estimated from the rotational speed of the cell body. The rotating cell body was simplified as the rotating cylinder. Under this condition, the rotational torque (T) is calculated by the equations,

$$T = \gamma\omega \tag{5.7}$$

$$\gamma = \frac{\frac{1}{3}\pi\eta L^3}{\ln(L/2r) - 0.66} \tag{5.8}$$

where γ : rotational drag coefficient, ω : rotational speed, η : viscosity of the environmental solution, L : length of the cylinder, r : radius of the cylinder [28].

The rotational torque was calculated by equations (5.6) and (5.7) where η : 0.001 Pa·s (pure water), L : 3 μm and r : 0.5 μm. As a result, the rotational torque was calculated ~2.8×10^3 / ~2.5×10^3 / ~2.3×10^3 pN·nm at ~6.9 Hz / ~6.3 Hz / ~5.6 Hz, respectively. Therefore, it was shown that the rotational torque of the flagellar motor could be controlled in 10^2 pN·nm orders by the local environmental control system with micro-nano dual pipettes. This value is hundredth part of the maximum rotational torque of the magnetic-actuated artificial flagella reported by L. Zhang *et al.* [29]. This shows that the driving force of the flagellar motor can be controlled more finely.

References

[1] Darnton, N., Turner, L., Breuer, K., Berg, H.C.: Moving Fluid with Bacterial Carpets. Biophys. J. 86, 1863–1870 (2004)

[2] Behkam, B., Sitti, M.: Effect of Quantity and Configuration of Attached Bacteria on Bacterial Propulsion of Microbeads. Appl. Phys. Lett. 93, 223901 (2008)

[3] Martel, S., Mohammadi, M., Felfoul, O., Lu, Z., Pouponneau, P.: Flagellated Magnetotactic Bacteria as Controlled MRI-trackable Propulsion and Steering Systems for Medical Nanorobots Operating in the Human Microvasculature. Intl. J. Robotics Research 28(4), 571–582 (2009)

[4] Martel, S., Felfoul, O., Mathieu, J.-B., Chanu, A., Tamaz, S., Mohammadi, M., Mankiewicz, M., Tabatabaei, N.: MRI-based Medical Nanorobotic Platform for the Control of Magnetic Nanoparticles and Flagellated Bacteria for Target Interventions in Human Capillaries. Intl. J. Robotics Research 28(9), 1169–1182 (2009)

[5] Weibel, D.B., Garstecki, P., Ryan, D., DiLuzio, W.R., Mayer, M., Seto, J.E., Whitesides, G.M.: Microoxen: Microorganisms to Move Microscale Loads. Proc. Natl. Acad. Sci. 102, 11963–11967 (2005)

[6] Macnab, R.: Flagella and Motility. In: Eschericia Coli and Salmonella, pp. 123–145. American Society for Microbiology, Washington D.C. (1996)

[7] Yorimitsu, T., Homma, M.: Na+-Driven Flagellar Motor of Vibrio. Biochim. Biophys. Acta 1505, 82–93 (2001)

[8] Aldridge, P., Hughes, K.T.: Regulation of Flagellar Assembly. Curr. Opin. Microbiol. 5, 160–165 (2002)

[9] Kojima, S., Blair, D.F.: The Bacterial Flagellar Motor: Structure and Function of a Complex Molecular Machine. Int. Rev. Cytol. 233, 93–134 (2004)
[10] Asai, Y., Kojima, S., Kato, H., Nishioka, N., Kawagishi, I., Homma, M.: Putative Channel Components for the Fast-Rotating Sodium-Driven Flagellar Motor of a Marine Bacterium. J. Bacteriol. 179(16), 5104–5110 (1997)
[11] Dean, G.E., Macnab, R.M., Stader, J., Matsumura, P., Burks, C.: Gene Sequence and Predicted Amino Acid Sequence of the motA Protein, a Membrane-Associated Protein Required for Flagellar Rotation in Escherichia coli. J. Bacteriol. 159(3), 991–999 (1984)
[12] Stader, J., Matsumura, P., Vacante, D., Dean, G.E., Macnab, R.M.: Nucleotide Sequence of the Escherichia coli motB Gene and Site-Limited Incorporation of Its Product into the Cytoplasmic Membrane. J. Bacteriol. 166(1), 244–252 (1986)
[13] Sowa, Y., Rowe, A.D., Leake, M.C., Yakushi, T., Homma, M., Ishijima, A., Berry, R.M.: Direct Observation of Steps in Rotation of the Bacterial Flagellar Motor. Nature 437, 916–919 (2005)
[14] Ying, L., Bruckbauer, A., Zhou, D., Gorelik, J., Shevchuk, A., Lab, M., Korchevb, Y., Klenerman, D.: The Scanned Nanopipette: a New Tool for High Resolution Bioimaging and Controlled Deposition of Biomolecules. Phys. Chem. Chem. Phys. 7, 2859–2866 (2005)
[15] Piper, J.D., Li, C., Lo, C.-J., Berry, R., Korchev, Y., Ying, L., Klenerman, D.: Characterization and Application of Controllable Local Chemical Changes Produced by Reagent Delivery from a Nanopipet. J. Am. Chem. Soc. 130, 10386–10393 (2008)
[16] Sánchez, D., Anand, U., Gorelik, J., Benham, C.D., Bountra, C., Lab, M., Klenerman, D., Birch, R., Anand, P., Korchev, Y.: Localized and Non-Contact Mechanical Stimulation of Dorsal Root Ganglion Sensory Neurons using Scanning Ion Conductance Microscopy. J. Neurosci. Methods 159, 26–34 (2007)
[17] Ying, L., Bruckbauer, A., Zhou, D., Gorelik, J., Shevchuk, A., Lab, M., Korchevb, Y., Klenerman, D.: The Scanned Nanopipette: a New Tool for High Resolution Bioimaging and Controlled Deposition of Biomolecules. Phys. Chem. Chem. Phys. 7, 2859–2866 (2005)
[18] Honda, S., Terabe, S.: Capillary Electrophoresis -Bases and Facts-. Kodansha Ltd., Tokyo (1995) (in Japanese)
[19] Darnton, N., Turner, L., Breuer, K., Berg, H.C.: Moving Fluid with Bacterial Carpets. Biophys. J. 86, 1863–1870 (2004)
[20] Behkam, B., Sitti, M.: Effect of Quantity and Configuration of Attached Bacteria on Bacterial Propulsion of Microbeads. Appl. Phys. Lett. 93, 223901 (2008)
[21] Weibel, D.B., Garstecki, P., Ryan, D., DiLuzio, W.R., Mayer, M., Seto, J.E., Whitesides, G.M.: Microoxen: Microorganisms to Move Microscale Loads. Proc. Natl. Acad. Sci. 102, 11963–11967 (2005)
[22] Homma, M., Oota, H., Kojima, S., Kawagishi, I., Imae, Y.: Chemotactic Responses to an Attractant and a Repellent by the Polar and Lateral Flagellar Systems of Vibrio Alginolyticus. Microbiol. 142, 2777–2783 (1996)
[23] Lo, C.J., Leake, M.C., Pilizota, T., Berry, R.M.: Nonequivalence of Membrane Voltage and Iongradient as Driving Forces for the Bacterial Flagellar Motor at Low Load. Biophys. J. 93, 294–302 (2007)
[24] Fung, D.C., Berg, H.C.: Powering the Flagellar Motor of Escherichia coli with an External Voltage Source. Nature 375, 809–812 (1995)

[25] Sowa, Y., Rowe, A.D., Leake, M.C., Yakushi, T., Homma, M., Ishijima, A., Berry, R.M.: Direct Observation of Steps in Rotation of the Bacterial Flagellar Motor. Nature 437, 916–919 (2005)

[26] Silverman, M., Simon, M.: Flagellar rotation and the mechanism of bacterial motility. Nature 249, 73–74 (1974)

[27] Sowa, Y., Berry, R.M.: Bacterial flagellar motor. Q. Rev. Biophys. 41(2), 103–132 (2008)

[28] Howard, J.: Mechanics of Motor Proteins and the Cytoskeleton, pp. 106–107. Sinauer Associates, Sunderland (2001)

[29] Zhang, L., Abbott, J.J., Dong, L., Peyer, K.E., Kratochvil, B.E., Zhang, H., Bergeles, C., Nelson, B.J.: Characterizing the Swimming Properties of Artificial Bacterial Flagella. Nano Lett. 9(10), 3663–3667 (2009)

Chapter 6
Nanomanipulation System under Electron Microscope

6.1 Configuration of Nanomanipulation System

In this chapter, we introduce our constructed nanorobotic manipulation system inside electron microscopes. The nanomanipulation system is basically constructed from 1) observation (imaging) system, 2) end-effector, 3) manipulator with actuator or sensor, 4) driver/controller, and 5) interface system. As we mentioned in Chapter 1, the imaging system is one of the important feature to restrict the environments and to decide the applicability.

Based on the nanorobotic manipulation system, a nano laboratory has been proposed. The nanolaboratory is integrates the nanomanipulation system for manipulating nanomaterials with a nanoanalytical system and a nanofabrication system for measuring the properties of nanostructures and constructing nanodevices.

6.2 Nanorobotic Manipulation System Inside SEM

6.2.1 Design of Nanorobotic Manipulation System Inside SEM

We aim at constructing a nanorobotic manipulation system with sensing capacity for 3D nanomanipulation of nanometer scale objects by controlling the dielectrophoretic forces between the tool and objects and by modifying the van der Waals forces between the objects and substrate. In the system, an optical microscope or SEM is served not only as real-time observation device, but as position/force sensing system as well. Moreover, for facilitating the observation and manipulation, an AFM cantilever is applied and set vertically, and a rotary DOF is adopted into the system. These are different from the conventional AFM-type manipulators and other manipulators inside SEM.

Design and Kinematics of Micro-Nanomanipulator

We have developed three generations of nanorobotic manipulators (NRM) as shown in Figure 6. 1, Figure 6. 2 and Figure 6. 3 using modular design [1, 2]. The 1st nanomanipulator (NRM-I) has 2 units and 4 DOFs. It was installed inside a SEM. Unit 1 has 3 DOFs including two prismatic DOFs along X and Y axes and a rotary one θ around Y. Unit 1 remained unchanged in all three generations. Unit 2 has only one DOF in NRM-I, i.e., a Z-direction prismatic DOF [3, 4]. All of these DOFs are actuated with commercially available $Pb(Zr,Ti)O_3$ (PZT) based actuators.

The 2nd generation of nanomanipulators (NRM-II), in which a 6-DOF manipulator is introduced [5]. NRM-II was firstly mounted inside the same SEM as NRM-I, and then transplanted inside a field-emission SEM later to obtain better resolution. The NRM-III is a newly developed one by adding a XY stage with finer resolution actuated with piezotube based actuators and a Z-direction PZT into Unit 2—the whole unit is denoted as Unit 2', and a new manipulator Unit 4 [6]. Unit 4 is a 3-DOF parallel manipulator actuated with 3 Nanomotors and serves as an independent manipulator or common sample stage for other three manipulators. NRM-III works inside the FESEM mentioned above.

Unit 1 and Unit 2

We take modular design for all the three generations of nanorobotic manipulators. NRM-I, NRM-II and NRM-III has a common Unit 1, and NRM-I and NRM-II share the same Unit 2. Unit 2' can be regarded as a combination of Unit 2 with 3-DOF prismatic joints for fine motion. Hence, we firstly introduce the design and kinematics of Unit 1 and Unit 2.

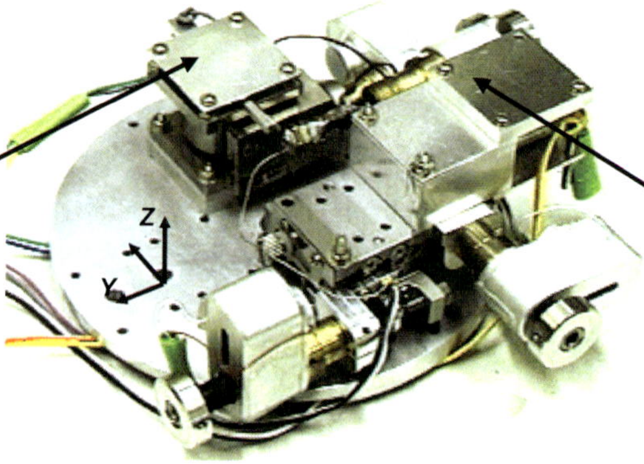

Fig. 6.1 Nanorobotic Manipulators (NRM-I: 4-DOF, 2-Unit)

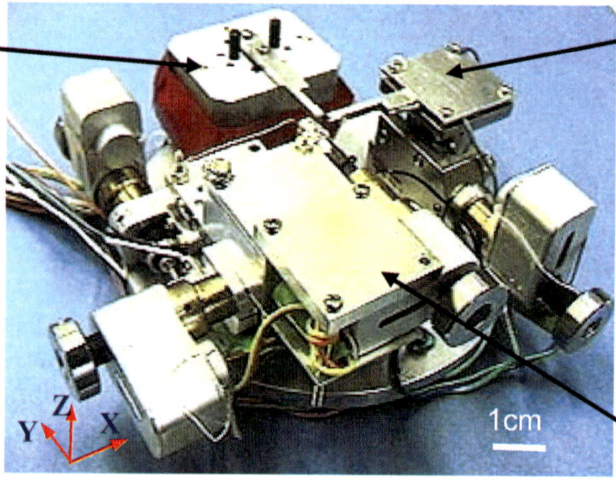

Fig. 6.2 Nanorobotic Manipulators (NRM-II: 10-DOF, 3-Unit)

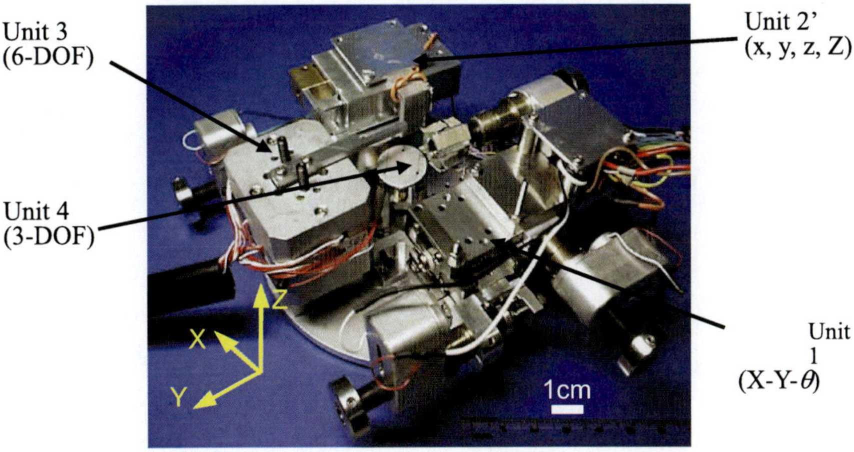

Fig. 6.3 Nanorobotic Manipulators (NRM-III: 16-DOF, 4-Unit)

Figure 6. 4 depicts the model of kinematics of these two manipulators, where J_i (i=1, 2, 3 and 1') are joints, whereas L_i (i=0, 1, 2, 3, 0' and 1') are links. Unit 1 is composed of joints J_1, J_2 (X-Y stage) and J_3 (a rotary one that can rotate around Y direction) for positioning a sample substrate and a cantilever side by side as two end-effectors. Unit 2 (Joint J_1', i.e. the Z stage) is designed for positioning another AFM cantilever. Unit 1 and Unit 2 have linear strokes 6mm in X, Y, Z directions and a 360 degree rotational one. The linear resolutions are 30nm (X, Y and Z stages) and that of the rotary one is 2mrad.

6.2.2 *Link Coordination of Nanorobotic Manipulation System Inside SEM*

Link coordinate systems are depicted according to Denavit-Hartenberg notation. Note that L_0 and L_0' are base links for units 1 and 2 and are fixed on their common base indicated with XYZ frame. The link parameters of the manipulators are listed in Table 6. 1, where i denotes the code number of joints, ai the link length, α_i the link twist, d_i the distance between links and θ_i the angle between links. Joint variables include d_x, d_y, θ and d_z.

Table 6.1 Link Parameters of Manipulators

i	α_{i-1}	a_{I-1}	d_i	θ_i
1	-90°	0	d_y	-90°
2	-90°	a_1	d_x	0°
3	90°	a_2	0	θ
1'	0°	a_1'	d_z	0°

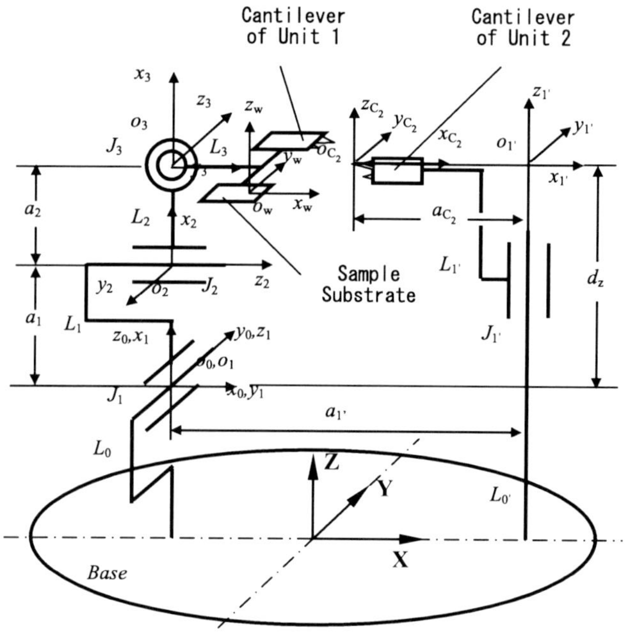

Fig. 6.4 Modeling of the Nanorobotic Manipulators

The homogeneous transform matrix of Units 1 and 2 in frame {0} are respectively

$$
{}^{0}T_{3} = \begin{bmatrix} S_{\theta} & C_{\theta} & 0 & d_{x} \\ 0 & 0 & 1 & d_{y} \\ C_{\theta} & -S_{\theta} & 0 & a_{1}+a_{2} \\ 0 & 0 & 0 & 1 \end{bmatrix} \tag{6.1}
$$

and

$$
{}^{0}T_{1'} = \begin{bmatrix} 1 & 0 & 0 & a_{1'} \\ 0 & 1 & 0 & 0 \\ 0 & 0 & 1 & d_{z} \\ 0 & 0 & 0 & 1 \end{bmatrix}, \tag{6.2}
$$

where $C_{\theta} = \cos\theta$ and $S_{\theta} = \sin\theta$.

Frame {w} is set for indicating the end point of objective workpiece either on the substrate or on the cantilever of unit 1. Let the coordinates of the origin point o_w of frame {w} be $({}^{3}x_{w}, {}^{3}y_{w}, {}^{3}z_{w})$ in the frame {3}, then

$$ {}^{0}T_{w}=\begin{bmatrix} C_{\theta} & 0 & S_{\theta} & d_{x}+{}^{3}x_{w}S_{\theta}+{}^{3}y_{w}C_{\theta} \\ 0 & 1 & 0 & d_{y}+{}^{3}z_{w} \\ -S_{\theta} & 0 & C_{\theta} & a_{1}+a_{2}+{}^{3}x_{w}C_{\theta}-{}^{3}y_{w}S_{\theta} \\ 0 & 0 & 0 & 1 \end{bmatrix}. \tag{6.3} $$

Similarly, let frame $\{C_2\}$ represent the tip point of the cantilever of unit 2, which is also the contact point of the tip and another terminal of the object, then

$$ {}^{0}T_{C_2}=\begin{bmatrix} 1 & 0 & 0 & a_{1'}-a_{C_2} \\ 0 & 1 & 0 & 0 \\ 0 & 0 & 1 & d_{z} \\ 0 & 0 & 0 & 1 \end{bmatrix}. \tag{6.4} $$

Eqs. (6.3) and (6.4) can be correlated with a transform matrix ${}^{w}T_{C_2}$ that indicates the transformation between two ends of the object.

$$ {}^{0}T_{w}\,{}^{w}T_{C_2}={}^{0}T_{C_2}. \tag{6.5} $$

Note that if the contact parts of object and the two end-effectors of Units 1 and 2 are point contact or can be regarded as ball joints, ${}^{w}T_{C_2}$ would include no rotation elements. In that case,

$$ \begin{cases} d_{x}=a_{1'}-a_{C_2}-{}^{3}x_{w}S_{\theta}-{}^{3}y_{w}C_{\theta}-{}^{w}x_{C_2} \\ d_{y}=-{}^{3}z_{w}-{}^{w}y_{C_2} \\ d_{z}=a_{1}+a_{2}+{}^{3}x_{w}C_{\theta}-{}^{3}y_{w}S_{\theta}+{}^{w}z_{C_2} \end{cases}, \tag{6.6} $$

where $({}^{w}x_{C_2}, {}^{w}y_{C_2}, {}^{w}z_{C_3})$ is the coordinate of O_{C_2} in the frame$\{w\}$.

It can be understood from Eq.(6.6) that for a given θ (a pose for making the observations readily under OM or inside SEM), the other joint variables of Units 1 and 2 can be determined. For 1D objects, e.g. thin cylinders such as carbon nanotubes, or 0D objects (orientations of terminals has no meaning), e.g. beads or ball-like objects, the relative position of O_{C_2} to O_w will determine the pose of the object in 3D space. That means the manipulators are applicable for 3D manipulations of 0D or 1D objects, which are just the commonly used nano building blocks.

The configuration of the sample substrate of Unit 1 and the cantilever of Unit 2 is shown in Figure 6. 5. In Unit 1, a rotary Picomotor is mounted above the X-Y stage actuated by two translatory Picomotors, and is used for mounting a silicon substrate, which is used for placing objects to be manipulated. Over the Si substrate, a thin film of aluminum is coated as an electrode for applying electrical field to generate dielectrophoretic force. Another electrode is wired to the cantilever of Unit 2. To make the tip of cantilevers and the Si plate insulated to each other, a thin polyimide film (thickness: 7.5μm) is bonded over the aluminum film. Cantilever of

Unit 1 is the same as that of Unit 2 but is set horizontally. As the two cantilevers are applied at the same time, it is also possible to measure the electronic properties of the object placed between them by wiring the cantilevers to measurement devices.

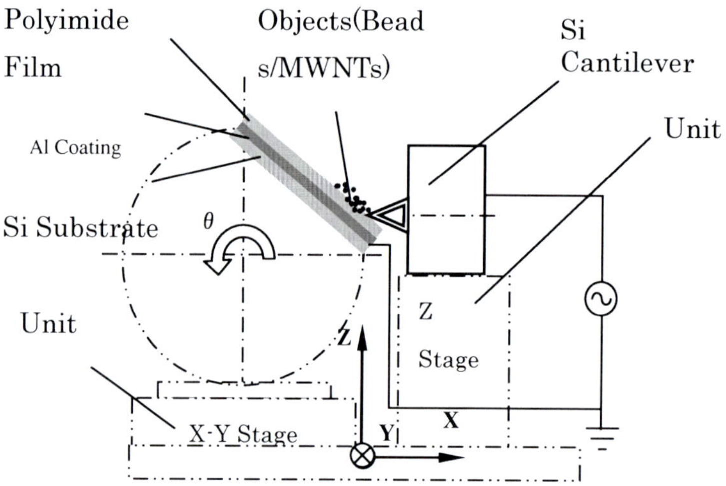

Fig. 6.5 Configuration of the Sample Substrate of Unit 1 and the Cantilever of Unit 2

Unit 3 is designed to compensate the stepping motion of Units 1 and 2, and also for multiple-probe manipulations. The inside view of Unit 3 is depicted in Figure 6. 6. The manipulator used 8 PZTs for getting bi-direction 6-DOF micro movements.

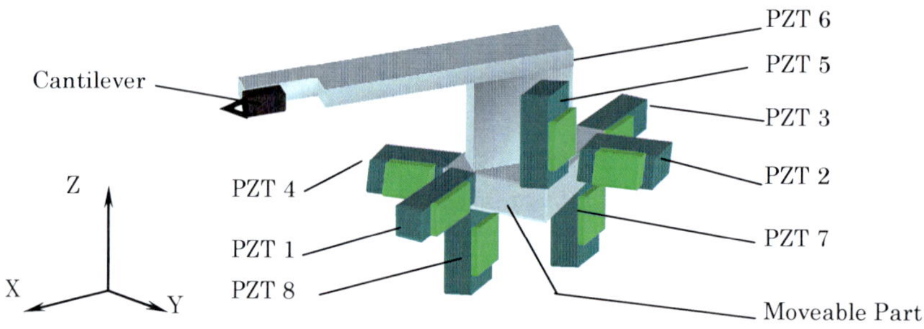

Fig. 6.6 Inside View of Unit 3

The working principle of Unit 3 is shown in Figure 6. 7 and Table 6. 2. Eight PZTs are applied both as actuators and sensors (superposition of the PZTs), and each can extends in z_i (i=1,2,…,8) direction. To get a translation along with the +X direction, PZTs 2 and 3 will be used as actuators, while 1 and 4 as position feedback sensors. Exchanging 2-3 with 1-4 pairs, a –X motion will be obtained. This table shows the realizations of all DOFs and PZTs involved. The purpose for doing so is to gain a larger working space (bi-direction actuation, e.g., the above mentioned +X and –X actuation) and a higher resolution (by close-loop control) without increasing the volume of the manipulator too much so as to place it into the SEM.

Table 6.2 Superposition of PZTs both as Actuators and Sensors in Unit 3

PZT No.	1	2	3	4	5	6	7	8
+X	S	S	A	A	-	-	-	-
-X	A	A	S	S	-	-	-	-
+Y	A	S	S	A	-	-	-	-
-Y	S	A	A	S	-	-	-	-
+Z	-	-	-	-	S	A	S	A
-Z	-	-	-	-	A	S	A	S
$+\alpha$	-	-	-	-	A	-	S	-
$-\alpha$	-	-	-	-	S	-	A	-
$+\beta$	-	-	-	-	-	S	-	A
$-\beta$	-	-	-	-	-	A	-	S
$+\gamma$	S	A	S	A	-	-	-	-
$-\gamma$	A	S	A	S	-	-	-	-

Note: "A"—Actuator, "S"—Sensor, "-"—Unused; α, β, and γ are the rotary directions around X, Y and Z.

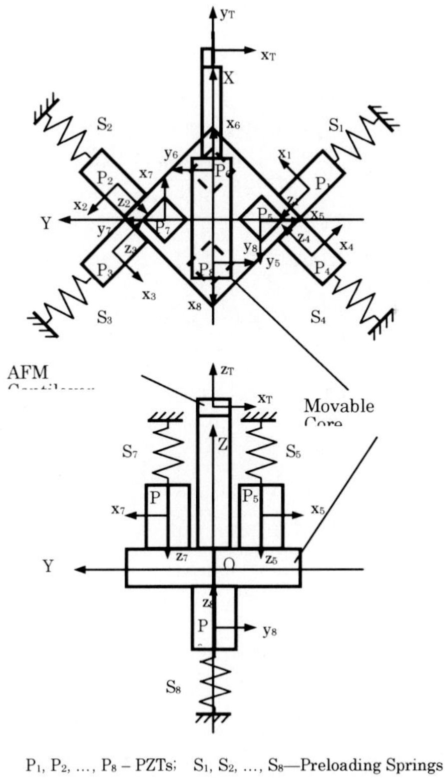

Fig. 6.7 Working Principle of Unit 3

Figure 6. 8 shows a SEM image of the tip of Unit3 with the tip of Unit 2 and sample (MWNT rod) in near position. Such a configuration is important for multi-probe manipulations.

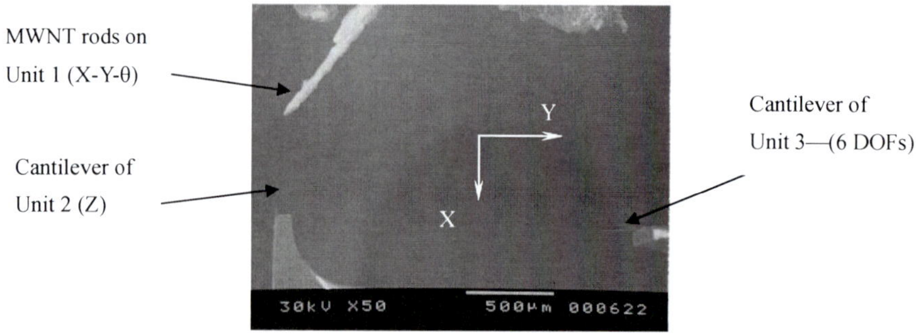

Fig. 6.8 Configuration of the Cantilevers and Sample-Stage of the Nanorobotic Manipulator NRM-II

To construct nanostructure in 3-D space, an active substrate is needed. It is also needed for 4-terminal (active) measurement. For these purposes, a Unit 4 as shown in Figure 6. 9. It is designed. A 3-RSP parallel mechanism is selected for the compact volume copied. Nanomotors are ideal selections for actuators in this schematic because it can provide several mm-stroke with a specified 0.01nm-resolution. The size of the Nanomotors is only ϕ 3.5×19mm. For 3-RSP scheme, the ball joint will be the most difficult matter to deal with. We used a gold wire instead of a necked rod because the large stroke causes the later too easy to be broken.

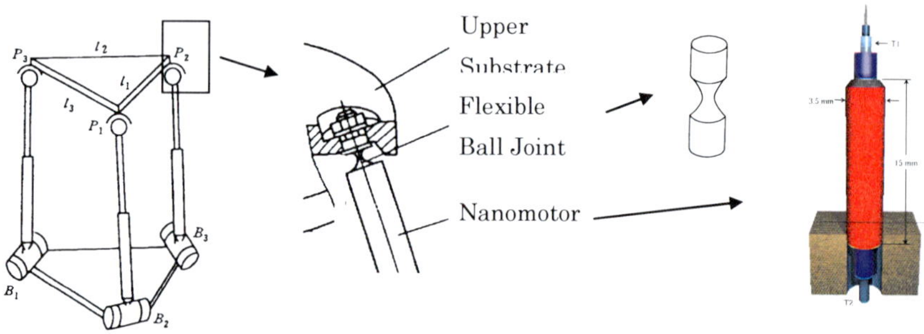

Fig. 6.9 Unit 4—a Parallel 3-RPS Manipulator

For getting a real-time observations of manipulations, the whole set of manipulators are mounted under an OM or inside a SEM/FESEM with a secondary electron detector, which has a relative large vacuum chamber. The resolution of the SEM is specified as 4 nm , whereas that of the FESEM is 1.5 nm. However, the real time video resolution is a factor of two or three lower. All of the wires are connected through the isolated vacuum feed-through passing through the SEM/FESEM chamber wall. All of the components and cables of the manipulators are properly shielded from the SEM/FESEM observation region to minimize image distortion from charging effect.

It can be found that the manipulators meet all of the basic requirements for nanomanipulations. It has a total of 16 DOFs (NRM-III) and four units with three to four cantilevers, the working space is $18\times18\times12mm^3$ with a 360º rotation, the coarse resolutions are 30nm and 2mrad. It is amounted under OM or inside the vacuum chamber of a SEM/FESEM, so the real time observations for the manipulations are realized and the forces between the cantilever tips and samples or that between samples and the substrate can also be measured with the SEM/FESEM images. The specifications including that for fine motions will be introduced later.

6.2.3 Configuration of Control System of Nanorobotic Manipulation System Inside SEM

The control system for the nanorobotic manipulation system is shown in Figure 6. 10. Generally, it uses typical mono-lateral master-slave control. The application of piezolevers provides the potential to realize bi-lateral control, but presently mainly for getting the force information between tools and objects.

The system can be divided into four parts including a multi-axis controller for actuators in Units 1 and 2, two LAN based Nanomotor controllers for the X-Y stage in Unit 2' and Unit 4, a PZT controller for Unit 3 and the z-direction PZT in Unit 2', and a controller for dielectrophoresis. Coarse motions generated by actuators can be controlled with a control panel, and potentially to be controlled with GPIB board.

Fine motions can be realized by taking proper actuation frequency of driving voltages and a scaling parameter provided by the controller. The open-loop control can provide nanometer resolution as to be shown in the next section.

Bulk PZT actuators do not suffering climbing motions but the hysteresis, which should be compensated by close-loop control. Two methods have been presented. The first is by using the PZTs in Units as both actuators and sensors, the so-called superposition of PZTs as have been shown. The second is to use strain gauges attached to the PZTs to form close loop. However, neither of them is full closed. The real time observation with microscopes provides a way to determine the final motion of the manipulators. If visual feedback is realized, it is possible to get 2-DOF (X and Y) closed for Unit 3. It is extremely difficult to control the motion in the direction along electron beam. Although focus depth (working distance) can provide rough information, it is hardly enough for nanometer positioning. Presently, a trial and correction process is taken by comparing the depth difference of the objects. It might be a better way by the combination of SPMs with the manipulators for getting higher vertical resolution.

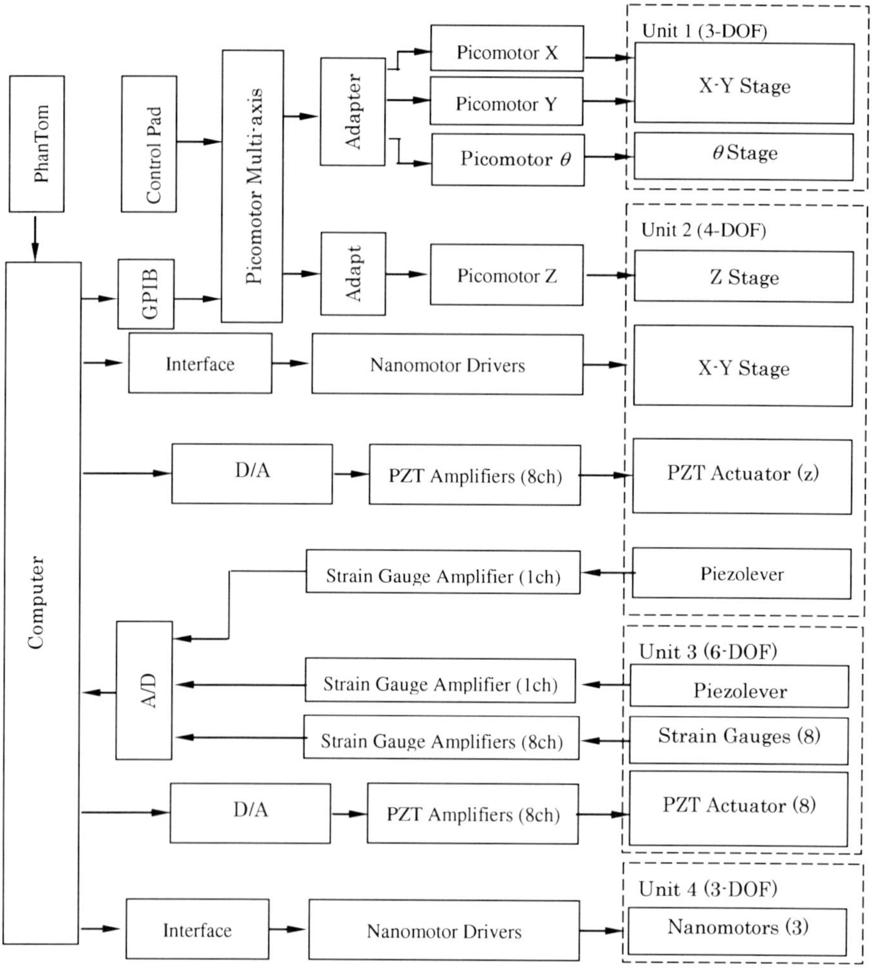

Fig. 6.10 Schematic Diagram of Control System

6.3 Hybrid Nanorobotic Manipulation System Inside SEM/TEM

"Hybrid nanorobotic manipulation system", which is designed for the nanomanipulations inside a scanning electron microscope (SEM) and a transmission electron microscope (TEM), was proposed [7]. We construct the SEM manipulators with 8 degrees of freedom (DOF), 3 units, to equip TEM manipulator for the effective TEM sample preparation. The TEM manipulator is designed with 3 DOF manipulator actuated by 4 multi-layer piezoelectric actuators and a 3 DOF passively driven TEM sample stage. High resolution transmission images of TEM are readily used for measurement and evaluation of samples. The passively driven

TEM sample stage is pre-manipulated by the SEM manipulator for coarse positioning within the working area of TEM manipulator. This methodology is named as hybrid nanorobotic manipulation so as to differentiate it from those only with an exchangeable specimen holder.

For the nanoanalytical system, the electron microscopes play one of the most important roles. Their high resolution image or various analytical techniques provide us the real-time change of samples with nanometer scale resolution. Various types of AFM cantilevers are used for the small-force measurement probe.

Hybrid nanorobotic manipulation system integrates a TEM nanorobotic manipulator (TEM manipulator) into SEM nanorobotic manipulator (SEM manipulator) is shown in Figure 6.11 of the basic idea of the configuration. The TEM manipulator can be set inside a SEM and a TEM for the sample preparation. It can also mount a passively driven sample stage. The stage is pre-manipulated by SEM manipulator for sample preparations and position setting inside the SEM. Figure 6.12 shows the scale of manipulated objects through SEM and TEM manipulations. The working space of a TEM manipulator is also much smaller than that of SEM manipulator with normally used actuators. If the actuation mechanism is installed in the outside of sample chamber, it is high cost and limited at the reconstruction of a TEM. On the other hand, manual adjustment can be reached with only ~100 μm positing resolution of a sample. Hence, it is difficult to adjust the position of sample within the working area of TEM manipulator manually. Here, we propose a sample stage passively driven by SEM nanorobotic manipulators for coarse positioning adjustment of TEM samples. This methodology is named as hybrid nanorobotic manipulation so as to differentiate it from those only with an exchangeable specimen holder. The schematic diagram of the overall of hybrid manipulation system is as shown in Figure 6.13. The system realizes an efficient sample preparation with wide enough working space of SEM manipulators, and high enough resolution to identify nano-scale objects with a TEM.

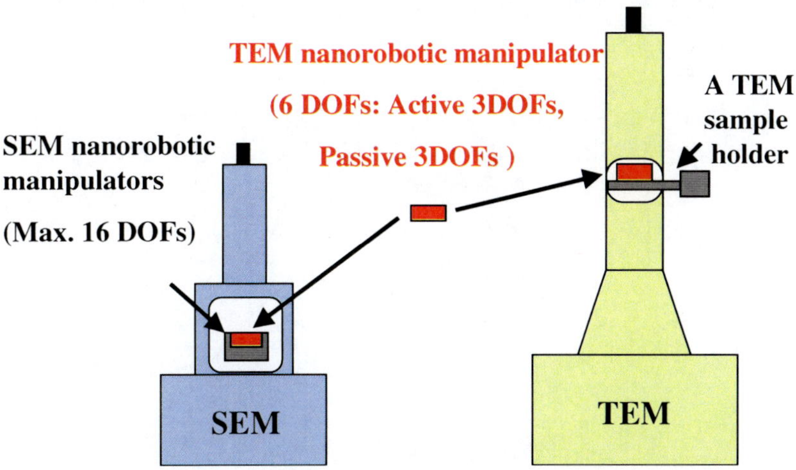

Fig. 6.11 Configuration of hybrid nanorobotic manipulation system

In this work, a FE-SEM and a TEM are used. Their specifications are listed in Table 6. 3. The resolution of TEM is ten times higher than that of FE-SEM; it is sub-nano order. On the other hand, the observation space of FE-SEM is approximately 10000 times larger than that of TEM as depicted in Figure 6. 12. Hence, TEM is well suited to measure and evaluate nano-scale objects, and SEM is preferred for setting manipulators and samples for efficient sample preparation. The detail specifications of hybrid nanorobotic manipulation system are summarized in Table 6. 4.

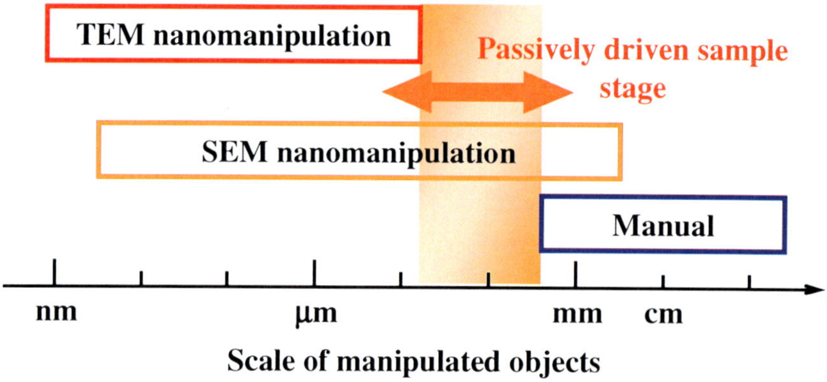

Fig. 6.12 Scale of manipulated objects through TEM and SEM manipulations

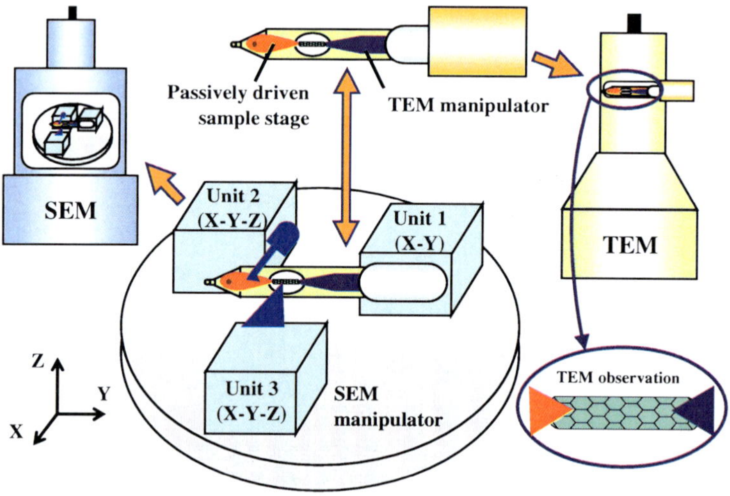

Fig. 6.13 Schematic diagram of hybrid nanorobotic manipulation system

Table 6.3 Specification of FE-SEM and TEM

Electron Microscope	TEM	FE-SEM
Acceleration Voltage	35 ~ 200 kV	0.5 ~ 30 kV
Magnification	×1,000 ~ ×600,000	×10 ~ ×500,000
Resolution	0.24 nm (lattice interval) ,0.45 nm (point-point interval)	1.5 nm
Observation Space	Φ2 mm × ± 0.3 mm	70 mm × 50 mm × 3~40 mm

Table 6.4 Specifications of hybrid nanorobotic manipulation system

Items	Specifications
TEM Nanorobotic Manipulator	
DOFs	3 DOFs (X, Y, Z), Passive 3 DOFs (X, Y, Z)
Actuators	4 multi-layer piezoelectric devices
Work. Space	~ 8 μm × ~ 60 μm × ~ 60 μm
SEM Nanorobotic Manipulator	
DOFs	Unit1: 2 DOFs (X, Y), Unit2: 3 DOFs (X, Y, Z), Unit3: 3 DOFs (X, Y, Z), Total: 8 DOFs
Actuators	6 Picomotors™, (New Focus Cop., Driving Force: 22 N, Max. Speed: 72 mm/s, Unit1, Unit2,Uni3-Z) 2 Nanomotors™ (Klocke Nanotechnik Cop., Driving Force: 0.03 N, Max. Speed: 2 mm/s, Unit3-X,Y)
Work. Space	~ 16 mm × ~ 16 mm × ~12 mm
Positioning Resolution	~ 30 nm (Unit 1, Unit2, Unit3: Z), ~2 nm (Unit3: X, Y)
TEM	
Acc. Volt.	35 ~ 200 kV
Resolution	0.24 nm (lattice-lattice), 0.45 nm (point-point)
Obs. Space	ϕ2 mm × ± 0.3 mm
FE-SEM	
Acc. Volt.	0.5 ~ 30 kV
Resolution	1.5 nm (at Acc. Volt.: 15 kV)
Obs. Space	70 mm × 50 mm × 3~40 mm

Scanning Electron Microscope (SEM) Manipulator

We reconstructed to be 8 DOF with 3 units for integration of the TEM manipulator as shown in Figure 6. 14. The characteristics of SEM manipulators include wide enough working space and multi-probe with multi-unit for preparing samples. In this paper, The installation of SEM manipulator is shown in Figure 6. 15. Unit 1 is used for driving TEM manipulator and the fixation of samples. Unit 2 is used for driving the passive sample stage of TEM manipulator, which will be described in Section II-A. Unit 3 is used for the fixation of samples. The specifications of SEM manipulators are listed in Table 6. 4.

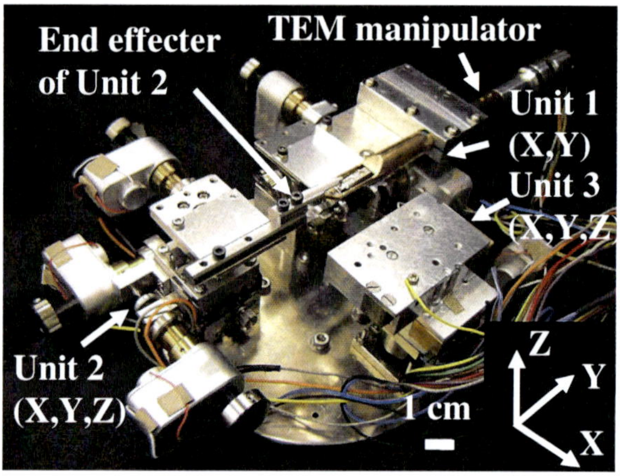

Fig. 6.14 SEM nanorobotic manipulators

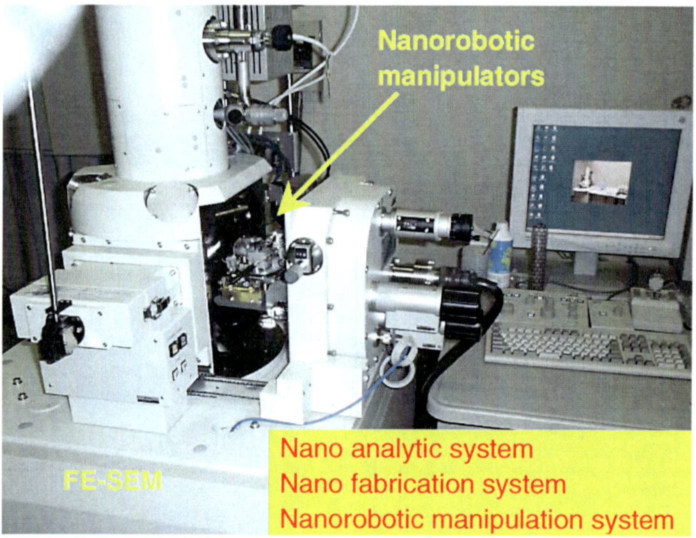

Fig. 6.15 Installation inside SEM Chamber

Transmission Electron Microscope (TEM) Manipulator

Figure 6. 16 shows 3-D schematic diagrams of the TEM nanomanipulator. It consists of 4 multi-layer piezoelectric devices, two 2-hinge beams, a plate for sample fixation, and a base. All parts are made of aluminum to avoid the magnetic field produced by processing. The two hinge beams are used for the translational movements in X and Z direction as shown Figure 6.16 (b) and (c).

Figure 6. 17 shows the overview of the constructed TEM nanomanipulator. The TEM manipulator can be installed inside a normal type side entry TEM sample holder, which has an inner diameter 7 mm as shown in Figure 6. 17. It is set on the side entry type TEM sample holder as shown in Figure 6. 18. The tip part is able to be removed for setting on SEM manipulator. The overview of used TEM is shown in Figure 6. 19, and the installation of TEM manipulator inside it is shown in Figure 6. 20. The specifications of TEM manipulator are listed in Table 6. 4.

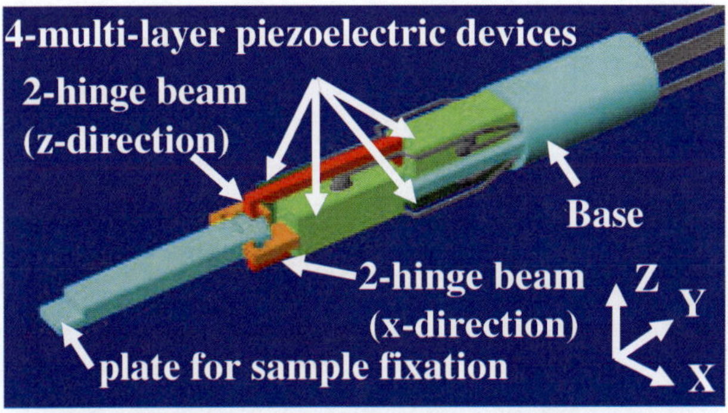

Fig. 6.16 3D image of a TEM nanorobotic manipulator

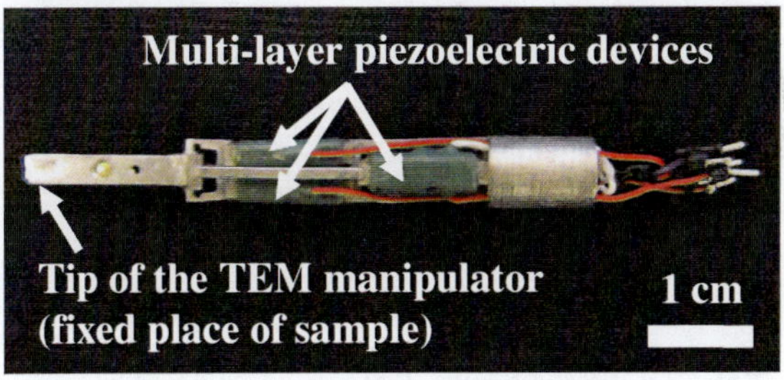

Fig. 6.17 TEM nanorobotic manipulator (Top view)

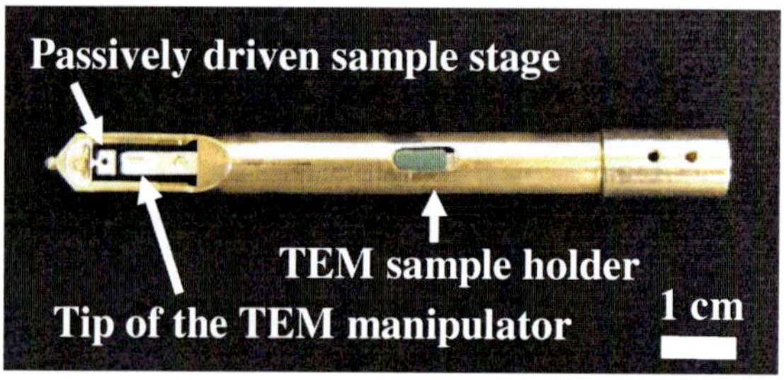

Fig. 6.18 TEM manipulator inserted inside a TEM sample holder

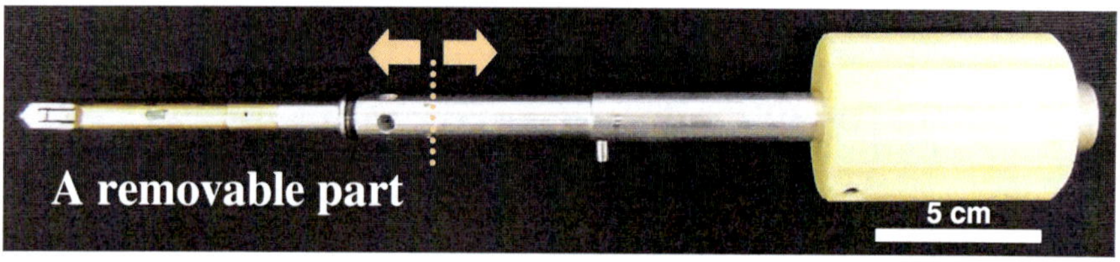

Fig. 6.19 TEM sample holder installed the TEM manipulator

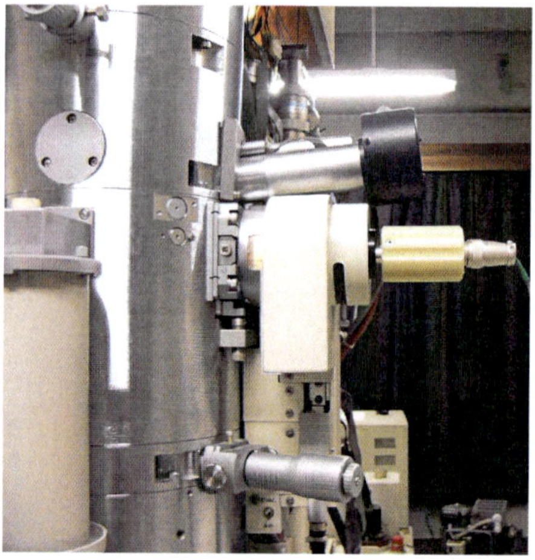

Fig. 6.20 Installation of TEM manipulator sample holder

Passively Driven Sample Stage

The space of a TEM sample chamber is quite narrow and strictly limited due to the principle of a TEM; high vacuum environment, short distance between an upper and a lower pole-piece, short working distance, and so on. These space limitations cause the difficulty of construction multi-DOF manipulation systems inside a TEM sample chamber. As shown in Figure 6.1, the working space of a TEM manipulator is also much smaller than that of SEM manipulator with normally used actuators. If the actuation mechanism is installed in the outside of sample chamber, it is high cost and limited at the reconstruction of a TEM. On the other hand, it is difficult to adjust the position of sample within the working area of TEM manipulator manually, because the manual adjustment has a positing resolution of only ~100 μm. Here, we propose a sample stage passively driven by SEM nanorobotic manipulators for coarse positioning adjustment of TEM samples. Classification of the passive driven mechanism is as follows (Figure 6. 21).

The types (a) Friction type, (b) Screw type, and (c) Plastic deformation type are relatively easy to be constructed and used as disposable. The types (d) Elastic deformation type, (e) Electrostatic force type, and (f) Magnetic force type have relatively high endurance, however their complex mechanism are not easy to be constructed with enough high positioning resolution. In this work, type (c) is applied for the mechanism of the passively driven stage from the aspects of easy fabrication. The hinge beams of a TEM sample stage are plastically deformed by SEM manipulator to be set within the working area of TEM manipulator.

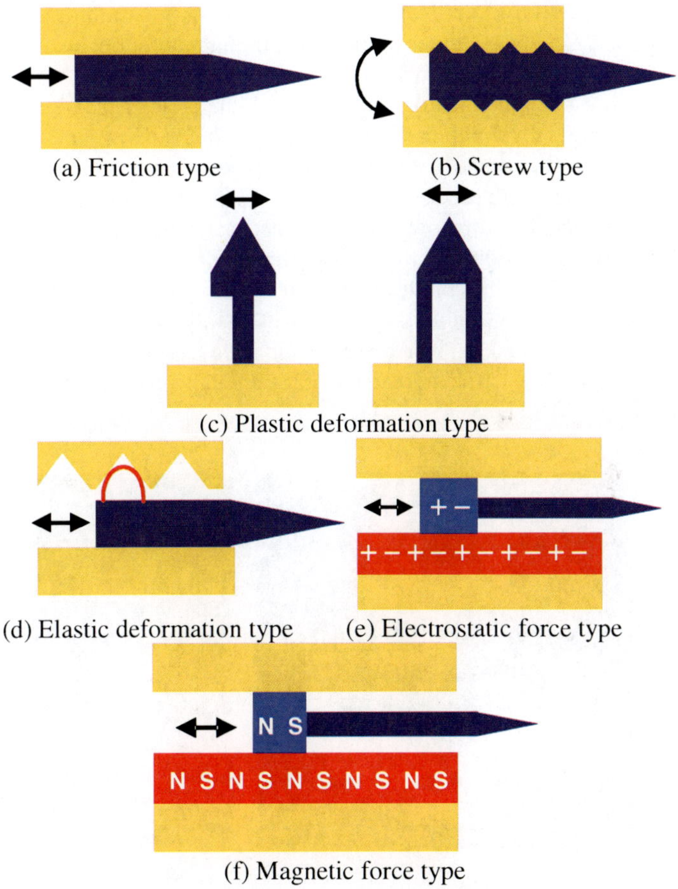

Fig. 6.21 Positioning mechanisms of passively driven sample stage

Figure 6. 22 shows the schematic diagram of constructed passively driven sample stage. Samples are fixed at the tip of sample stage. The sample stage has two holes; one is used for driving by SEM manipulators and the other is used for fixation on a TEM sample holder. The thin two beams and a plate are driven in X, Y, and Z directions, for X-direction by bent of two beams, for Y-direction by tensile and compressive bent of two beams, and for Z-directions by twist bent of a plate and beams. Figure 6. 23 shows the configuration of passively driven sample stage and an end-effecter of SEM manipulator.

To avoid the magnetic field produced by processing, it is needed to use nonmagnetic metal. We have always used aluminum, but a relatively large elastic deformation is not suitable for precise operation; it is cause of the error of stage movement at plastic deformation. Hence, indium and lead are used as low strength material. Figure 6. 24 shows the typical example of plastic deformation of indium passively driven sample stage at each movement. On the average of ten times ~20 μm movements, the positioning resolutions are ~12 μm in X direction (positioning error: ~6 μm), and ~15 μm in Y direction (positioning error: ~5 μm). These values are ~ 5 and ~2 times higher positioning resolution, and ~14 % and ~3 % low positioning error comparing with aluminum stage.

By using low strength material, such as indium or lead, it is easy to set the position of sample with small elastic deformation effects. However, endurances of these type stages are extremely low at fixation hole; after several time usages, the stages cannot be used with precision. Hence, in this paper, hybrid stage is proposed, which is constructed with low strength material, like lead as only plastic deformation part, and copper as the other parts. TEM sample stage. Figure 6.18 shows the movement of the sample stage with 10 step movements. The positioning resolutions of the passive sample stage is ~1 μm by moving ~25 μm in X direction, and ~1 μm by moving ~10 μm in Y direction, on the average. This result shows high precision positioning, ~1 μm, is realized. It is enough precision for setting TEM sample positions.

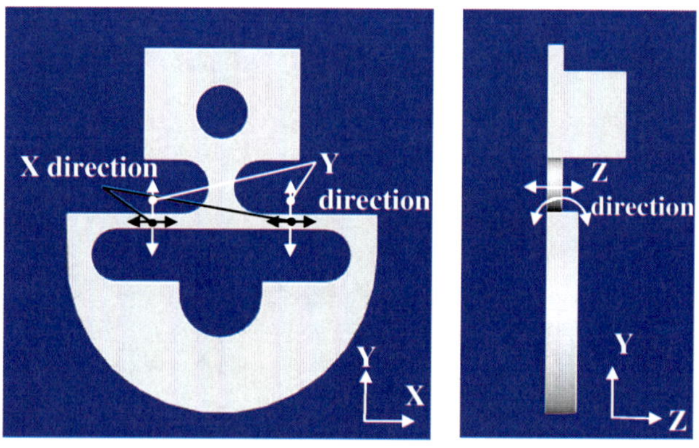

Top view Side view

Fig. 6.22 Schematic diagram of a passively driven sample stage

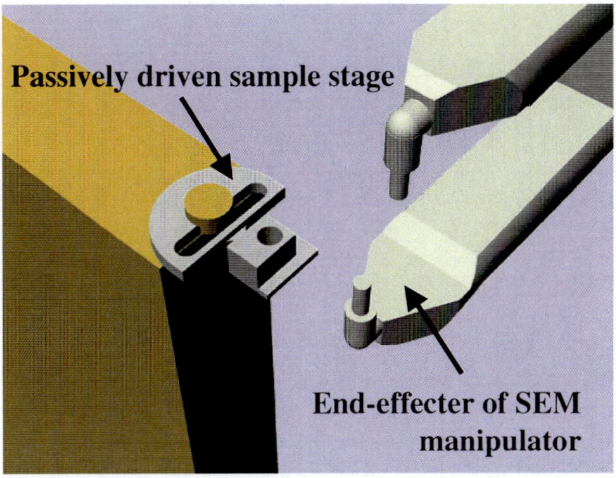

Fig. 6. 23 Schematic diagram of a passively driven sample stage and an end-effecter of SEM manipulator

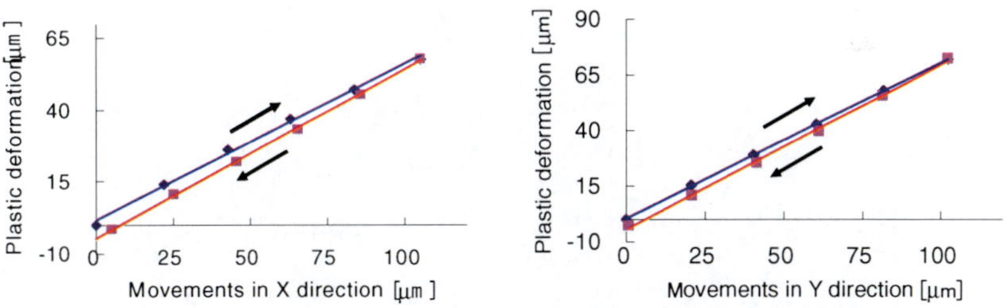

(a) X direction (b) Y direction

Fig. 6.24 Plastic deformation of indium passively driven sample stage

6.4 Nanorobotic Manipulation System Inside E-SEM

As explained in Chapter 3, the Environmental-SEM (E-SEM) can realize semi-wet (or dropwise condensation condition: high-humility condition) observation of samples with high chamber pressure (10 up to 2700 Pa). The capability to observe single biological cells in a native state using E-SEM is tremendous to overcome the limitation of conventional SEM which is limited in high vacuum environment (~1 × 10^{-3} to $^{-5}$ Pa). The observation by ESEM can provide morphological information of the cells, however, the other properties of the cells such as mechanical, electrical and chemical remain mystery.

We proposed a nanomanipulation system inside E-SEM mainly for biological sample manipulation using various nanotools to realize single cell nanosurgery system as explained in detail in Chapter 8 [8, 9, 10, 11, 12, 13, 14, 15, 16]. In order to clarify the other properties of the cells, the ability to physically manipulate the cells would be critical and necessary. Once the cells can be physically approach, interesting and new information on cells' properties, such as mechanical and electrical, could be revealed. Therefore, the characterization of the single cells requires some tools of manipulation. Since, the size of the biological cells is expected to be in the range of 3 – 6 μm, precise and high resolution manipulation system is required.

6.4.1 Design of Nanorobotic Manipulation System Inside E-SEM

The constructed system is shown in Figure 6. 25. The nanomanipulator consists of three units which can provide a total of seven degree-of-freedom (DOFs). Unit 1 has 3 DOFs, i.e. all linear translational motions in X, Y and Z axes. Unit 2 has 4 DOFs, i.e. three linear translational movements in X, Y and Z axes, and one rotational movement in Z axis. Unit 3 has no DOF and acting as a sample holder for the Peltier cooling stage and biological samples. All of these DOFs are actuated with commercially available PZT-based actuators.

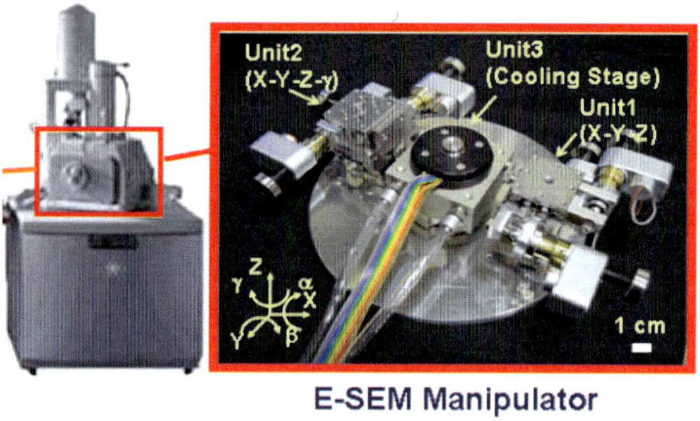

Fig. 6.25 The image of nanomanipulator system inside ESEM

Table 6.5 Specification of ESEM nanomanipulation system

Environmental Scanning Electron Microscope (ESEM)	
Vacuum Mode	E-SEM Mode (10–2600 Pa) Low Vacuum Mode (10–130 Pa) High Vacuum Mode (~10^{-4} Pa)
Acc. Voltage	0.2 ~ 30 kV
Resolution	3.5 nm (E-SEM Mode) 15 nm (Low Vacuum Mode) 3.5 nm (High Vacuum Mode)
Obs. Space	150 mm × 150 mm × 65 mm
Max. Obs. Area	f0.5 mm (E-SEM Mode) f18 mm (Low and High Vacuum Mode)
Detectors	SED, RED

Nanomanipulator	
DOFs	Unit1: 4 DOFs (X-Y-Z-g), Unit2: 3 DOFs (X-Y-Z) Total: 7 DOFs
Actuators	7 Picomotors™, (Unit1, Unit2)
Work. Space	~ 16 mm × ~ 16 mm × ~12 mm × ±5°
Positioning Resolution	~ 30 nm (Unit 1, Unit2)
Cooling Stage	Unit3 (Cooling water temp. ± 20 ° C)

Nanoprobe	
Multi-shapes	•Sharp Tip (0.02-2 N/m spring constants) •Flat Tip (0.2-2 N/m spring constants) •Nanoneedle (0.2, 0.8, 2 N/m spring constants) (< 200 nm diameters)

6.4.2 *Link Coordination of Nanorobotic Manipulation System Inside E-SEM*

An Integrated ESEM-Nanomanipulator System

Having explaining characteristics of ESEM system and the design and kinematics of nanomanipulator system, both systems are now ready for integration. ESEM, which has an image resolution up to 3.5 nm, is an excellent imaging instrument for observing surface characteristics of biological samples. Furthermore, nanomanipulator system, which has a linear resolution of 30 nm, is an excellent actuator for acting like a 'hand' for the operator to touch and characterize the physical properties of biological sample. By attaching a nanoprobe or a nanosensor on the substrate at frame t, several property characterizations of single cells, especially mechanically and electrically. Table 6.5 summarized the detailed specifications of an integrated ESEM nanomanipulation system.

The following kinematics derivation of the nanomanipulator is based on the modified Denavit-Hartenberg (D-H) notation.

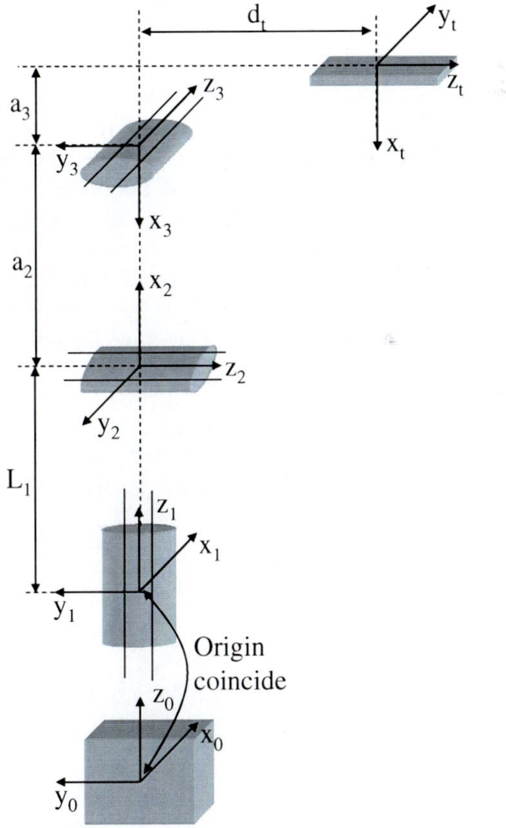

Fig. 6.26 Kinematic parameters and frame assignments for Unit 1 nanomanipulator

Table 6.6 Link Parameters for Unit 1 Nanomanipualtor

i	α_{i-1}	a_{i-1}	d_i	θ_i
1	0	0	d_1	0
2	90	0	d_2	90
3	90	a_2	d_3	180
t	90	$-a_3$	d_t	0

Modeling of the Unit 1 nanomanipulator is shown in Figure 6. 26. From the link frame assignments of Figure 6. 26, it can be seen that frame *0* and frame *1* are coincident. The frame 1 has a translational movement in the + Z_0 axis away from frame 0. Frames 2, 3, and *t* will be moved relative to the movement of frame *1*. Frame *2* has a translational movement in $|Y_0|$ axis and its movement will drag frames *3* and *t* together without affecting frame *1*. Frame 3 has a translational movement in $|X_0|$ and only drags frame t during its movement without affecting the position and orientation of frames *1* and *2*. Frame t is a tool frame.

The link parameters of Unit 1 nanomanipulator is shown in Table 6. 6, where i denotes the code number of joints, a_i the link length, α_i the link twist, d_i the distance between links and θ_i the angle between links. It can be seen from Table 6. 6, the link parameter of L_1 is not listed. The absence of L_1 is occurred because we were unable to define its values based on the standard definition of the D-H parameters, i.e. . a_i, α_i, d_i and θ_i. In order to overcome this issue, we intentionally added the parameters of L_1 in the transformational matrix of frame 2 with respect of frame 1, i.e. ${}^1T_2$ as will be explained in the following paragraph. By doing this slight modification of ${}^1T_2$, we are able to include L_1 in the kinematics derivation of Unit 1.

By defining new symbols for sine and cosine as s and c, respectively, the homogeneous transformations for each link are described below,

$$
{}^0T_1 = \begin{bmatrix} c\theta_1 & -s\theta_1 & 0 & a_{i-1} \\ s\theta_1 c\alpha_{i-1} & c\theta_1 c\alpha_{i-1} & -s\alpha_{i-1} & -s\alpha_{i-1}d_1 \\ s\theta_1 s\alpha_{i-1} & c\theta_1 s\alpha_{i-1} & c\alpha_{i-1} & c\alpha_{i-1}d_1 \\ 0 & 0 & 0 & 1 \end{bmatrix} = \begin{bmatrix} 1 & 0 & 0 & 0 \\ 0 & 1 & 0 & 0 \\ 0 & 0 & 1 & d_1 \\ 0 & 0 & 0 & 1 \end{bmatrix} \tag{6.7}
$$

$$
{}^1T_2 = \begin{bmatrix} c\theta_2 & -s\theta_2 & 0 & a_{i-1} \\ s\theta_2 c\alpha_{i-1} & c\theta_2 c\alpha_{i-1} & -s\alpha_{i-1} & -s\alpha_{i-1}d_2 \\ s\theta_2 s\alpha_{i-1} & c\theta_2 s\alpha_{i-1} & c\alpha_{i-1} & c\alpha_{i-1}d_2 + L_1 \\ 0 & 0 & 0 & 1 \end{bmatrix} = \begin{bmatrix} 0 & -1 & 0 & 0 \\ 0 & 0 & -1 & -d_2 \\ 1 & 0 & 0 & L_1 \\ 0 & 0 & 0 & 1 \end{bmatrix} \tag{6.8}
$$

$$ {}^{2}T_3 = \begin{bmatrix} c\theta_3 & -s\theta_3 & 0 & a_{i-1} \\ s\theta_3 c\alpha_{i-1} & c\theta_3 c\alpha_{i-1} & -s\alpha_{i-1} & -s\alpha_{i-1} d_3 \\ s\theta_3 s\alpha_{i-1} & c\theta_3 s\alpha_{i-1} & c\alpha_{i-1} & c\alpha_{i-1} d_3 \\ 0 & 0 & 0 & 1 \end{bmatrix} = \begin{bmatrix} -1 & 0 & 0 & a_2 \\ 0 & 0 & -1 & -d_3 \\ 0 & -1 & 0 & 0 \\ 0 & 0 & 0 & 1 \end{bmatrix} \tag{6.9} $$

$$ {}^{3}T_t = \begin{bmatrix} c\theta_t & -s\theta_t & 0 & a_{i-1} \\ s\theta_t c\alpha_{i-1} & c\theta_t c\alpha_{i-1} & -s\alpha_{i-1} & -s\alpha_{i-1} d_t \\ s\theta_t s\alpha_{i-1} & c\theta_t s\alpha_{i-1} & c\alpha_{i-1} & c\alpha_{i-1} d_t \\ 0 & 0 & 0 & 1 \end{bmatrix} = \begin{bmatrix} 1 & 0 & 0 & -a_3 \\ 0 & 0 & -1 & -d_t \\ 0 & 1 & 0 & 0 \\ 0 & 0 & 0 & 1 \end{bmatrix}. \tag{6.10} $$

As can be noticed from ${}^{1}T_2$, the parameter of L_1 is added with the element in the third row and fourth column position in the transformational matrix. It can be acknowledged later on after the completion of the kinematics derivation, that this addition of L_1 in ${}^{1}T_2$ gives satisfactory kinematics behaviors of Unit 1.

Finally, the product of all four link transformation is,

$$ {}^{0}T_t = {}^{0}T_1\,{}^{1}T_2\,{}^{2}T_3\,{}^{3}T_t = \begin{bmatrix} r_{11} & r_{12} & r_{13} & p_x \\ r_{21} & r_{22} & r_{23} & p_y \\ r_{31} & r_{32} & r_{33} & p_z \\ 0 & 0 & 0 & 1 \end{bmatrix} = \begin{bmatrix} 0 & 1 & 0 & d_3 \\ 0 & 0 & -1 & -d_2 - d_t \\ -1 & 0 & 0 & a_2 + a_3 + d_1 + L_1 \\ 0 & 0 & 0 & 1 \end{bmatrix}. \tag{6.11} $$

By evaluating the position and orientation of frame t as respect to frame 0 as described by ${}^{0}T_t$, the correctness of the kinematics derivation of Unit 1 can be confirmed.

$$ {}^{0}T_t = \begin{bmatrix} \overset{\text{A}\,\downarrow}{0} & \overset{\text{B}\,\downarrow}{1} & \overset{\text{C}\,\downarrow}{0} & d_3 \\ 0 & 0 & -1 & -d_2 - d_t \\ -1 & 0 & 0 & a_2 + a_3 + d_1 + L_1 \\ 0 & 0 & 0 & 1 \end{bmatrix} \begin{matrix} \leftarrow \text{D} \\ \leftarrow \text{E} \\ \leftarrow \text{F} \\ \end{matrix} \tag{6. 12} $$

6.5 Hybrid Microscope

The E-SEM provides a direct observation of water-containing samples with nanometer high resolution by specially built secondly electron detector. However, these scanning electron microscopes can provide us the surface imaging. The TEM can provide the interior imaging, however it needs HV environment. The hybrid microscope can provide the nano-scale imaging by E-SEM and interior/florescent imaging by OM with water-containing biological specimen. Optical microscope

(OM) is widely used for biological specimen analysis and imaging. It provides the interior image of transmitted biological specimen, and also the fluorescent image of protein or dying materials. However, its limitation of its imaging resolution prevents for nano-scale evaluation, caused by the diffraction limit of optical wavelength.

We proposed a novel type microscope called as hybrid microscope which is combined the optical microscope and E-SEM [17, 18, 19, 20, 21]. The hybrid microscope can provide the nano-scale imaging by the E-SEM and interior/florescent imaging by OM with water-containing biological specimen. Figure 6. 27 shows the configuration of hybrid microscope schematically. The the DOFs of nanomanipulators are extended as 13 DOFs with 3 units as shown in Figure 6. 28. The list of characteristics of the hybrid microscope is summarized in Table 6. 7 comparison with other microscopes for conventional biological imaging. It is needed to control the position of biological specimen to observe by OM and E-SEM continuously. Our E-SEM nanorobotic manipulators inside the E-SEM can be realized a direct observation of water-containing samples with nanometer high resolution by specially built secondly electron detector. The evaporation of water is controlled by the low sample temperature and high sample chamber pressure.

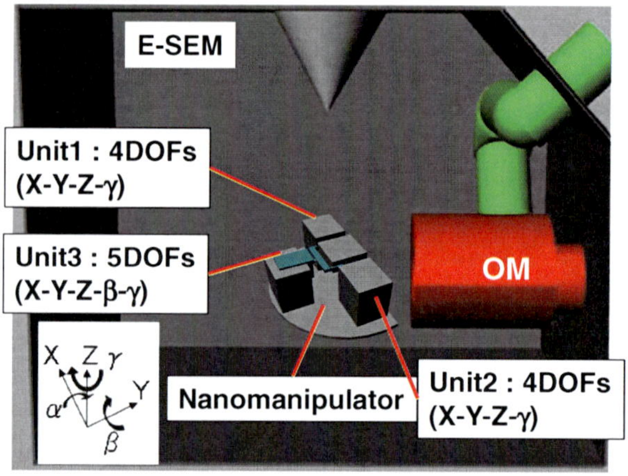

Fig. 6.27 Schematic of hybrid microscope which is combined with OM and ESEM

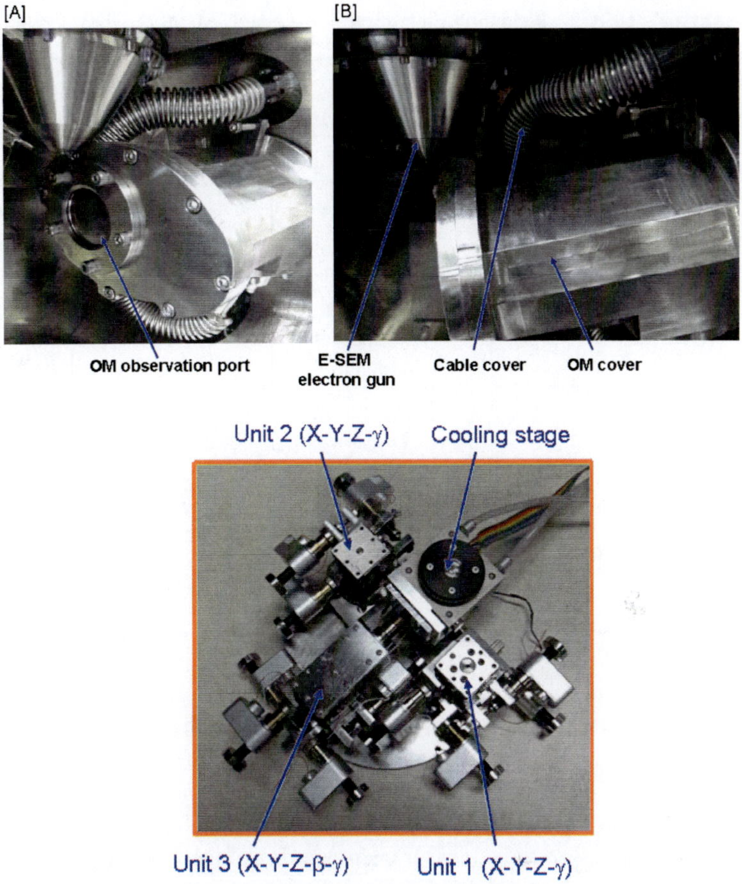

Fig. 6.28 Nanomanipulation System inside Hybrid Microscope, Built up cover inside E-SEM chamber. ([A] Side view, [B] Front view) and overview of constructed nanomanipulators.

Table 6.7 Characteristics of Hybrid Microscope (OM: Optical Microscope, AFM: Atomic Force Microscope, SEM: Scanning Electron Microscope, E-SEM: Environmental-SEM)

Types of Microscope	Imaging Resolution	Imaging Features	Imaging Environment	Fluorescent Imaging
OM	~ 200 nm	Interior Imaging	Water/Air	Good
AFM	~ 1 nm	Surficial Imaging	Water/Air/ High Vacuum	Difficult
SEM	~ 1 nm	Surficial Imaging	High Vacuum	Difficult
E-SEM	~ 3.5 nm	Surficial Imaging	Water - Containing	Difficult
TEM	~ 0.1 nm	Interior Imaging	High Vacuum	Difficult
Hybrid Microscope	~ 3.5 nm	Interior and Surficial Imaging	Water - Containing	Good

6.6 Nano-tool Exchanger System under Hybrid Microscope

The nanomanipulation system has multi-degrees of freedom and multi-units. The "nanotools" proposed to realize flexible and complex nano-scale stiffness measurement, adhesion force measurement, cutting, injection and so on. One of the difficulties of nanomanipulation system inside electron microscope (EM) is consumption of exchange time after setting the tools and samples. It needs time to evacuate the pressure of sample chamber by vacuum pump to observe using electron beams by opening the chamber. To observe under high-vacuum condition (around 10^{-3}~10^{-4} Pa), several ten minutes to several hours are required depending on the size of chamber and vacuum pump system. Moreover, the setting is manually operated, because those tools or samples are very small size and brittle. But, in some cases, some mistakes are easily happened such as hitting the tools with stage, missing or losing the samples by electro-static charge and so on.

As previous work, the cantilever tip of atomic force microscope (AFM) was replaced by the nanobits which are fabricated by electron beam lithography and standard silicon processing [22]. This method is one of effective ways to replace the tip in nanoscale, but they did not apply to solve the problem to exchange for nanomanipulation in the EM.

We proposed a novel nanotool exchanger system (NTExS) to realize the tool exchange without opening the sample chambers [23]. Figure 6. 29 shows the schematic of proposed nanotool exchange system. This system mainly consists of nanotool adapter, nanotool holder, and nanotool attachment for various types of nanotools. At first, the nanotools are fixed in the nanotool holder with nanotool adapter. The nanotool adapter is designed to hold the various types of nanotools. To apply the nanotool to the sample, the nanotool attachment is used to fix the nanotool as an end-effector of nanomanipulator with nanotool adapter. Through the nanotool exchanger system, it is not needed to open the sample chamber to exchange the nanotools and to evacuate the sample chamber pressure.

The nanotool exchanger system is composed with three parts, 1) "nanotool adaptor", 2) "nanotool attachment", and 3) "nanotool holder".

1) The "nanotool adaptor" is used to fix different types nanotools. This adaptor has a mechanism to fix the nanotool attachment in a same manner.

2) The "nanotool attachment" has a function to attach nanotool to use as an end-effector of manipulator through nanotool adaptor. This attachment is considered to be a two cased for using nanotool adaptor or attaching nanotool directly.

3) The "nanotool holder" is used for to arrange the different multiple nanotools. This holder is useful to exchange the nanotool effectively by the nanotool attachment. It is needed to fix the nanotool adaptor or nanotool, but to release them when the nanotool attachment is fixed the nanotools.

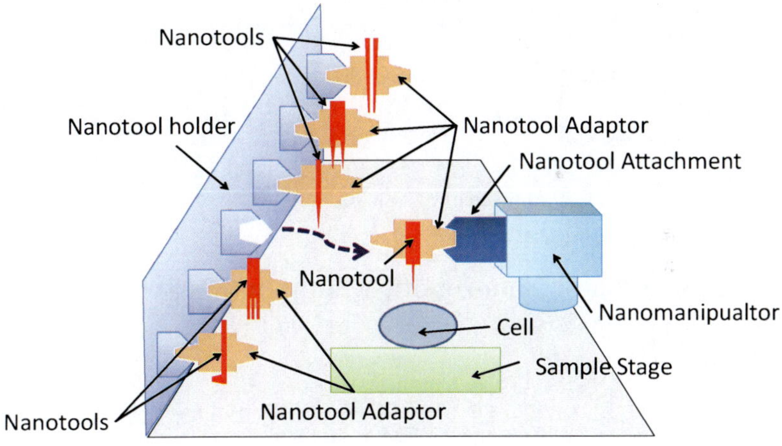

Fig. 6.29 Schematic of Nanotool Exchanger System

Figure 6. 30 shows the flow chart of nanotool exchanger system consisted with the three parts. The new nanotools are attached to the nanotool holder. To exchange new nanotools, the nanotool adaptor is used to fix for different nanotool in a same manner. It is possible to fix the nanotools by nanotool attachment directly. By selecting the target nanotools, the nanotool attachment is used to attach a new nanotool as end-effector of manipulator. Then we can use the new nanotool for the application for single cell surgery for example.

By using the nanotool for the application in long time, it is wearied and deteriorated. In some cases, it is needed to change to do the different task by different nanotools. In these cases, we need to change the nanotool newly. At first, the nanotool is collected by the nanotool holder from the nanotool attachment. Normally, the nanotool is expected to use as disposable devices. Then, the new nanotool is selected and fixed to the nanotool attachment as explained in the paragraph up.

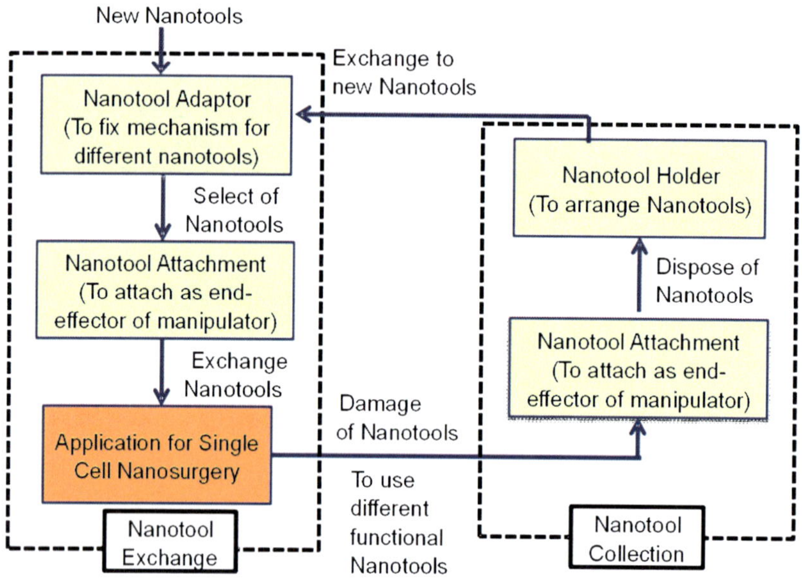

Fig. 6.30 Flow-chart of Nanotool exchanger system

6.7 Automation of Nanorobotic Manipulation System Inside E-SEM

The flow chart of the automatic control is shown in Figure 6. 31. Firstly, a real-time image was got from the ESEM system. Secondly, the position of the endeffector and the target cell was recognized based on image processing. The difference of these two position values is taken as the signal to control the movement direction of the nanomanipulator. Thirdly, the end-effector moved to the target cell automatically driven by the nanorobotic manipulator according to the position information. After the movement, the updated image was captured again to calculate the new position of the end-effector.

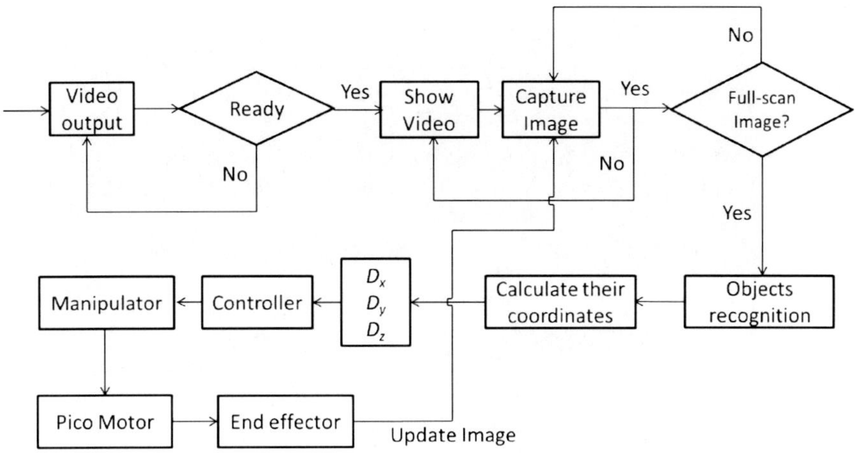

Fig. 6.31 Flow chart of the automatic manipulation system

Image Capture/Processing

The key part of the automatic control is based on the real time images. Therefore, the first step is to catch and save the ESEM image. Here, we used two methods for the image capture, i.e. manually saving and auto saving with a given frequency. The manually saving is used when we set the template for the end-effector, and the automatic capture method is used during the movement of the end-effector.

The first step for image processing is to reduce the noise. There are two primary forms of noise in the images: noise due to the statistical nature of electron collision and emission, and noise due to the SEM detector and signal processing electronics. The number of electrons hitting the specimen and then the corresponding emission of secondary electrons is inherently a statistical process. This results in shot noise extending up to high frequencies in the image. Shot noise is usually assumed to be both zero mean and independent. Noise can also originate from stray high-energy backscattered electrons colliding with either the detector or the chamber walls and emitting additional secondary electrons.

Generally, the ESEM system can eliminate these noises by itself using the hardware and software design. However, it is hard to deal with the noise from the outside, including the movement of the motor, and the external voltage potential inside ESEM chamber. For example, the periodical noise is appeared when we using the oscillation knife to cut the single cell, which is considered to be caused by the piezo actuator. Therefore, a median filter is used to reduce the noise in ESEM image as shown in Figure 6. 32.

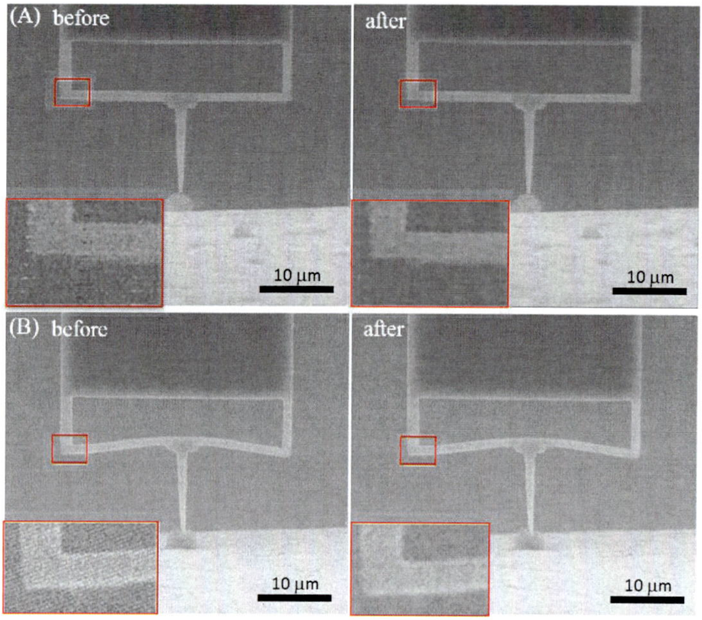

Fig. 6.32 Images comparison before and after the image filter. (A) and (B) are two examples at different experiment, the left images are the original images and the right images are the ones after filter. The left-bottom shows the enlarge image of part of the nano knife.

The second step for imaging processing is to find out the interesting regions. In our case, we are interest in two parts, i.e. single cell and the endeffector for single cell analysis.

The endeffector is identified by the template matching technique. In the nano knife indemnification, the tip of the nano knife was taken as a template due to its obvious shape. It was demonstrated that the nano knife can be identified exactly. Next, the position of the nano knife tip can be obtained based on the image pix.

Yeast cell usually kept a semicircle shape inside ESEM due to its adhesion to substrate, and the size and shape of cells are slightly different from each other. It hard for the definition what kind of cell do we want to analysis. Meanwhile, the interesting points of the cell for researcher are also different according to the analysis purposes. For example, the interesting area is the top part when we want to evaluate the cell wall's stiffness and the interesting area should change to the side part when we want to evaluate the shear adhesion force of the cell to substrate. Therefore, the single cell which will be used for analyzed is decided and chosen by the researcher. The program was set to recognize and save the position of the cell once people point out the cell and click the mouse on it. By using this human-robot corporation methodology, the analysis system can work more efficient.

Picomotor Control

One picomotor driver can only drive one picomotor each time. In order to increase the movement speed, the movements of the nanomanipulator to three directions are controlled by three picomotor drivers respectively. Therefore, the endeffector driven by the three motors can move to the target directly.

There are two kinds of movement mode for the picomotor, i.e. step mode and speed mode. The picomotor will move a certain distance with a given speed at the step mode. On the other hand, the picomotor will move continuously at a certain speed without stop until setting the stop flag to true.

We designed two types of movement, i.e. the manually control at step mode and the automatic control at speed mode. The manually control is used for the cell manipulation purpose and the automatic control is used for the end effector positioning. In the automatic movement control, the distance between the endeffector and the cell is used as the control signal to set the stop flag value. The stop flag will be true when the distance is zero. Therefore, the picomotor will move to the single cell automatically until the distance becomes zero.

The movement speed of the picomotor ranges from 0 to 20 μm/s. As shown in Figure 6. 33, in order to reduce the movement time, the movement speed is set to the maximum value 20 μm/s when the distance between the end effector and the single cell is far than 30 μm. On the other hand, to prove the movement accuracy, the speed was set to 1 μm/s when the distance is less than 5 μm. The movement speed reduces linearly when the distance is between 30 μm to 5 μm.

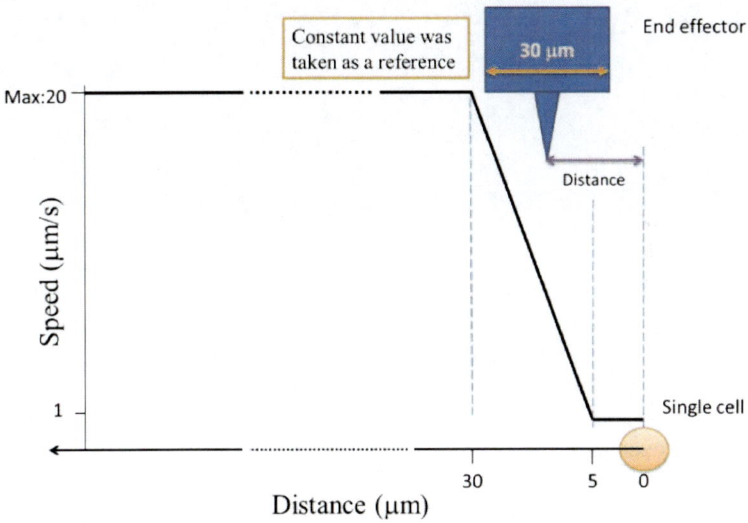

Fig. 6.33 Speed control of the picomotor based on the distance between the end effector and the single cell

The distance between the end effector and the single cell is calculated based on the image pix. First, the width of the cantilever beam is taken as a scale bar 30 μm. the number of pixel is counting by image processing. Second, the distance value of each pixel is calculated. Finally, the distance between the end effector and the single cell can be got in real time.

Experiment Result

The automatic movement of the end effector to singe cell was demonstrated by taking the single cell cutting as an example. The nano knife was fixed to the nanorobotic manipulator, and the cell was arranged on the substrate inside ESEM. The real time ESEM video was captured and then the position of the objects was used as the control signal. The automatic movement of the nano knife was controlled based on real time image processing.

Figure 6. 34 shows the captured image during the automatic movement of the endeffector. The tip of the nano knife was recognized and marked with a green circle. The motors were controlled based on the position information. It can move to the target cell by itself. Form 0 s to 40 s, the nano knife moves to the first target cell automatically. Next, we choose another cell (target cell 2) for testing. It can be seen that the nano knife can move to the second cell automatically. The total time for the whole experiment is around 70 s.

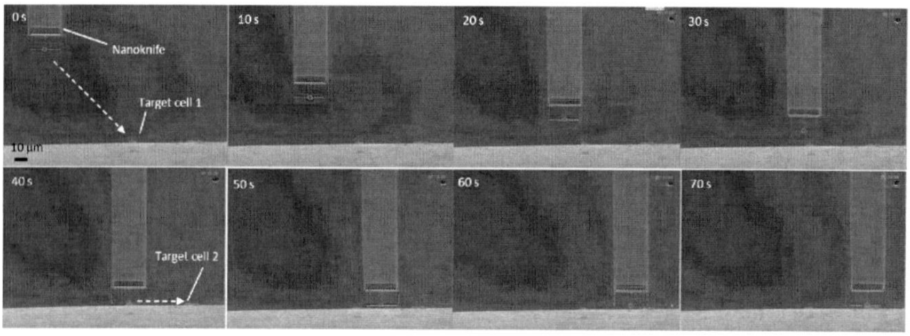

Fig. 6.34 Application of the automatic manipulation system for single cell cutting

After that, single cell cutting experiment can be performed once the nanoknife tip touches the single cell, as described in chapter 6. Therefore, we only need to focus on the single cell cutting analysis with the supporting of this automatic manipulation system.

Single cell analysis is important for the biological field. ESEM system combining with the nano manipulator allows the real time single cell analysis at nanometer scale. In this chapter, an automatic analysis system was presented. The nanomanipulation system with endeffector allows the automanipulation. Therefore, the movement can be realized automatically and we only need to focus on the import part. This system allows us to perform the single cell analysis at a relative high speed. It would benefit the research in the biological and medical fields.

This research shows the primary work of the automation system. Only the target recognition and the auto movement were realized up to now. There are still lots of works need to be considered, such as the path planning, control algorithm and so on. The cell array inside ESEM is also an interesting topic. We have shown the possibility to arrange the cell in a simple way based on the "coffee ring" theory. However, how the parameters, such as the evaporation speed, the structure of the chip, influence the final structure is still not clear. It would be very interesting if we can explain the mechanism and arrange the cell in any shape. This automatic manipulation method gives us a new insight for the single cell analysis, though it still has lot problems required to be solved. I believe that further study in this area will benefit both the engineering and biological field.

References

[1] Fukuda, T., Arai, F., Dong, L.X.: Assembly of Nanodevices with Carbon Nanotubes through Nanorobotic Manipulations. Proc. of the IEEE 91(11), 1803–1818 (2003)

[2] Fukuda, T., Arai, F., Dong, L.X.: Nanorobotic Systems. Int. Journal of Advanced Robotic Systems 2(3), 264–275 (2005)

[3] Dong, L.X., Arai, F., Fukuda, T.: 3D Nanorobotic Manipulation of Nano-Order Objects Inside SEM. In: Proc. of the 2000 Int. Symposium on Micromechatronics and Human Science, Nagoya, Japan, pp. 151–156 (2000)

[4] Dong, L.X., Arai, F., Fukuda, T.: 3-D Nanorobotic Manipulations of Nanometer Scale Objects. J. of Robotics and Mechatronics 13(2), 146–153 (2001)

[5] Dong, L.X., Arai, F., Fukuda, T.: Three-dimensional Nanorobotic Manipulations of Carbon Nanotubes. Journal of Robotics and Mechatronics 14(3), 245–252 (2002)

[6] Dong, L.X., Arai, F., Fukuda, T.: Destructive Constructions of Nanostructures with Carbon Nanotubes through Nanorobotic Manipulation. IEEE/ASME Transactions on Mechatronics 9(2), 350–357 (2004)

[7] Nakajima, M., Arai, F., Dong, L.X., Matsuura, H., Fukuda, T.: Measurement of a Bending Modulus of a Nanotube through Hybrid Nanorobotic Manipulation System inside SEM and TEM. Transactions of the Japan Society of Mechanical Engineers, Part C 71(704), 1349–1354 (2005) (in Japanese)

[8] Ahmad, M.R., Nakajima, M., Kojima, S., Homma, M., Fukuda, T.: In-situ Single Cell Mechanics Characterization of Yeast Cells using Nanoneedles inside Environmental-SEM. IEEE Transactions on Nanotechnology 7(5), 607–616 (2008)

[9] Ahmad, M.R., Nakajima, M., Kojima, S., Homma, M., Fukuda, T.: The effects of cell sizes, environmental conditions and growth phases on the strength of individual w303 yeast cells inside esem. IEEE Transactions on Nanobioscience 7, 185–193 (2008)

[10] Ahmad, M.R., Nakajima, M., Kojima, S., Homma, M., Fukuda, T.: Nanoindentation Methods to Measure Viscoelastic Properties of Single Cells using Sharp, Flat and Buckling Tips inside ESEM. IEEE Transactions on Nanobioscience 9(1), 12–23 (2010)

[11] Ahmad, M.R., Nakajima, M., Kojima, S., Homma, M., Fukuda, T.: Buckling Nanoneedle for Characterizing Single Cells Mechanics inside Environmental SEM. IEEE Transactions on Nanotechnology 10(2), 226–236 (2011)

[12] Ahmad, M.R., Nakajima, M., Kojima, S., Homma, M., Fukuda, T.: Buckling Nanoneedle for Characterizing Single Cells Mechanics inside Environmental SEM. IEEE Transactions on Nanotechnology 10(2), 226–236 (2011)

[13] Shen, Y., Nakajima, M., Ahmad, M.R., Kojima, S., Homma, M., Fukuda, T.: Effect of ambient humidity on the strength of the adhesion force of single yeast cell inside environmental-SEM. Ultramicroscopy 111(8), 1176–1183 (2011)

[14] Shen, Y., Ahmad, M.R., Nakajima, M., Kojima, S., Homma, M., Fukuda, T.: Evaluation of the single yeast cell's adhesion to ITO substrates with various surface energies via ESEM nanorobotic manipulation system. IEEE Transactions on Nanobioscience 10(4), 217–224 (2011)

[15] Ahmad, M.R., Nakajima, M., Kojima, M., Kojima, S., Homma, M., Fukuda, T.: Nanofork for Single Cells Adhesion Measurement via ESEM-Nanomanipulator System. IEEE Transactions on Nanobioscience 11(1), 70–78 (2012)

[16] Ahmad, M.R., Nakajima, M., Kojima, M., Kojima, S., Homma, M., Fukuda, T.: Instantaneous and Quantitative Single Cells Viability Determination Using Dual Nanoprobe Inside ESEM. IEEE Transactions on Nanotechnology 11(2), 298–306 (2012)

[17] Nakajima, M., Hirano, T., Kojima, M., Hisamoto, N., Homma, M., Fukuda, T.: Direct Nano-injection Method by Nanoprobe Insertion based on E-SEM Nanorobotic Manipulation under Hybrid Microscope. In: Proc. of the 2011 IEEE Int. Conf on Robotics and Automation (ICRA 2011), pp. 4139–4144 (2011)

[18] Nakajima, M., Akimoto, H., Hirano, T., Kojima, M., Hisamoto, N., Homma, M., Fukuda, T.: Biological Specimen Viability Analysis by Hybrid Microscope Combined Optical Microscope and Environmental-SEM. In: Proc. of 2010 IEEE Nanotech. Materials and Devices Conf (IEEE-NMDC 2010), pp. 315–320 (2010)

[19] Nakajima, M., Hida, H., Shen, Y., Kojima, M., Sato, K., Fukuda, T.: Multi-slicing of C. elegans Tissue using Micro-Nanocutting Probe based on Nanomanipulation. In: Proc. of the 2012 12th IEEE Conf. on Nanotechnology, IEEE-Nano 2012 (2012)

[20] Nakajima, M., Hirano, T., Kojima, M., Hisamoto, N., Homma, M., Fukuda, T.: Direct Nano-injection Method by Nanoprobe Insertion based on E-SEM Nanorobotic Manipulation under Hybrid Microscope. In: Proc. of the 2011 IEEE Int. Conf. on Robotics and Automation (ICRA 2011), pp. 4139–4144 (2011)

[21] Nakajima, M., Hirano, T., Kojima, M., Hisamoto, N., Nakanishi, N., Tajima, H., Homma, M., Fukuda, T.: Local Nano-injection of Fluorescent Nano-beads inside C. elegans based on Nanomanipulation. In: Proc. of the 2012 IEEE/RSJ Int. Conf. on Intelligent Robotics and Systems (IROS 2012), pp. 3241–3246 (2012)

[22] Rajendra Kumar, R.T., Hassan, S.U., Sardan Sukas, O., Eichhorn, V., Krohs, F., Fatikow, S., Boggild, P.: Nanobits: customizable scanning probe tips. Nanotechnology 20, 395703 (2009)

[23] Nakajima, M., Kawamoto, T., Hirano, T., Kojima, M., Fukuda, T.: Nanotool Exchanger System based on E-SEM Nanorobotic Manipulation System. In: Proc. of the 2012 IEEE Int. Conf. on Robotics and Automation (ICRA 2012), pp. 2773–2778 (2012)

Chapter 7
Measurement/Manipulation/Assembly of Carbon Nanotubes under FE-SEM/TEM

7.1 Application Fields of Nanomanipulation System under FE-SEM/TEM

Nanofabrication as a key technology can improve the applications of nanotechnology. In recent years, CNTs have been proposed as a basic building block for a new generation of nanoelectronic and mechanical systems. Previous researches showed that CNTs can be used as linear [1] and rotational nanobearings[2], mass conveyors [3], field emitters [4, 5], AFM probes [6], nanotweezers [7], nanoposition sensors [8] and so on. All of those applications are based on as-grown CNTs without any change in the mechanical structure of CNTs except some chemical decoration [9]. In fabrication, manipulation or assembly of nanotubes, the length of CNTs is an important factor and will influence the function and configuration of the nanostructures and nanodevices. Therefore, a method of defining the length of nanotubes precisely is indispensable. The applications of CNTs in special structure or shapes can be significantly improved by a method capable of making an irreversible bending of CNTs. In addition of these structure manipulation methods, also an effective welding technique is essential for the development of CNT based nanostructures and devices. The reliability and efficiency of all these nanofabrication methods can be improved, if they can be completed inside the same equipment, which decrease the risk of destruction of the structure and electronic properties by minimizing the need for the opening of the vacuum chamber. This would also make the combination of the methods more effective which is needed if more complicated structures are manufactured.

7.2 Mechanical Evaluation of Carbon Nanotubes

There are lots of theoretical and experimental approaches about the elastic modulus of CNTs. Typical results are summarized in Table 7.1. From these results, the elastic modulus is almost 1 TPa. This value is the same order of magnitude as that of the basal plane of graphite 1.06 TPa. Hence, normally it is assumed to be ~1 TPa.

From theoretical aspects, J. P. Lu considered to be almost same values with different diameters [10]. On the other hand, E. Hernadez et al. presented that they are clearly different with diameters [11]. From experimental aspects, p to the present, there is no perfect synthesis process, with which uniform CNTs can be produced. In this chapter the calibration method of the elastic modulus of as-grown CNTs are proposed on experimental aspects. All CNTs are synthesized by arc discharge method in this chapter.

Table 7.1 Elastic modulus of CNTs (previous research works)

No.	Papers	Type of CNTs	Young's Modulus	Methods
1	[10]	SWCNTs, MWCNTs	0.97 - 1.11 TPa	Empirical force constant model
2	[11]	SWCNTs	1.22 - 1.25 TPa	By nonorthogonal tight-binding formalism
3	[12]	SWCNTs	0.94 - 1.18 TPa	Ab initio calculation
4	[13]	MWCNTs	~0.1 - ~ 1.0 TPa	Electromechanical resonance
5	[14]	MWCNTs	Ave. 1.28 ± 0.59 TPa	Bending with AFM
6	[15]	MWCNTs	Ave.1.8 TPa (0.40 – 4.15 TPa)	Thermal vibration
7	[16]	MWCNTs	0.27 - 0.95 TPa	Tensile-loading by AFM cantilever using "Nanostressing stage"
8	[17]	MWCNTs	Arc-discharge MWCNTs: Ave.0.81 TPa (0.67 - 1.10 TPa)	Measuring the change in deflection as a function of the change in load by contact mode AFM
			Annealed MWCNTs: Ave.0.755 TPa (0.60 – 0.83 TPa)	
			Catalytic MWCNTs: Ave. 0.27 TPa (0.12 - 0.50 TPa)	
9	[18]	SWCNTs	Ave. 1.25 TPa	By observation of their room-temperature vibration in a TEM
10	[19]	SWCNT ropes	~1 TPa	Loading by AFM tip
11	[20]	SWCNT ropes	0.32 ~ 1.47 TPa Ave. 1.002 TPa	Mechanical response under tensile load

Note: Elastic modulus along the basal plane of graphite: 1.02 – 1.06 TPa [10, 13].

Measurement of Elasticity of Carbon Nanotube by Buckling Process

To demonstrate the effectiveness of the hybrid nanorobotic manipulation system, the elasticity of a MWCNT is measured inside a TEM with *in situ* force measurement using an AFM cantilever. The multi-walled carbon nanotubes are synthesized by arc discharge method.

Figure 7.1 shows the analytic model for the measurement of the elastic modulus of a CNT by buckling process. Figure 7.1 (a) and (b) are the continuous buckling process of a CNT. Assuming a CNT to be continuum model, from Euler's formula and force balance relations, the following equations can be obtained,

$$\begin{cases} W_1 = \dfrac{\pi^2 E_{CNT} I_{CNT}}{(c l_1)^2},\ W_2 = \dfrac{\pi^2 E_{CNT} I_{CNT}}{(c l_2)^2} \\ \delta F = k\delta \\ \delta F = \dfrac{h_2}{(cl_2)} W_2 - \dfrac{h_1}{(cl_1)} W_1 \end{cases} \tag{7.1}$$

where W_1 and W_2 are the buckling forces applied to a CNT, δF is the difference of reaction force by an AFM cantilever, k is the spring constant of an AFM cantilever δ is the deflection of an AFM cantilever, E_{CNT} is the Young's modulus of a CNT, l_1 and l_2 are the distance between two ends of the buckled CNT, and I_{CNT} is the geometrical moment of inertia of a CNT. In this equation, l_1 and l_2 are corrected by the multiplication of c: 0.8, which is recommend by Structural Stability Research Council (SSRC) [21]. From these equations, the Young's modulus of a CNT E_{CNT} is as follows,

$$E_{CNT} = \frac{64k\delta}{(d_{oCNT}{}^4 - d_{iCNT}{}^4)\pi^3 (h_2/(c l_2)^3 - h_1/(c l_1)^3)} \tag{7.2}$$

where d_{oCNT} and d_{iCNT} are is the outer and inner diameter of a CNT. The values of a_1, a_2, h_1, h_2, δ, d_{oCNT}, and d_{iCNT} are measured from TEM images, I_1, and I_2, are calculated from these equations.

$$\begin{cases} I_1 = \sqrt{a_1{}^2 + h_1{}^2} \\ I_2 = \sqrt{a_2{}^2 + h_2{}^2} \end{cases} \tag{7.3}$$

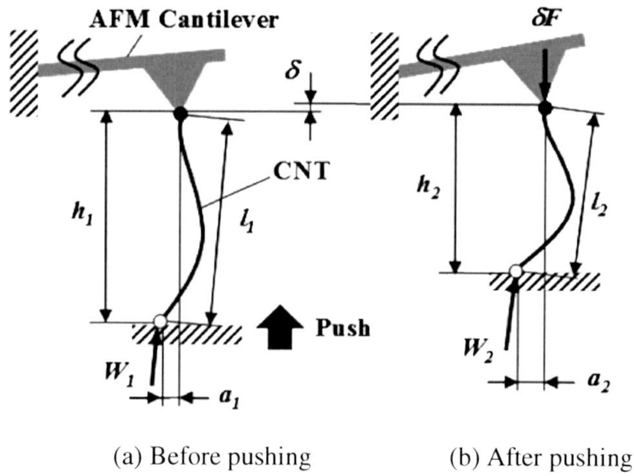

Fig. 7.1 Schematic diagram of a buckled carbon nanotube for measurement of its bending modulus

At first, a CNT is picked up on an AFM cantilever (spring contant k: 0.03 N/m) from bulk samples through electron-beam- induced deposition: EBID. The AFM cantilever is set on the passively driven sample stage. It is positioned within the working area of TEM manipulator (described in Chapter 2). Typical result is shown in Figure 7.2. The gap between the tip of an AFM cantilever and a tungsten probe is reduce to ~4 μm from ~150 μm.

Figure 7.3 shows two sequential TEM photographs during buckling a MWCNT (d_{oCNT}: 14.3 nm, d_{iCNT}: 3.2 nm). One end of a CNT is fixed on an AFM cantilever by EBID inside a SEM. The other end is adhered on a tungsten needle probe by van der Waals forces. From these images, the parameters are measured as a_1: 0.447 μm, a_2: 0.318 μm, h_1: 2.36 μm, h_2: 2.10 μm, and δ: 84.0 nm. The values of l_1, and l_2 are calculated as 2.41 μm, and 2.12 μm. Hence the elastic modulus of this CNT E_{CNT} is calculated to be $1.23\ ^{+0.09}_{-0.08}$ TPa. This value is the same order as that ~1 TPa of the theoretical analysis and other experimental results (as mentioned in Table 7. 1). Comparing with the *in situ* SEM measurement at same experimental parameters (maximum image resolution: 1.5 nm, JSM-6500F), -33.1% ~ +58.2 % relative error is caused by the measurement error on diameter of a CNT. It is not possible to measure the inner diameter of a CNT by SEM, hence 0.2 % relative error is caused d_{iCNT}: 0 in the equation (4.2). From the *in situ* measurement inside a TEM, we can measure the inner diameter of a CNT and improve the accuracy in measuring the Young's modulus of a CNT.

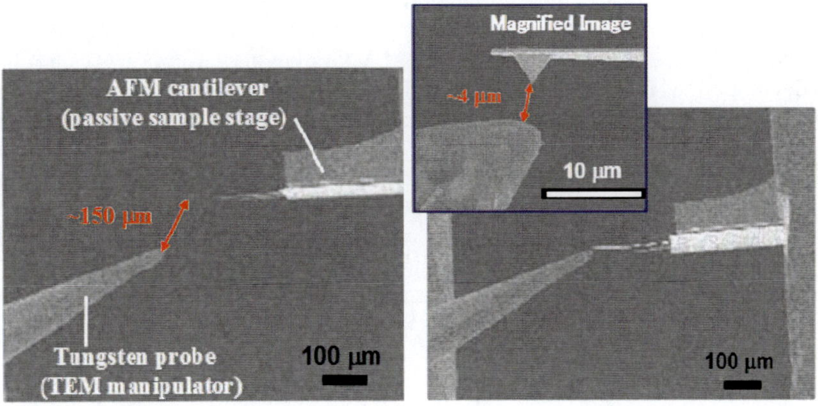

(a) Before positioning (b) After positioning

Fig. 7.2 Positioning of TEM sample through passively driven sample stage

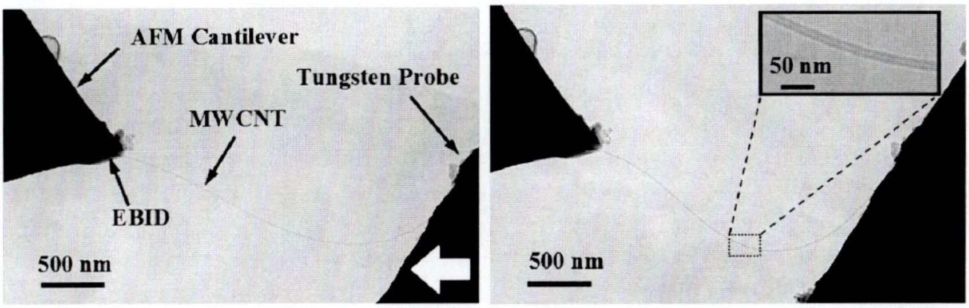

(a) Before pushing (b) After pushing

Fig. 7.3 Sequential TEM photographs during buckling of a multi-walled carbon nanotube

7.3 Deposition Using Carbon Nanotube Emitters

Electron-beam-induced deposition (EBID) has been demonstrated as an additive nanolithography technique for nanostructuring and nanosoldering [22]. Focused ion beam can also be used to deposit nanostructures, but its resolution (>50 nm) is lower than that of EBID (1-10 nm) [23]. Deposition with CNT emitters with pump oil as precursors has shown a remarkable improvement in the specific growth rate. For the fabrication of scanning tunneling microscope tips, nanowires in a circuit or electrodes with EBID, conductive deposits are requested. In previous research, the contents of metal are shown in Table 7. 2 [24, 25] with different precursors. For getting the higher contents of metals, process via heating substrate [26] or using precursors without carbon and oxygen has been demonstrated. To improve the growth rate of conventional EBID caused by the single-beam nature in a FESEM or similar apparatus, we have presented parallel EBID by using CNTs as emitters [27] due to their superior field emission properties. However, the deposits were insulating. Field emission not only from MWNTs but also from SWNTs has been demonstrated. MWNTs are comprised of 2 to several tens of concentric graphitic layers. The outer diameters can be several nanometers to tens of nanometers, and their length is generally several micrometers. SWNTs are much thinner, generally 0.4 to >3nm in diameter, while their length is usually several micrometers. A high aspect ratio and small radius of curvature at their tips are realized in carbon nanotubes. The geometric characteristics of nanotubes, together with excellent chemical stability and high mechanical strength, are advantageous for using as field emitters. Most studies have focused on using SWNTs and MWNTs as field emission electron sources for cathode ray tube [28], flat panel displays [29], gas discharge tubes, surge protector [30], X-ray [31], and microwave generators. The shape modifications of nanotubes have succeeded in the assistance of EBID deposits on a deformed tube, which can find application in nano devices such as nanotube scissors [32].Some examples have shown the applications and opportunities, which arise from the employment of the described deposition technique. Nanometer-sized dots at distances of a few nm are the elements to build SED electronics. The high resolution of the deposition technique provides a tool for fabrication of Coulomb-blockade devices. A high-resolution stitching procedure employing an in-field deposited mark and image processing in a search field, located in the overlapping field region of subsequent address fields for stitched patterns allows subsequent layers to be overlaid with an accuracy of ±2 pixelsteps. This can be as small as 5 nm if a step of 2.5 nm is used in the pattern address field [33]. Tips grown by EBID can be added to commercial SFM tips to improve tip radius and aspect ratio. With this improved probe pictures of surfaces can be obtained containing fewer tip-induced artifacts [34].Special tip-geometries render nanometer metrology of deep trenches or steep surface steps possible. Another application is additive lithography of fine patterns, e.g., as etch mask for further dry processing. The corrugation of the edges of deposited lines has been measured in a TEM at 400 keV electron energy. The peak-to-peak edge definition of deposited patterns is better than 2 nm. This illustrates the superior resolution of EBID compared to organic resist systems. Selecting precursor material and deposition conditions allows to directly deposit the desired features with freely adjustable geometry. One example of such components is electronic devices such as resistors or connecting wires. Dielectric materials as obtained at very low beam currents from the investigated precursors are used to grow Fresnel lenses. Zone plates deposited on the head end of optical fibres may accomplish improved coupling to a wave-guide. Another application is production of X-ray lenses [35]. This application emphasizes the advantage of EBID over conventional lithography in freedom in choosing any kind of substrate. Fabricating electrically conducting deposits and field-emitting tips enables the construction of vacuum microelectronic devices such as switches, microtubes and miniaturized electron optical columns. And more, EBID is used to fabricate thermal probe [36].

Using the emitter of the FESEM, three dimensional structures such as super tips from micro to nano scale have been fabricated [37]. For instance, the smallest radius of the tip of commercial available W-probes is larger than 1μm, which is not convenient for manipulating nano scale objects. Therefore, we deposit super tip with EBID at the tip of W-probe. The diameter of the probe can be reduced to 50 nm. The length and conductivity of the tip can be selected according to specific purpose.

Figure 7.4 (a) shows the tip of tungsten probe viewed from the top before deposit. Figure 7.4 (b) and (c) shows a super-probe deposit at the tip of tungsten. The super-probe has 2.22 micrometer length and 0.44 micrometer diameter. For measurement of super-probe strength, we use an AFM cantilever to touch the super-probe. It has been found that it is destructed easily from its root connecting to the tungsten probe like kinds of weak materials. The cause is considered that the deposit is not crystal but amorphous and includes a amount of amorphous carbon, revealed by following experiments. Although it has these characteristics, it has wide applications such as tip for atom force microscopy, a nano scale thermal probe which can be used in two modes: passive (the temperature is monitored a it is scanned across the surface) and active (the probe induces heat flow to the sample which depends on the thermal conductivity of the substrate) [12], a field electron emission source [38] or three dimensional nano structure like Smith-Purcell electron optics [39] and so on. It has other important properties that it can be form arbitrary parallel pattern for increasing the rate of produce.

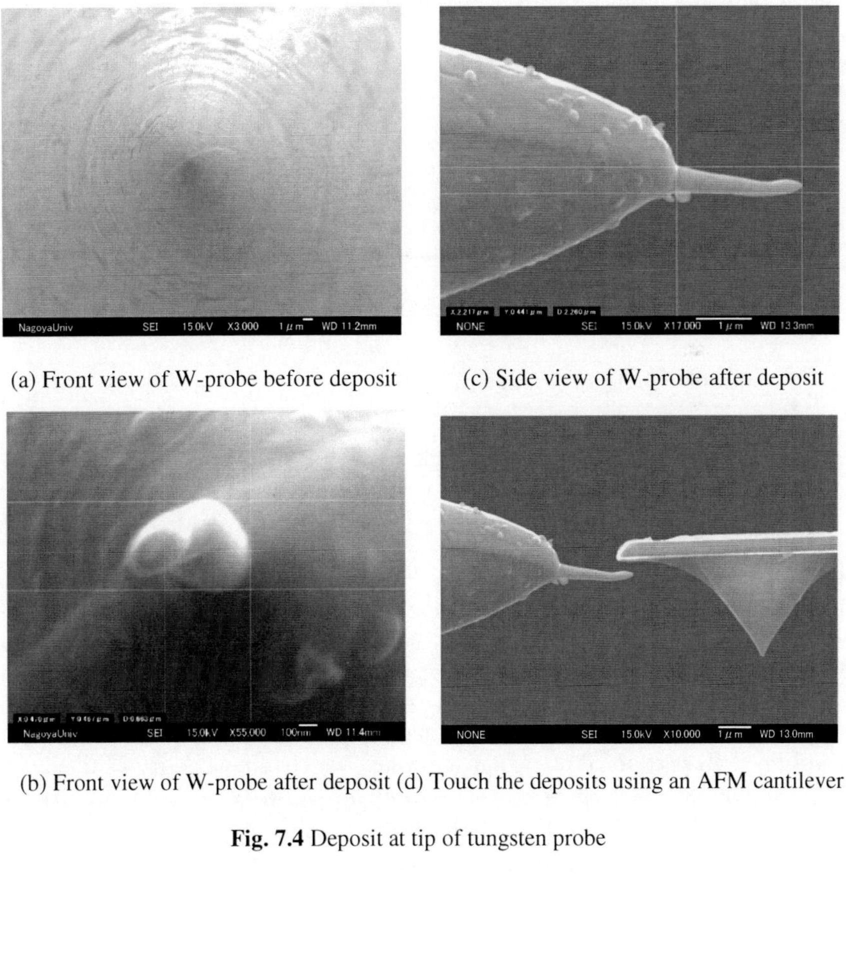

(a) Front view of W-probe before deposit (c) Side view of W-probe after deposit

(b) Front view of W-probe after deposit (d) Touch the deposits using an AFM cantilever

Fig. 7.4 Deposit at tip of tungsten probe

Table 7.2 Contents of metal at conventional studies

Precursors	Deposit	Percentage (at.%)	Reference
$CpPtMe_3$	Pt	14%	[40]
$Mo(CO)_6$	Mo	7%	[41]
$W(CO)_6$	W	55%	[42]
WF_6	W	66%	[43]
$Me_2Au(tfac)$	Au	45%	[44]

Deposit without Introducing Precursors

The compositions of deposits without any precursors have been analyzed by energy dispersive X-ray spectrometer (EDS). The percentage of the compositions is shown in Table 7.3. Here, gold and silicon are background. Deposit can be formed without introducing precursors, because the residual gas molecules such as pump oil exist in vacuum chamber. The deposit normally is called a contamination. The deposition conditions are as follows. The pressure of specimen chamber is 3.0×10^{-4} Pa, the acceleration voltage is 15 kV and the field emission current is 3×10^{-10} A.

Table 7.3 Compositions in deposits without introducing precursors

Compositions	keV	Mass%	Error%	Atomic quantity%
C	0.277	39.80	0.51	54.47
O	0.525	34.12	1.06	35.06
Si	1.739	16.52	0.29	9.67
Au	2.121	9.56	1.47	0.80

Tungsten Hexacarbonyle ($W(CO)_6$) as Precursors

The compositions of the deposit are identified through EDS. In the case of using $W(CO)_6$, Table 7.4 shows the percentage of the compositions. The prominent element energy peak is consistent with tungsten. The element W has almost 56.38 wt.% (9.80 in at.%) that can explain why the deposit is conductive. Besides tungsten, carbon and oxygen are also detected. A bit of gold was detected because deposition was performed on a gold substrate. The conditions of deposition are as follows. The pressure of specimen chamber is 5.0×10^{-3} Pa, the acceleration voltage is 15 kV and the field emission current is 3×10^{-10}A.

Table 7.4 Compositions in deposits from $W(CO)_6$

Compositions	keV	Mass%	Error%	Atomic quantity%
W	1.774	56.38	1.19	9.80
C	0.277	20.69	0.40	55.02
O	0.525	17.15	0.93	34.24
Au	2.121	5.79	1.89	0.94

Dimethyl-Gold ($Au(CH_3)_2(C_5H_7O_2)$) as Precursors

The element spectrum of deposit by precursor of dimethyl-gold was analyzed and the percentage of compositions is presented in Table 7. 5. There, deposition is performed on a mesh of copper. The mass of the gold up to 56.87 wt.% (12.15 at.%) is detected and oxygen and carbon are also included at depositions. In the two cases using $W(CO)_6$ and dimethyl-gold as precursors, both of the carbon and oxygen are involved in deposition. One of the reasons that a part of organometallic material cannot be decomposed is that they immediately adsorb to the substrate. Another cause is a little part of the organometallic materials decomposes into tungsten carbide, e.g. W_xC_y.

Table 7.5 Compositions in deposits from dimethyl-gold

Compositions	keV	Mass%	Error%	Atomic quantity%
Au	2.121	56.87	0.62	12.15
Cu	8.040	21.42	1.06	14.19
C	0.277	18.96	0.22	66.41
O	0.525	2.76	0.45	7.26

Growth Rate of Deposits

According to the theoretical analyses by Koops, the monolayer growth rate through EBID in a steady state is:

$$R = v \cdot N_O \cdot \frac{(g \cdot F / N_O) \cdot q \cdot f}{(g \cdot F / N_O) + 1/\tau + q \cdot f} \tag{7.4}$$

where v is the volume occupied by a dissociated molecule or its fractions, N_o the molecule density in a monolayer, g the sticking coefficient, F the molecular flux density arriving on the substrate, q the cross section for dissociation of the adsorbed molecules under electron bombardment, f the electron flux density, and τ the lifetime of the molecules.

Note that the growth rate increases when the electron flux f and/or the molecular flux F are increased. In practice these parameters are limited, e.g., by electron optical constraints, the vapor supply and differential pumping system, or simply the fact that only a limited vapor pressure can be tolerated above the sample before significant electron scattering starts. Under these conditions, and note that the influence of the term including the lifetime τ of the molecules can be made negligible, e.g., by cooling the sample, the maximum rate obtainable for any given electron flux is

$$R_{\max} = v \cdot N_O \cdot q \cdot f \tag{7.5}$$

and choosing

$$q \cdot f = g \cdot F / N_O \tag{7.6}$$

yields 50% of the maximum obtainable growth rate at supply F.

Growth rates of the EBID using $W(CO)_6$ and dimethyl-gold as precursors were measured by depositing on an atomic force microscope (AFM) cantilever with different period. The time of each deposit from above is 1min, 2 min, and 3 min, respectively as shown in Figure 7.5. The growth rate is about 17.6 nm/min by using $W(CO)_6$ and is about 65 nm/min in average by using dimethyl-gold. It is also found that the deposition rate and the diameter of deposit depend on the beam current of FESEM emitter. Large current brings out high vertical growth rate but large diameter.

It is also found that the deposition rate and the diameter of deposit depend on the beam current of FESEM emitter. Large current brings out high vertical growth rate but large diameter. Figure 7.6 shows the depositions at different beam current of 0.01, 0.03, 0.1 nA and every dots deposit time is 3 minute.

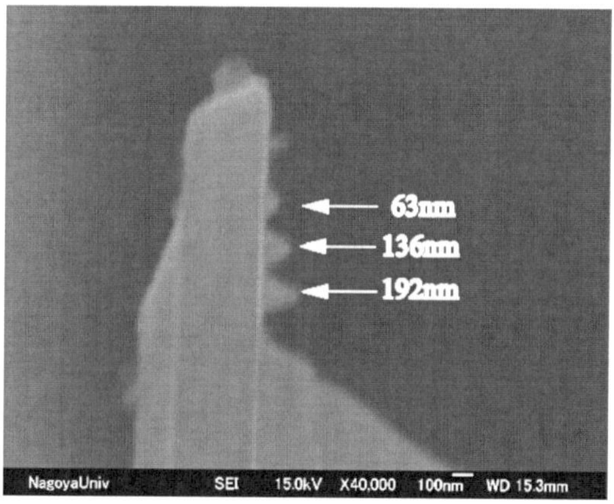

Fig. 7.5 Deposit with in different time interval

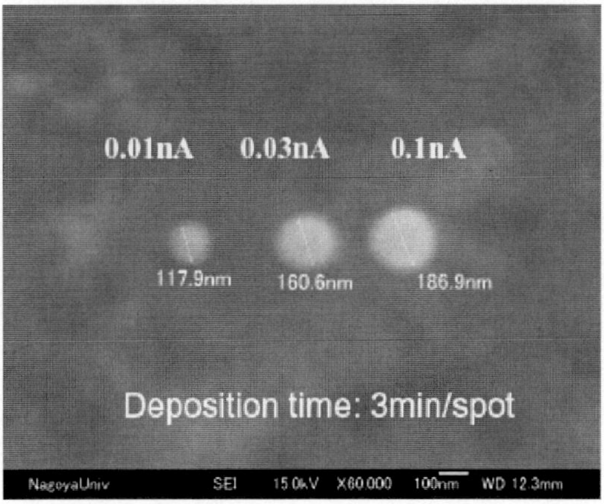

Fig. 7.6 Deposited spots under different emission currents

Resistivity of Deposits

If EBID is used for circuit repairing, device manufacturing, or prototype wiring, low resistivity deposit will be requested. The deposition of conductive or insulating structures can be realized through selecting different precursor material. For example, the deposit is insulate when the pump oil (contamination) is selected as precursor, while it is conductive when the organometallic compound, e.g. $W(CO)_6$ or dimethyl-gold, is selected. It can be proved by following experiments. The conductivity of deposit has been measured by using 2-probe method with the electrical circuit as show in Figure 7.7. Two electrodes are prepared by cutting the

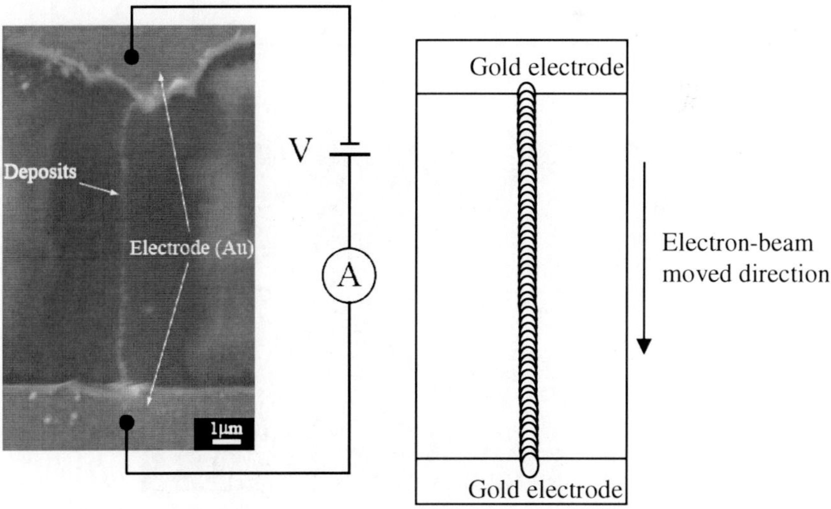

Fig. 7.7 Electric circuit for resistance measurement

gold film on a Si_3N_4 substrate with deep-ultraviolet laser (DUV). While moving the stage, a nanowire is deposited by EBID with $W(CO)_6$ or dimethyl-gold between the two gold electrodes.

Result of the experiment reveals that the deposition is insulating when using pomp oil, and with 50 kΩ resistance when using $W(CO)_6$. Since the dimension of the deposit is 10 μm in length and 200×100 nm^2 in section area, its resistivity is obtained to be about 0.01 Ω·cm. In the case of the dimethyl-gold, the resistivity is about 0.008 Ω·cm.

Electron-Beam-Induced Deposition with Carbon Nanotube Emitters

It has been demonstrated that CNTs are well defined field emission source and can work at low bias. By using CNT-emitters, the resolution of the deposits can be increased and its principle is shown in Figure 7.8. For producing nanotube emitters, a CNT can be soldered on the tip of a tungsten probe or an AFM cantilever. The surface of tungsten probe is easily oxidized in air that will increase the contact resistance. To remove an oxide layer, discharge fabrication is effective. Figure 7.9 (a) shows a typical example.

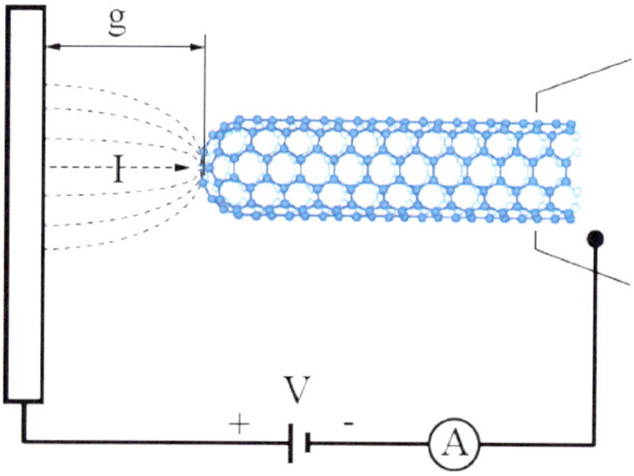

Fig. 7.8 Setup of field emission experiment

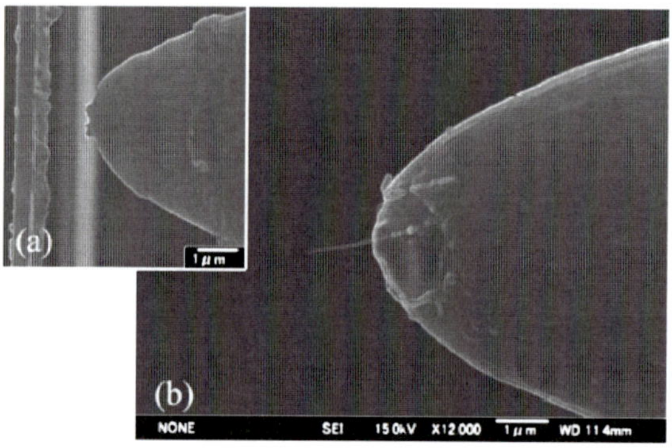

Fig. 7.9 (a) Removal of an oxide layer by discharge fabrication (b) CNT-emitter on top of the W-probe after discharge fabrication

EBID without Precursors

Although EBID is an effective way for nanolithography and is expected to have a higher resolution than its counterpart: focused ion beam deposition for the fabrication of 3D nanostructures, the low deposition rate and high cost hinder its applications. Although multi-exposure electron-beam [45] can improve its productivity, it provides no solution for the problem of cost because it still involves field-emission scanning electron microscope (FESEM) or similar equipments that equipped with expensive thermal electron-filaments. Parallel EBID with cheaper field emitter array will be a synthetic solution for the above problem, but no works have been demonstrated so far.

On the other hands, carbon nanotubes (CNTs) have been demonstrated as well-defined cold cathodes [46], because they can emit high current density under very low bias. Although intensive works have been done on the investigation of the field emission properties of CNTs [4] and their applications in flat-display [47], there has no demonstration yet showing the feasibility of EBID with CNT emitters. Because of the possibility to fabricate well-aligned CNT-arrays [48,49], they may also find applications in parallel EBID, which will lead to broader application of EBID.

There have been some demonstrations on conventional 2D lithography by using CNT tips [50,51], but as have been shown by Koop, conventional lithography is not effective for the fabrication of 3D nanostructures. Here we show the feasibility of EBID through individual CNT emitters, and present parallel EBID with CNT arrays as a 3D large-scale nanofabrication technique.

The experiment system of CNT emitter for EBID is shown in Figure 7. 10. In which, a CNT is mounted on an AFM cantilever and is placed above an Au-coated Si-substrate. The CNT is used as cathode, while the substrate as anode. The gap L between the tip of the CNT and the substrate can be adjusted by mounting the cantilever and substrate onto unit 1 and unit 2 of the nanorobotic manipulation system. The nanorobotic manipulation system also provides the possibility to generate a deposit pattern through EBID onto the substrate by servo control of the substrate. The deposit pattern is achieved by magnetic servo control of e-beam in a conventional EBID system.

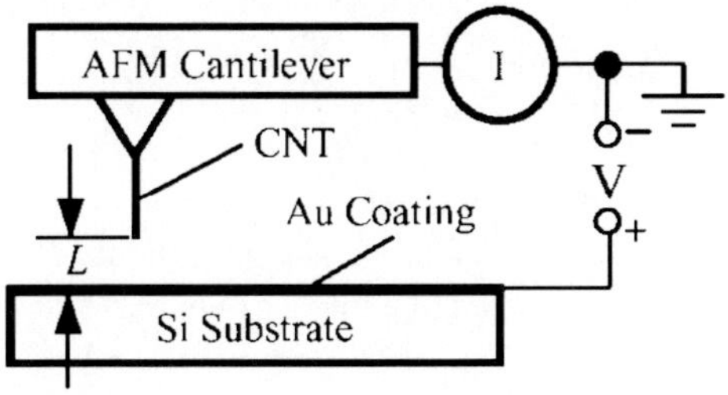

Fig. 7.10 Configuration of CNT-emitter EBID system

Nanorobotic manipulators provide the possibility to select CNTs with proper diameters so as to obtain high enough emission current, to assemble CNT electron emitters and to adjust the field by positioning the CNT above the substrate with a proper gap.

The nanorobotic manipulators are mounted inside a FESEM for obtaining real-time observation, for the preparation of CNT emitters, and for the comparison of EBID with CNT emitters and conventional EBID in the same circumstance. A typical CNT emitter is shown in Figure 7.11. It was prepared through nanoassembly of a multi-walled carbon nanotube (MWNT) onto a commercially available Au-coated Si_3N_4 AFM cantilever as field emitter. The inset shows the MWNT mounted with conventional EBID deposit.

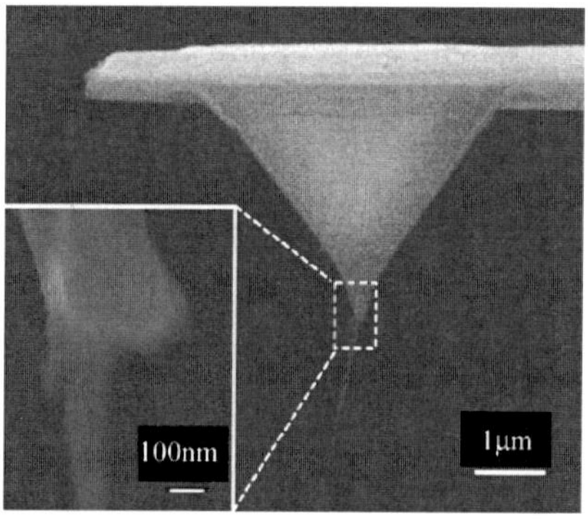

Fig. 7.11 A CNT emitter, the inset shows deposit with EBID

Typical precursors for EBID include WF_6, carbonyls, styrene, pump-oil, etc, and deposits are respectively W, metal carbides, carbonaceous, contamination, and so on. Here we apply pump oil (chamber contamination) inside the FESEM to simplify the problem of gas supplying.

Figure 7. 12 (a) shows the configuration of the MWNT emitter and the substrate before EBID, where the dimensions of the MWNT are Φ28 nm×2.7 μm and the gap L=377 nm. By applying a DC bias V=120 V (Field 318.3 V/μm) between the nanotube and the substrate, field emission current I=2 μA occurred (Current Density: 6.1×10^8 A/m^2). Figure 7. 12 (b) shows the deposit on the substrate after 20min deposition. The height of the deposit is about 70 nm, and the diameter of the deposit is 204 nm. The growth rate R_T is calculated as 0.17 nm/s. The CNT tip after deposit became thicker than that before the deposition.

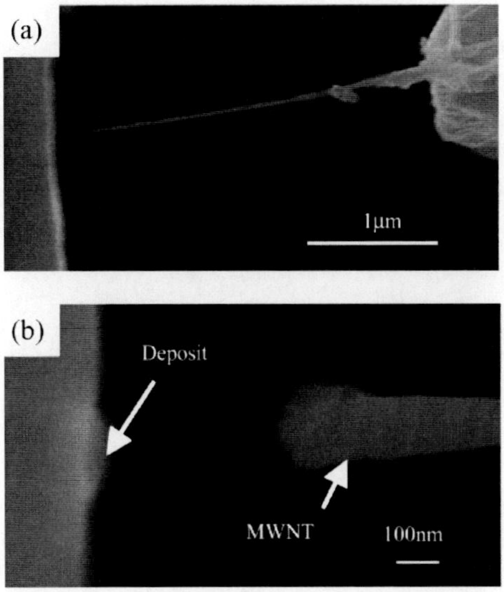

Fig. 7.12 EBID through MWNT electron emitter: (a) before EBID and (b) after EBID

Analyzing Compositions of Deposit

Compositions of the deposited films via CNT-emitters are analyzed with EDS and the result shows that the mass of tungsten is nearly 80.67 wt.% (24.97 at.%) among all elements of the deposit. Other compositions include carbon and oxygen as shown in Table 7.6. It demonstrated that conductive nanostructures have been fabricated via EBID with CNT emitters. EDS analysis of the deposits reveals the composition of tungsten as using $W(CO)_6$ is about 2 times richer than that of conventional EBID using SEM.

Table 7.6 Compositions in deposits from $W(CO)_6$

Compositions	keV	Mass%	Error%	Atomic quantity%
W	1.774	80.67	1.47	24.97
C	0.277	12.82	0.52	60.74
O	0.525	3.80	1.12	13.51
Au	2.121	2.72	2.65	0.78

Figure 7. 13 shows the deposition growth process. The molecular of $W(CO)_6$ adsorbed on surface or near to the substrate will be dissociate by irradiation of electron beam. Two types molecular will be formed: one is solid molecular, such as W or W_xC_y, another one is gas molecular, CO or CO_2 or O_2. The gas molecular will be exhaust by vacuum pump. The ionized W or W_xC_y will adsorb on the substrate to form deposits. This process also can be occurred on the deposits surface and continued.

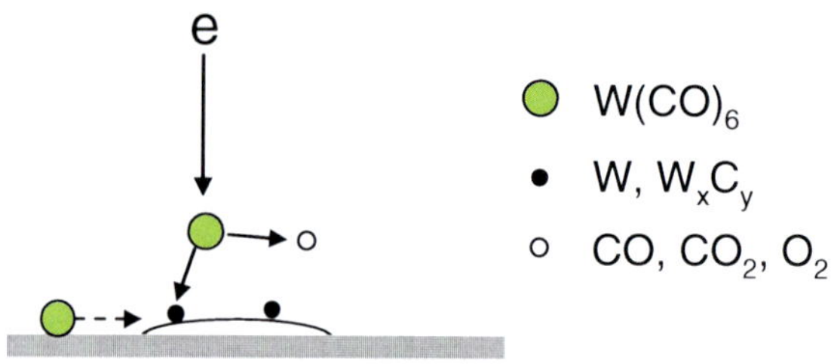

Fig. 7.13 Deposition growth process

Observation of Electron Beam Pattern with Field Emission Microscopy Inside a Scanning Electron Microscope

Field Emission Microscopy (FEM) has been extensively used in the past 60 years in surface science [52, 53, 54]. In such an experiment a phosphor screen is used as the counterelectrode and the obtained pattern reflects the spatial distribution of the emitted current. Since the emitted current depends critically on the local field amplification and the work-function, FEM has been used with great success for diffusion and adsorption/desorption studies as well as to determine the workfunction from different surfaces. FEM is often used in combination with field ion microscopy (FIM), which can be performed with the same setup but with a positive applied voltage under partial pressure of an imaging gas (He or Ne) [20, 21]. The gas atoms are ionized by the high local electric field at surface steps and salient atoms, accelerated by the potential difference and impact on the phosphor screen. FEM and FIM are of great interest since they offer a direct way of visualizing the spatial distribution of the emitted electrons as well as the atomic structure of the cap. These expectations have not quite been fulfilled yet for nanotubes, even though several groups performed such experiments with sometimes contradictory findings.

Saito et al. obtained FEM patterns from closed MWNTs and observed single bright spots with dimensions that suggested that the emission proceeds from the whole cap of MWNTs and not from a few atoms only [55]. Ring patterns with a darkcenter were detected on open MWNTs and were interpreted as originating from the edges of the graphene layers of the open MWNT [56]. SWNTs patterns were similar to those of closed MWNTs and were assigned to SWNT ropes [57]. On purified SWNT films, Zhu et al. found ring patterns that were interpreted as coming from nanotube ends because of the circular symmetry [58]. Ma et al. observed circular arcs and rings on films of open-ended fibers that were attributed to emission from the open edges [59]. Dean et al. measured FEM on SWNTs at room temperature as well as at 600°C [60]. At room temperature, the patterns show one to four lobes that are typical of, and behaved like, molecules adsorbed on the tube cap. These patterns disappeared above 600°C, and fine-structured patterns with sometimes five or sixfold symmetry were detected. Dean et al. argue that these patterns represent the electronic distribution of the emitting states at the nanotube cap and not the atomic structure of the cap.

Lovall et al. tried to image the cap of a SWNT by FIM using He as the imaging gas [61]. The resulting pattern was composed of ring structures consisting of several features located around the circumference of the ring with two or three additional faint features located inside the innermost ring. The image is compatible with the edge atoms of an open SWNT, and the observed SWNT was tentatively identified as a tube of 28° chiral angle and 2.1 nm diameter.

In order to measure the electron lens effects, we created a FEM inside a SEM so as to obtain the electron beam pattern. The FEM consists of a charge-coupled device (CCD) camera, a viewing port and a fluorescent screen. A fluorescent screen with 32 mm diameters is created on indium tin oxide (ITO) glass deposited with fluorescent material (ZnO). The ITO glass can be used as an anode, and the electron beam can penetrate it for recoding beam patterns. It is set on a stage of scanning electron microscope and is perpendicular to the emitters. A port is modified to the viewing port on the wall of FESEM. So the field emission electron beam pattern can be recorded with a CCD camera behind the fluorescent screen. An electrostatic lens is created with circle shaped copper thin film, on the center of which a hole is milled by a focused ion beam as shown in Figure 7. 14.

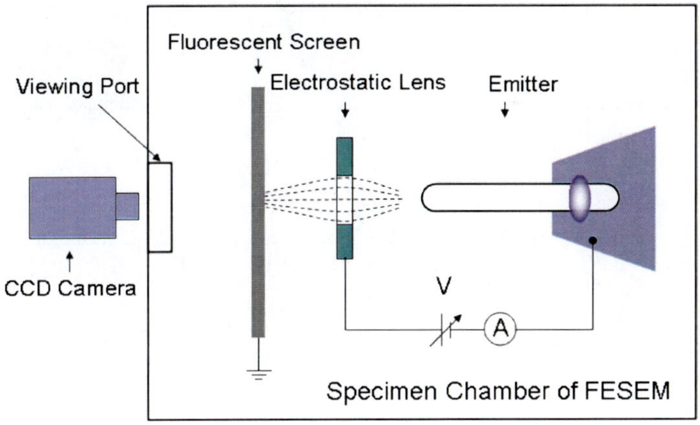

Fig. 7.14 A field emission microscope system inside a FESEM

A MWNT bundle with 50 μm diameter and 650 μm length which was fabricated by arc discharge method as emitters was fixed on a copper probe with conductive adhesive tape and set on a manipulator unit with 3 degrees of freedom.

An electrostatic lens, with a 100 μm diameter hole, was set perpendicular to the emitter and fixed on another 3 degrees of freedom manipulator unit. The hole and the emitter are on the same axis. All of these equipments are inside a vacuum chamber of FESEM with typical pressure of 10^{-4} Pa during observation. The distance was kept at 200 μm between the emitter and the lens.

Figure 7.15 shows the field emission property of such MWNT bundle when a voltage from 0 to 1440 volt was applied. The emission current did not appear until the applied voltage increased to 580 V. From about 700 V, we can observe the electron beam pattern with fluorescent screen. As shown in this figure, emission current rises immediately to several tens of microampere. In order to prevent burning of the emitter, the applied voltage is limited to 1440 V. The emission current lacks stability due to amorphous carbon present on the CNT bundle and noise of energy supply, among others. Stability can be improved if these problems are dealt with. In lager scale of emission current, the degradation of CNTs influences the stability of currents. Another important cause is that residual gases in the vacuum chamber also influence field emission when they are adsorbed to the emitter surface.

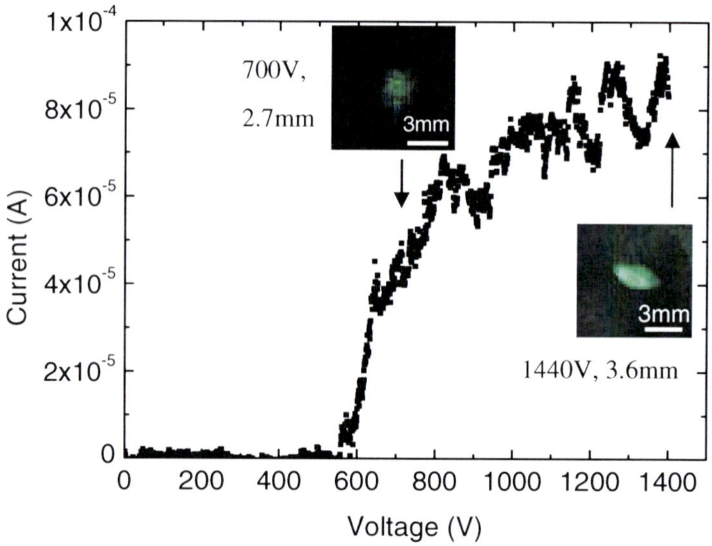

Fig. 7.15 Field emission property of a MWNT bundle

We increased the electrostatic lens effects through nanorobotic manipulation. In system, the center of lens is zero point. The emitter is moved back and forth by 500 μm. The electron beam pattern was obtained by fluorescent screen. Figure 7.16 (a)-(e) show the electron beam diameter reduction during emitter-lens distance increase from 200 to 700 μm at 1440 V, while Figure 7. 16 (f)-(j) show electron beam diameter enlargement during emitter-lens distance decrease.

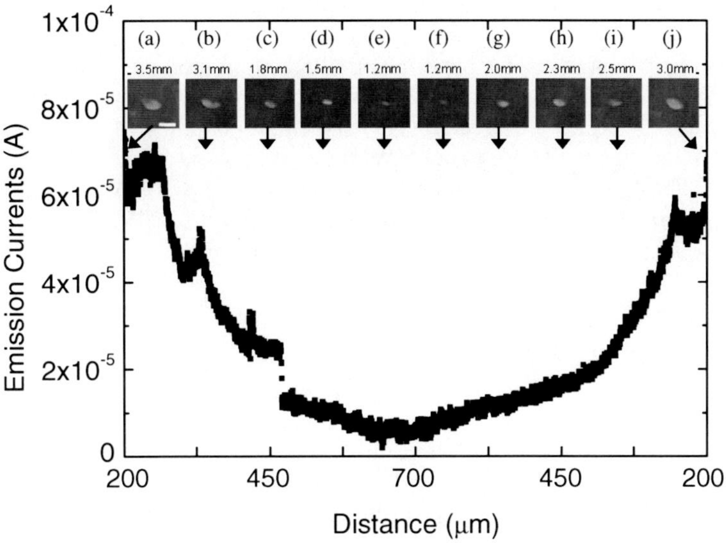

Fig. 7.16 Field emission change when emitter is moved by nanorobotic manipulator (Bar: 3 mm)

Electron-Beam-Induced Nanowire Growth from Cathode

Nanomaterials such as nanowires [62], carbon nanotubes (CNTs) [63] and nanobelts [64, 65] are becoming increasingly attractive for nanotechnology and have been proposed as the basic building blocks for a new generation of nanoelectronic and mechanical systems. Especially, field emission properties of CNTs have attracted many researchers and have been used in generating photoluminescence [66] and as emitter in field emission display [67], electron-beam-induced deposition (EBID) [68], and nanoposition sensor [69]. Some field emission based nanotube applications, in which emitters work at higher emission current region, face degradation of carbon nanotube emitter [70]. The CNT is shortened when emission current over the saturation is applied. The main reason can be considered to be joule heat. This is an important problem in field emission application based on CNTs. J. T. L. Thong et al. reported that nanowires were grown from a cold-field-emission tip in the presence of a precursor [71]. They argued that the field-emission current has to be limited to 100 nA or less for obtaining single nanowires. Carbon nanowires growth on the tip of a CNT was reported by Y. G. Wang and co-workers [72]. They measured current-voltage characteristic of the nanotube-nanowire that showed a nonlinear relation between the current and applied bias voltage due to the saturated sp^3 bonds formed at the junction. The field assisted nanowire growth has been used for atomic force microscope probe [73, 74] and electrical interconnects [75]. We propose that nanowire can be used for modifying the degraded emitters.

But normally, a tree-like nanowire can randomly grow at the tip of the emitter during electron field emission, using CNT as emitters and tungsten hexacarbonyl ($W(CO)_6$) as precursors. This paper reports success in growing a single nanowire by controlling the distance between the emitter and substrates with a nanorobotic manipulator. We also characterized nanowire properties with transmission electron microscopy (TEM). The MWNT is 1.2 μm in length and 20 nm in diameter. The gap between the tungsten probe and the tip of MWNT was set to 2 μm. To test the CNT, we varied the voltage from 0 V to 160V. The emission of current began from about 120 V and reached 7 μA at 160 V.

Tree-Like Nanowires

Several tree-like nanowires with different heights were grown from the tip of a CNT. It took 30 min to obtain the height shown in Figure 7. 17. A sprout structure can be found at the tip of each nanowire. The nanowire is 20 nm in diameter. We also found that the MWNT tip became thicker than that before field emission. The diameter changes from 20 to 40 nm. We characterized the nanowire composition by TEM energy dispersive X-ray spectrometry (EDS) at four points shown in Table 7.7.

The radius of curvature of the tip is less than 5 nm in diameter. Tungsten was detected and was 1 at.% among the nanowire. Carbon (49 at.%), oxygen (47 at.%) and some backgrounds were also detected. We surmise that the structure is

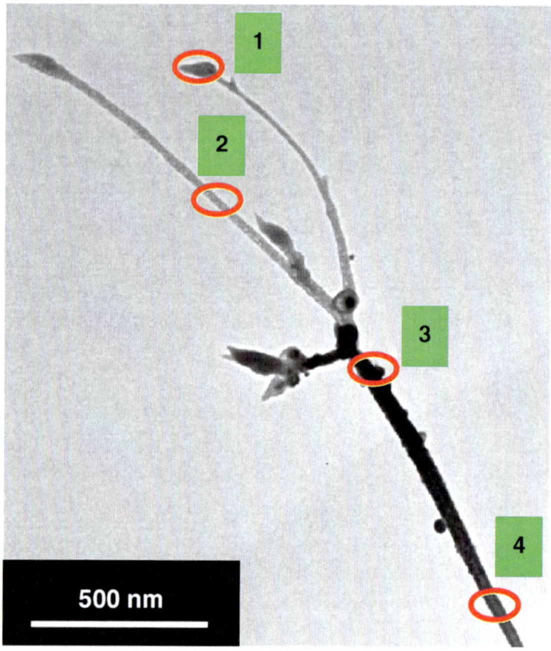

Fig. 7.17 Components analysis with TEM EDS at 4 points

Table 7.7 Compositions in each point 1, 2, 3 and 4

Point	W (at.%)	C (at.%)	O (at.%)
1	1	49	47
2	5	-	94
3	6	-	93
4	6	38	54

nanowires covered with a carbon layer. 5 at.% W and 94 at.% O were detected at the central nanowire. On the surface and nearly at CNT tip, tungsten is 6 at.% and oxygen is 93 at.%. Tungsten balls were found and some amorphous carbon is around the balls. It was found that a half of that ball, the rest was hidden under the CNT. The part of CNT near the substrate has 6 at.% W, 54 at.% O and 38 at.% C.

Nanowires Growth Mechanism

The TEM image of tungsten nanowires clearly presented the geometric structure of nanowires as shown in Figure 7. 18. Two long wires and short third one were grown from CNT tip. By examining the contact part between the nanowire and the CNT, we can see some spheres on the CNT surface. Their sizes are different as shown in red circle. The positive ion of tungsten is separated from $W(CO)_6$ during electron beam emission (Figure 7. 18 (a)). At the beginning of the electron emission and for several seconds, thin tungsten film is coated on the surface of the CNT (Figure 7. 18 (b)). CNT is then heated to higher temperature, above 2000 K, induced by field emission current [76]. The film was condensed and became a sphere. (Figure 7. 18) Two spheres nearly merge together causing the mass translation [77].

At the end of the nanowire, as shown in dashed circumference, some half spheres were found. We suppose that the nanowire grows from tungsten sphere because the field strength at surface of sphere is larger than that at the surface of CNT. The number and grown position of nanowires is random. At the tip of the nanowire, it has a bud structure like a plant and keeps a stronger electron field. (Figure 7. 18 (d)) The width of the bud is wider than the nanowire, but its tip is less than 5 nm. All of the buds have the same orientation directed forward the anode. This is due to the electrostatic force attraction. Electrons are emitted continually to decompose $W(CO)_6$ and the tungsten positive ion are attracted the foot of the bud to keep the nanowire growing. The relation between diameter of nanowires and currents is not clearly understood. This is because the behavior is more complicated than previously thought. The growth continues until the nanowire contacts the substrate or the emission is stopped.

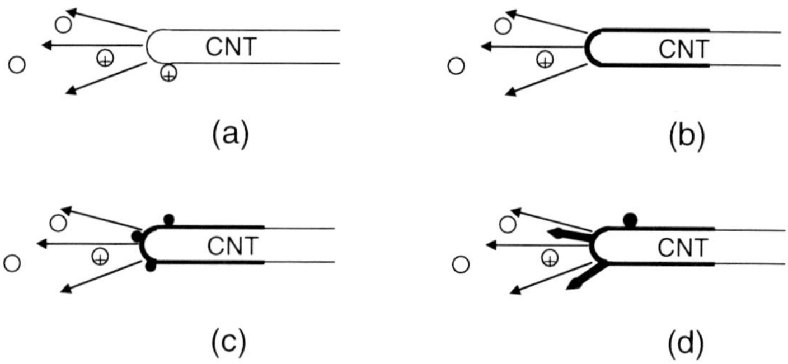

Fig. 7.18 Nanowire growth process

Single Nanowires Growth Controlled with a Nanorobotic Manipulator

The nanowire is randomly grown on the tip of a CNT and the number of nanowires cannot be estimated. Unfortunately, these nanowires are not suitable to be used as emitters for field emission-based applications such as EBID [7], nanoposition sensor [8], because the tree-like structures will influence the resolution and stability of those nanodevices. We propose that by changing the distance between emitter and substrate, the growth of nanowires can be controlled. Our concept is that when we draw the emitter far from the anode during electron emission, the longest nanowire at the tip can grow continually, because the electron field at the tip is stronger than the other parts of nanowires. As a result, the electron field emission on other nanowires will be weaker as we move farther away. Therefore other nanowires will stop growing. Although multiple of nanowires will grow at the beginning, the result is that only one nanowire can grow continually at tip of CNTs. A CNT with 500 nm in length and 25 nm in diameter was prepared. The applied voltage was fixed at 65 volt and the interelectrode distance was 500 nm. The precursor gas $W(CO)_6$ was introduced in the vacuum chamber. From the start of electron emission, the emitter is moved back by 15 nm/s using the nanomanipulator.

After 5 minutes, we stop the power. A 500 nm nanowire grew from CNT tip. We start again the experiment, with the same conditions. After another 5 minutes, the nanowire grew up to 800 nm. The last 5 minutes result showed that the nanowire was 1100 nm in length. The TEM image shows that our method of controlling the distance between emitter and substrate is useful for a single nanowire growth. The nanowire length can be selected by controlling the growth time. The nanomanipulator was halted while the power was applied from 0 to 65 volt.

When CNT is used as an emitter in applications such as EBID or field emission displays, the emitter suffers degradation under large emission current. This problem can be solved with our method of nanowire growth. The nanowire can modify the degraded emitters. It also has good field emission property so that it can endure high emission current (0.1 μA) at low bias voltage (65 V).

We measured the conductivity of nanotube-nanowire with two point method. A single nanowire was grown from CNT until it came in contact with the anode. The applied voltage is 75 volt. When the nanowire contacted the anode, the current became stable and the emission stopped. The resistivity of nanotube-nanowire was about 7 Ωcm. The resistivity of nanotube-nanowire is higher than that of deposits on substrate described on section 4.1.5 (0.01 Ωcm), it can be considered that it caused by the conduct resistivity and lower composition of tungsten.

The peeling and sharpening of MWNTs through electrically driven vaporization presents a possible method for the removal of CNTs [78], but this method is difficult to control and needs an electrical contact to the CNT. Cutting or removal of CNTs on a nanometer or even an angstrom scale with high-energy electron beam inside a TEM was report by F. Banhart et al. [79, 80]. Although, by using a TEM, CNTs can be cut at high resolution, the small specimen chamber of TEM make nanofabrication and assembly difficult, which restrict the applicability of the method in the nanodevice fabrication. Also destructive fabrication technique is proposed for the cutting of CNTs, but unfortunately the destructive point is unknown and difficult to control, which makes the method unsuitable for precise nanomanufacturing. [81] The other mechanical cutting methods, like the cutting with an AFM [82] or a scanning tunneling microscope (STM) tip [83], are time consuming and therefore not very effective for complicated applications. Also the CNT etching via FIB is proposed, but the FIB could damage the rest of nanotube, which is undesired in nanofabrication. [84]

Suzuki et al. reported that the electron beam irradiation induce damages on CNTs under the low acceleration voltage of a scanning electron microscope (SEM). This method was applied in selective removal of nanotube with a beam accompanied with oxygen, as reported in [85]. As another SEM based method, in situ cutting CNTs with water vapor inside an environmental SEM has reported by Zettl's group. They showed mass loss of CNTs caused by water molecular. [86] Despite the variety of the cutting methods, none of these researches have explained why the CNTs are cut under low-energy electron beam. As we know, for removing a carbon atom by a knock-on collision, minimum incident electron energy of 86 keV is required. [87] The high-energy beam can be easily obtained inside a TEM, but in a SEM, the energy of electron beam is normally limited to 30 kV. Therefore, the cutting mechanism inside a SEM has to differ greatly from the one inside a TEM.

Reversible bending of CNTs with AFM was reported by M. R. Ralvo et al. [88]. They found that the carbon nanotube is a material with extraordinary strength through the bending experiments. The AFM method is restricted by the difficult fabrication of the suspended structure. Bending of MWNTs with manipulation inside a TEM was performed for the dynamical measurement of the electrical conductivity of CNT during the deformation. [89] Plastic deformation of double-walled carbon nanotubes was also induced by applying current over the nanotube. [90] The location of the kink is, however, difficult to define in these two methods. They are not capable of sharp angle formation either. W. H. Knechtel reported a reversible bending induced by electron beam inside TEM. [91] The bending was caused by van der Waals force between the CNT suspended on two gold lines and the substrate.

Bending can not only be used for building nanostructures but also for nano electronic devices. The electrical conductivity properties of a single-walled carbon nanotube with an intramolecular junction were measured in [92]. The results show that the metal-semiconductor junction behaves like a rectifying diode with nonlinear transport characteristics. The sharp angles can be made also by using FIB, but the method easily damages the sample, which is not favorable. [93]

Another fundamental technique for the manufacturing of NEMS is a precise peeling of a MWNT, which enables the production of nanobearings as the nested tubes are released to slide over the almost optimal shell surfaces [94, 95]. These nanobearings can be exploited in linear or rotating nanoactuators where their offer almost frictionless movements.

Techniques for high speed cutting and bending of CNTs by introducing oxygen gas into the vicinity of the sample will be introduced. In addition, the peeling of MWNT inside a SEM will be presented. Because both of the techniques can be conducted inside a SEM, the combination of the techniques is easy, which increases the efficiency of the techniques and makes them very suitable for nanodevice manufacturing. Also the size of the specimen chamber of a SEM enables the use of nanorobotic manipulator in the nanofabrication and assembly as discussed. Finally, the usefulness of these methods is demonstrated in *in situ* nanofabrication of a two (2D) and a three dimensional (3D) nanostructures based on CNTs.

7.4 3D Assembly of Carbon Nanotube

Cutting of Carbon Nanotubes Assisted with Oxygen Gas

The proposed cutting technique is depicted in Figure 7. 19. In the cutting, oxygen is introduced in the vicinity of a CNT and the CNT is exposed to an electron beam by using a spot mode of a SEM. The acceleration voltage of the beam is low (1~10 kV) which enables the electrons to interact with the carbon atoms of the CNT and makes them active. The active carbon atoms react with the oxygen gas according to the following reactions

$$O_2 + C \rightarrow CO_2 \tag{7.7}$$

$$O_2 + 2C \rightarrow 2CO \tag{7.8}$$

In these reactions the carbon atoms react with oxygen atom and forms carbon dioxide and carbon monoxide. The reaction energy is provided by low energy electron beam. It is noteworthy; that the cutting does not happen if either oxygen or electron beam is not involved. The necessity of the electron beam enables accurate positioning of the cutting point, because the cutting happens accurately in the point where the beam is located. The accuracy of the technique makes it a very useful to be used in nanomanufacturing applications. The method is also very effective and a single CNT can be cut rapidly. The accuracy and efficiency of the cutting method is tested in an experiment discussed in this session.

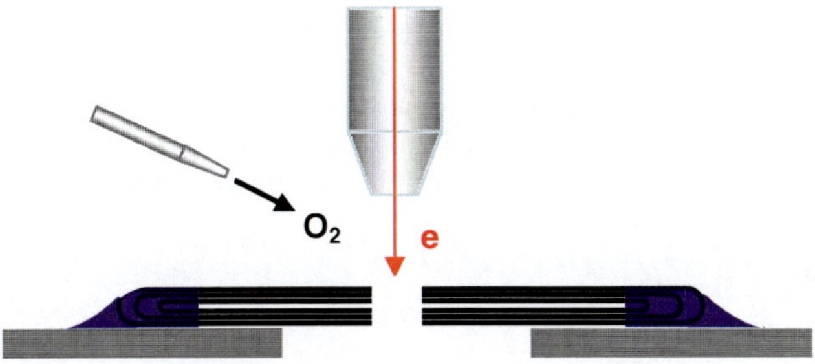

Fig. 7.19 The procedure of cutting a single CNT inside a SEM by using an electron beam assisted with oxygen gas

We show three sets of results for the cutting technique under different conditions. One is for irradiation of a CNT with a low-energy electron beam in the absence of oxygen. The second is by irradiation of a CNT with the introduction of oxygen gas when the gas nozzle was 1 mm from the samples. Finally, we show the cutting of a single CNT at high speed in the presence of oxygen and with the gas nozzle at 90 μm from the samples.

A Single Carbon Nanotube Exposed to the Electron Beam without Introducing Oxygen Gas

A single CNT was irradiated with an electron beam current of 6×10^{-10} A, an acceleration voltage of 1 kV and in the absence of oxygen. The vacuum in the specimen chamber was 2.7×10^{-4} Pa. The electron beam was irradiated in a spot mode in an area 500 nm away from the tip of the CNT. The results of irradiation after 0, 5 and 10 minutes. No variation in the appearance of CNT was observed.

A Single Carbon Nanotube Exposed to the Electron Beam with Introduced Oxygen Gas from a Distant Nozzle

Oxygen gas of 1 sccm was introduced to the specimen chamber, but the gas nozzle was 1 mm from the cutting samples. The total pressure of specimen chamber is 1.6×10^{-2} Pa. Irradiated point was far from 500 nm from the tip of CNT. Irradiation times are 0, 5, and 10 minutes. As a result, the CNT did not show any conspicuous change except for a bending of a little angle of 4 degrees after 10 minutes, but the nanotube was not cut.

Cutting of a Single Carbon Nanotube

A single CNT was cut by the electron beam in the presence of oxygen gas, as shown in Figure 7. 20. Cutting was performed using the same electron beam current and acceleration voltage as the two previous cases. The vacuum pressure was 1.6×10^{-2} Pa and the oxygen gas flow rate conditions were the same as previously. However, the nozzle was located at a distance of 90 μm from the CNT in this case. Figure 7. 20 (a) shows the CNT before cutting, Figure 7. 20 (b) shows the CNT after a length of 650 nm has been cut off and Figure 7. 20 (c) shows the CNT after removing a further 700 nm. Figure 7. 20 (d) shows the cutting of a CNT such that it has the same length as the one on the left. These experimental results demonstrate that the length of CNT can be precisely controlled by cutting using an electron beam, assisted with oxygen gas. The acceleration voltages and the beam currents that can cut CNTs in less than 1 minute are shown in Figure 7. 21. Cutting is easy and rapid at low-acceleration voltages and high beam currents.

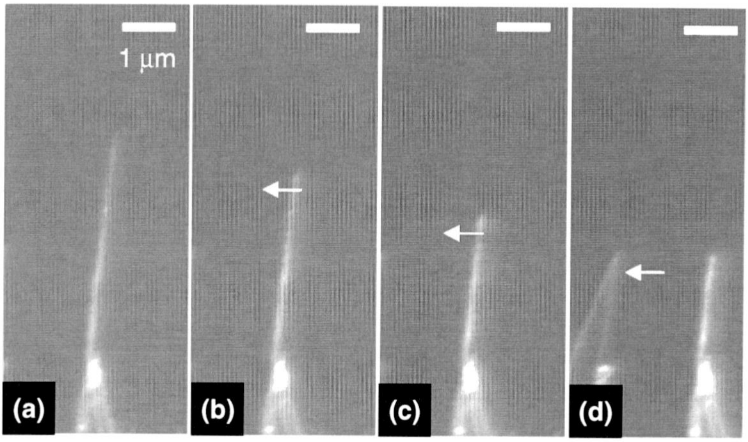

Fig. 7.20 (a) Before and (b-d) after cutting of a single CNT in less than 1 minute. (b) A length of 650 nm was cut off the 1st time. (c) A further 700 nm was cut off the 2nd time. (d) The CNT on the right after cutting has the same length as the one on the left. The arrows point cutting positions.

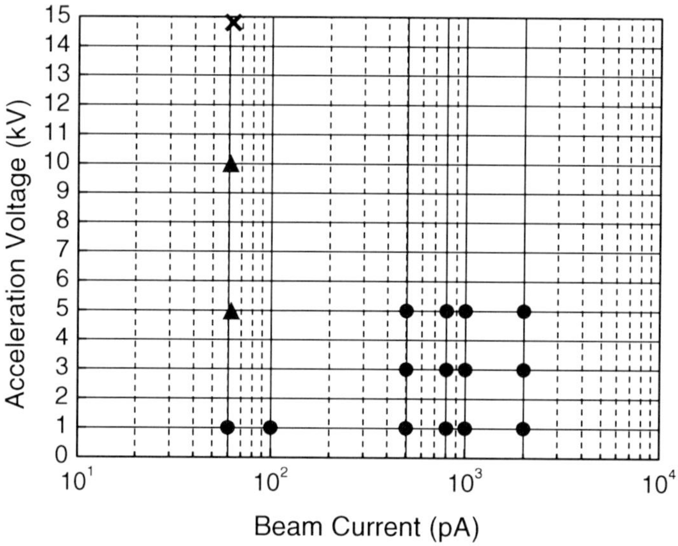

Fig. 7.21 Cutting of CNTs in less than 1 minute under various acceleration voltages and beam currents show by the black circles. ▲: it took more than 1 min for cutting. ×: no cutting even for 30 min irradiation.

For getting clear structures about cutting point, TEM images were taken before and after the cutting process using acceleration voltage of 200 kV. Figure 7.22 shows the nanotubes deposited on a TEM grid before and after cutting. Figure 7.22 (a) shows the nanotube suspended across a gap before cutting. The cutting was performed inside a SEM. Figure 7.22 (b) shows the CNTs after cutting and as can be seen, about 100 nm of material was removed. It can be also detected that the gap is larger than the beam spot size of 3~5 nm, and that the tip of CNTs after cutting is sharpened. The main reason for this is the drift of the beam drifts.

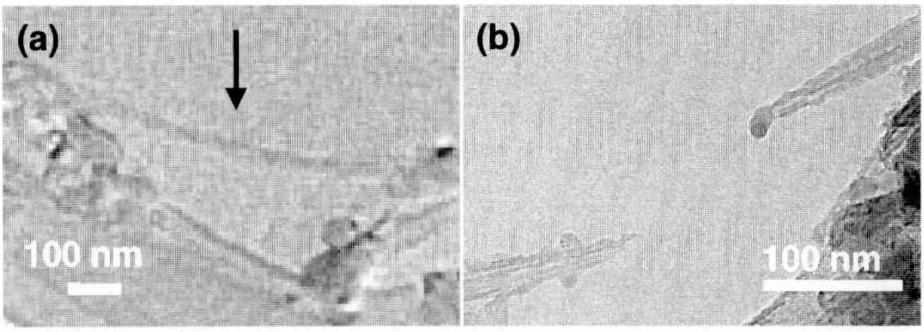

Fig. 7.22 TEM images of MWNTs (a) before and (b) after cutting. It was estimated that about 100 nm of material was removed.

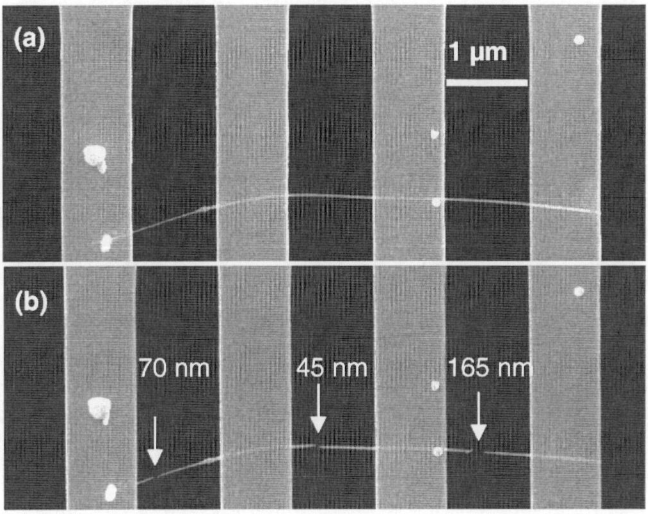

Fig. 7.23 (a) Before and (b-d) after cutting of a single CNT. (b) The cut off length are 70 nm, 45 nm and 165 nm, respectively.

In another cutting experiment, the electron beam with an acceleration voltage of 2 kV and a beam current 3×10^{-10} A is pointed into the desired cutting point by using the spot mode of the SEM. The electron beam was assisted with oxygen gas flow of 1 sccm, flowing from a nozzle sited 500 µm from the sample. A single

CNT was picked up by an AFM cantilever and then put it on the substrate. Figure 7. 23 (a) shows the CNT before cutting. As a result, the CNT was cut by the electron beam at three points, in about 2 min at the left two points and 3 min at the right point as can be seen from Figure 7. 23 (b). The cut off length are 70 nm, 45 nm and 165 nm, respectively. According to the results, the cutting system can produce the smallest removable length of about 45 nm.

Mechanism of Cutting CNTs with Electron Beam and Oxygen Gas

As shown by the experimental results, two conditions are required for CNT cutting. One is a low-energy electron beam damaging the CNT and another is the presence of oxygen gas that reacts with carbon molecules to form CO and CO_2. The cutting process requires a low-energy electron beam probably because the ionization of carbon molecules is dependent of the energy of the primary electrons. The ionization efficiency is a maximum when the irradiation beam energy is three or four times larger than the core-electron binding energy. The core-electron binding energy of carbon is 290 eV. [96] Since the irradiation beam energy is 870 - 1160 eV, which allows us to obtain the maximum ionization efficiency readily. Secondary electron yields are obtained at a maximum of 300 - 1100 eV. [97] Carbon molecules are easily excited and ionized under a low acceleration voltage (~1 kV). We have only demonstrated these phenomena at acceleration voltages from 1 to 30 kV because the SEM cannot be operated at less than 1 kV. The second experiment reveals described in section 3.5.2 that the oxygen gas has an uneven distribution even though the nominal pressure in the specimen chamber was 10^{-2} Pa. For cutting CNTs in less than 1 minute, the gas nozzle needs to be close to the sample. In our case, the gap between the CNTs and the gas nozzle was less than 100 μm. If this gap is increased the amount of oxygen available to react with the CNTs will decrease, thereby increasing cutting time.

However, what is the influence of the beam irradiation-induced heating on the cutting of the CNTs? We can estimate the increase in temperature ΔT induced by electron beam irradiation at the point mode from the following equation. [98]

$$\Delta T = 0.48 \times E_0 I_p / kd \tag{7.9}$$

Where E_0 is the acceleration voltage, I_p is the irradiation current, k is the thermal conductance of CNTs and d is the diameter of the beam. In our experiment, $E_0 = 1$ kV, $I_p = 0.6$ nA and $d \approx 3$ nm. The value of k is 3500 $W \cdot m^{-1}K^{-1}$. [99] ΔT is then calculated to be 0.027 K. The ΔT will increase according as the $E_0 I_p$ rising. But the experimental data shows that cutting cannot occur even under larger $E_0 I_p$. Therefore, the increase in temperature is sufficiently small that its influence on the cutting of CNTs can be ignored.

In conclusion, we have demonstrated that CNTs can be cut in less than 1 minute under a low acceleration voltage assisted by the presence of oxygen gas inside a field emission scanning electron microscope. Oxygen gas was regulated by a mass flow controller and introduced with a gas nozzle. It is found that although the total pressure in the specimen chamber reached 10^{-2} Pa, cutting at high speeds only occurred in an area close to the gas nozzle. We assume that the cutting of a CNT requires a low-energy electron beam, since the ionization efficiency of carbon is high at low primary electron energies. The electron beam induced heating can be ignored during the cutting process since it is small. We expect that the in situ cutting technique demonstrated here will be widely used for rapid prototyping of CNT nanodevices.

Bending of Carbon Nanotubes Assisted with Oxygen Gas

The presence of oxygen gas in the vicinity of the CNT can be used also for the bending of CNT. The only changes needed in the process conditions are the increase of the acceleration voltage or the receding of the oxygen gas nozzle from sample, or reducing the irradiation time. Figure 7.24 shows the schematic of the experimental setup for bending a CNT.

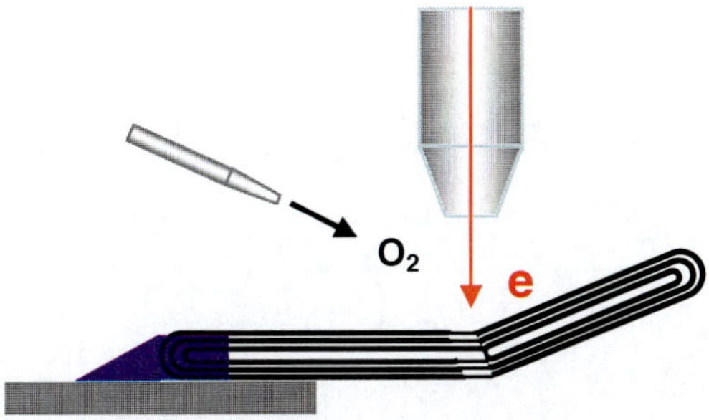

Fig. 7.24 A schematic of the experimental setup using an electron beam for bending a nanotube assisted with oxygen gas

Bending of a Single Carbon Nanotube

The bending procedure was demonstrated in the following experiment. In the experiment, the gas nozzle was setup 170 μm distance from the sample. The acceleration voltage and the irradiation current were 2 kV and 3×10^{-10} A, respectively, and the oxygen gas flow was 1 sccm. An individual CNT was irradiating in spot beam mode for 2 min or 30 s at three points, and the results are shown in Figure 7. 25. Figure 7. 25 (a) shows the carbon nanotube before bending. In Figure 7. 25 (b) a point on the CNT was irradiated 2 min and a clear angle can be observed in the respective point. Irradiation was continued for 30 s on the same point which increased the angle as can be seen from the Figure 7. 25 (c). The experiment was repeated in the other location of the CNT and after 2 min irradiation approximately same angle was bended in the CNT as in the first experiment, as shown in Figure 7. 25 (d). From the Figure 7. 25 (e), it can be seen that the 30 s additional irradiation was able to increase the angle as in the first experiment. The additional 2 min irradiation increased the angle forming an over 90° angle, as shown in Figure 7. 25 (f).

The experiment clearly shows that the CNT can be bended accurately by using the proposed technique. The technique enables accurate determination of the bending point and the bending angle. The bending process can be executed with the same equipments and experiment settings as the cutting, discussed earlier.

Some theoretical calculations about the bending of CNT have been reported in [100-103]. Based on these the bending of CNTs is assumed to have three typical configurations: rippling [104], buckling [105] and pentagon-heptagon mode [106]. The rippling mode shows a wavelike structure on the inner arc of CNTs and smooth curvature on the outer arc, appearing always on slightly bended thick nanotubes. From the rippling mode CNTs can return to their initial shapes after removing the bending stress. The buckling mode will occur on thin nanotubes or on large deformations in which the compression or stretching is over the elastic limit of CNT. In the third mode, pentagon and heptagon nets will appear in hexagonal carbon lattice in pairs. The mechanism of bending in the experiment is more likely pentagon-heptagon induced deformation. This is because no mechanical stress is involved, but the carbon-carbon bonds of hexagonal carbon lattice are destructed and part of carbon molecules are removed by the oxygen molecules.

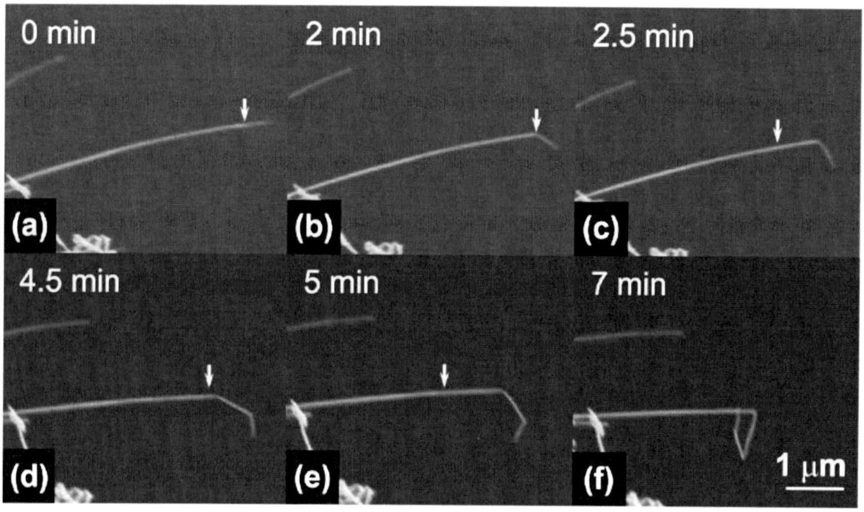

Fig. 7.25 Bending of a CNT for 2 min or 30 s at three points, respectively. The arrows show the irradiation point at each image.

Bending Direction

The CNTs were bent toward to up and down as shown in Figure 7. 26. We consider that the bending direction depends on the beam irradiation position. As shown in the insets of Figure 7. 26, when the beam position shifts from the center axis to a side, then the result is that the CNT will be bent and turn to that side. If the beam position is just on the center axis, the CNT will bend toward the direction of the electron beam flux. The bending of the tubes is a natural cause of the shrinkage on one side only.

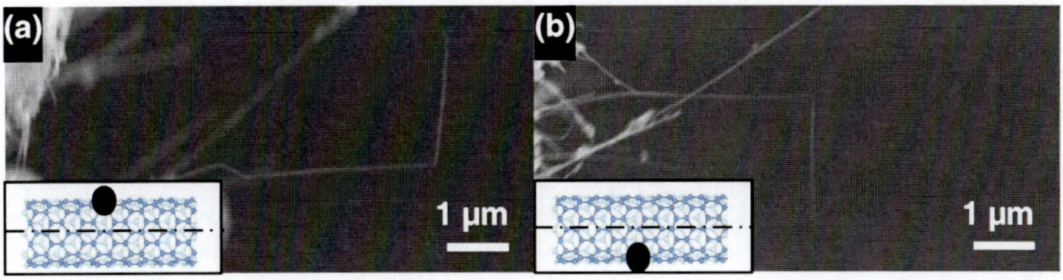

Fig. 7.26 CNTs were bent toward to up (a), and down (b). The both beam irradiation positions are away from the center axis of CNTs.

Peeling of Carbon Nanotubes Assisted with Oxygen Gas

The proposed peeling technique is closely related to the cutting technique, because also in this technique oxygen is introduced in the vicinity of the MWNT and the activation of carbon atoms happens by using electron beam. Therefore, the peeling happens through the same reactions as the cutting. Instead of the spot mode of the electron beam used in cutting, in the peeling a scanning mode is used. The "burning" of the CNT is restricted only on the outermost shell because the oxygen is on the surface of the CNT and can react only with the outermost carbon atoms, not the inner ones. The peeling area can be accurately set according to the scanning area which makes the technique suitable for nanomanufacturing. If compared to cutting, the density of oxygen is normally set lower to prevent the inner layers injure and therefore the peeling takes more time than simple CNT cutting.

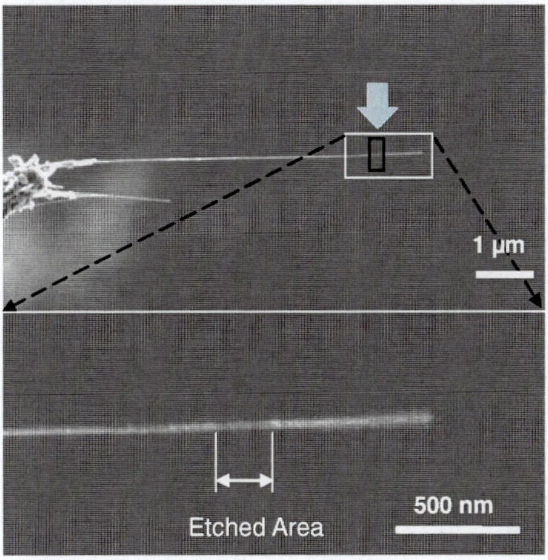

Fig. 7.27 Two SEM images of peeling MWNTs. In the upper image the area, exposed to electron beam, is marked as a black square. In the lower image a magnification of the upper image is shown, revealing the absence of the outer shells.

The peeling technique was tested by using the same experiment arrangements as in the cutting. During the experiment, the electron beam was set to a scan mode with a scan area of 0.8 × 0.2 μm. The acceleration voltage and the irradiation current were 2 kV and 3×10^{-10} A, respectively, and the oxygen gas flow was 1 sccm. The scanning area was exposed to the electron beam for 30 minutes and the results are shown in Figure 7.27. The peeled MWNT with the black box indicating the scanned area is shown in the upper figure and the magnification of the area of the white box in the lower figure. As can be seen, the diameter of the MWNT is smaller in the area exposed to the electron beam. This is because the outer shells of the MWNT are etched by the electron beam. Because the peeling of a CNT can be done by using the same kind of experiment device as the cutting and bending, these three techniques can be easily combined. This makes the techniques very suitable for nanodevice manufacturing as will be discussed next.

Assembly of Nanostructures with Electron-Beam-Induced Nanofabrication

Because of the great effect of the CNT length to the functionality of the nanostructure, the effective cutting techniques are indispensable for nanofabrication and nanostructure assembly. The usefulness and functionality of the proposed cutting technique is shown in the experiment where a three-dimensional nanostructure is created from precisely cut CNTs, cut by using the cutting technique and fixed by using the welding technique. The combination of the cutting, bending, peeling, welding techniques and the nanorobotic manipulation is demonstrated in assembly of a three-dimensional nanostructure.

Two-Dimensional Nanostructures

Figure 7. 28 (a) shows an array of eight tungsten dots deposited with FIB on a thin gold film coated substrate. The dots are 1 μm in diameter, 350 nm in height and have a 1.5 μm intervals. The side view of the dots is shown in Figure 7. 28 (b). The nanostructure was created by picking up a single CNT with an AFM cantilever and suspending it across two dots by using a nanorobotic manipulator as shown in Figure 7. 28 (c). The other end of a nanotube was then fixed by using EBID on the top of the dot and the nanotube was strained so that the tension was stronger than the adhesion between the CNT and the substrate. After that also the other and was welded and the excess nanotube were cut by using the cutting technique described in previous sections. The resulting stretched CNT is shown in Figure 7. 28 (d). The cutting point is indicated by the arrow in the figure. After that the rest CNT were assembled by applying the same procedure. The final result was the letters N and U being assembled with 6 CNTs based on eight tungsten dots as shown in Figure 7. 28 (e).

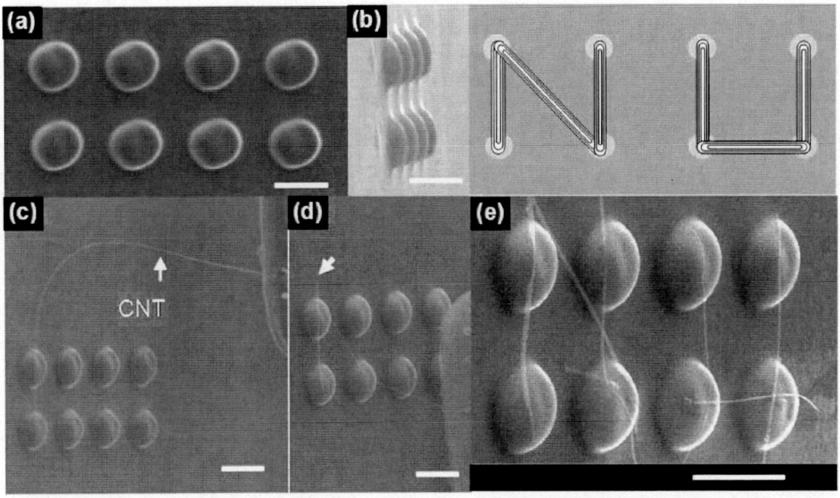

Fig. 7.28 Top view (a) and side view (b) of the tungsten dots deposit with FIB on gold coated substrate, a carbon nanotube picked up by an AFM cantilever and put on the tungsten dots (c), cutting process for separating CNT from substrate. Letters NU were assembled with 6 carbon nanotubes. Bars are 1 μm.

Three-Dimensional Nanostructures

A 3D nanostructure is constructed by the electron-beam-induced nanofabrication. The assembly progress of the structure is shown step by step in Figure 7. 29. Figure 7. 29 (a) shows a CNT picked up by an AFM cantilever and manipulated by the nanorobotic manipulator. The other end of the CNT was fixed on AFM cantilever surface by a tungsten deposit, produced by the proposed welding technique. The other end was set to touch the surface of another AFM cantilever. The proposed bending technique was applied on the CNT and, as can be seen from Figure 7. 29 (b), the CNT was bent at this point. The direction and angle of bending can be controlled by the manipulator.

The first bending was followed by another bend in other CNT point, as shown in Figure 7. 29 (c). The location and orientation of the CNT was changed by the manipulator and the second knick was set to touch the substrate as shown in Figure 7. 29 (d). Finally, the CNT was cut at third point shown in Figure 7. 29 (e) and the created 3D nanostructure was separated from the substrate one. As the result, a letter N was assembled in a CNT and stand on the substrate at two points as shown in Figure 7. 29 (f). The two points attach the structure on the substrate only by van der Waals force.

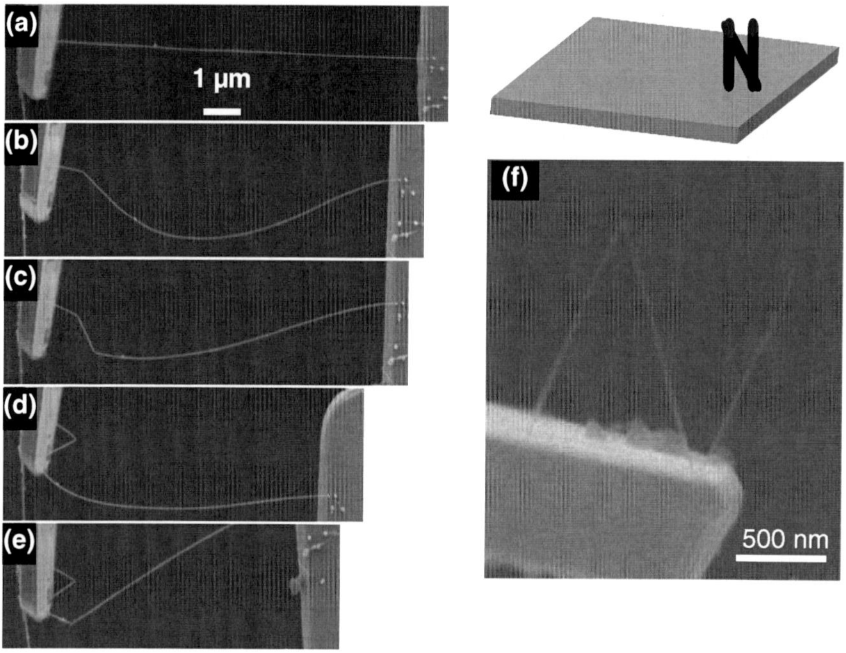

Fig. 7.29 Assembly of three dimensional nanostructure based on a CNT assisted with welding (a), bending (b)-(d), and cutting (e) techniques. The letter N stands on the substrate (f).

7.5 Nano-actuator Using Telescoping Carbon Nanotube

Nanoactuation of Telescoping Nanotubes

In this work, we report on the direct observation of electrostatic actuation of a telescoping MWCNT inside a TEM. L. X. Dong et al. proposed that the field emission from a one-end opened telescoping MWCNT is used for position feedback system, and they tried the experiment inside a SEM [107].However, no direct confirmation of its actuation was observed, because the charge up effects prevented the observation of the sample while applying the electric field. Here, we showed the direct observation of its actuation under applied electrostatic field inside a TEM. Analyses of electrostatic forces by finite element method (FEM) and van der Waals interactions are performed, and it is shown that the actuation force is mainly caused by electrostatic forces.

J. Cumings et al. demonstrated that pulling out the inner core of one-end opened type telescoping nanotube mechanically inside a transmission electron microscope (TEM) [108]. The inner core of telescoping nanotube is pulled out with van der Waals forces between its tip and probe. Interestingly, the inner core is automatically retracted inside the outer layers, when the connection is freed. It is clearly observed that the inner core is separated from outer layers with mainly van der Waals force interactions. A. M. Fennimore et al. demonstrated that MWCNTs are applied for the rotation axis of silicon chip as rotational actuators [42]. From the computational aspects, several researchers proposed that telescoping MWCNTs have possibilities for Gigahertz oscillator applications [109, 110, 111]. The van der Waals interaction with the inner core and outer layers are effectively used. A. Hansson et al. showed that the inner core of telescoping carbon nanotube has a distributed inter-shell conductance [112]. This shows the possibility of high precise sensing of inter-shell distance.

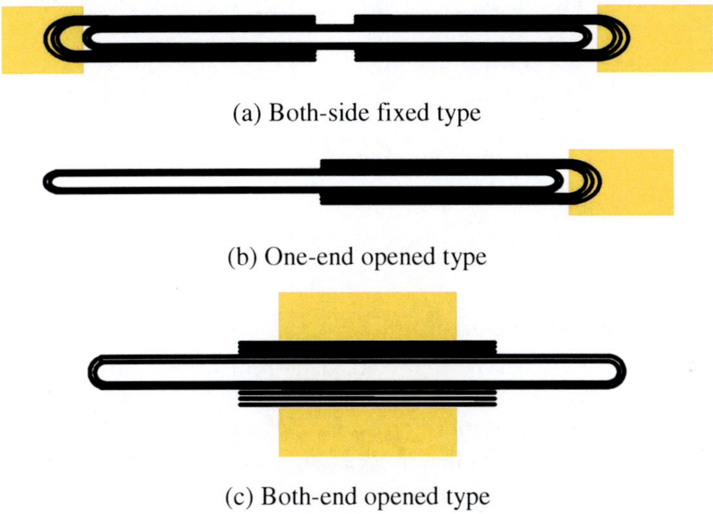

(a) Both-side fixed type

(b) One-end opened type

(c) Both-end opened type

Fig. 7.30 Schematic diagram of telescoping nanotubes

Modeling

The telescoping structure is fabricated by peeling off the outer layers of a MWCNT as shown in Figure 7. 30. We performed basic experiments of electrostatic actuation with a one-end opened telescoping nanotube. The schematic diagram of experimental setup of the electrostatic actuation of telescoping nanotube is shown in Figure 7.31. One-end of a MWCNT was fabricated by peeling off the outer layers as the telescoping structure. The tip of the telescoping nanotube is set to face the electrode with a gap G_{CNT} by the TEM manipulator. Electric voltage is applied between the telescoping nanotube and the electrode, and field emission current is measured. While the voltage is applied, electrostatic forces are generated at the tip of the telescoping nanotube. In the counter direction, friction force F_R and retracting van der Waals force F_W are acted. When the voltage is turned off, only the retracting van der Waals force F_W works on the inner core. Hence, the inner core is automatically retracted inside the outer layers[1]. The friction force is also acted on the counter direction impeding the motion of the inner core. When the friction force F_R is higher than retracting van der Waals force F_W, the inner core is stopped and cannot fully retract into the outer layers.

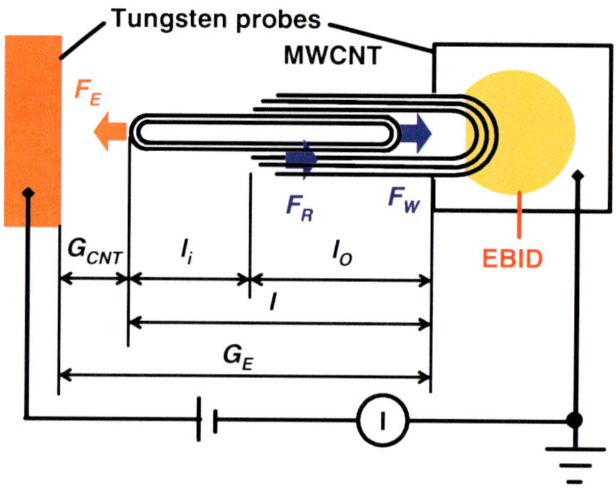

Fig. 7.31 Schematic diagram of electrostatic actuation of a telescoping MWCNT

Analysis of Electrostatic Forces

The friction force F_R has been investigated from theoretical and experimental aspects. From molecular static and dynamic investigations on the interlayer friction in a bi-walled CNT, the friction forces are calculated as ~0.5 to ~1.5 nN [113]. Experimentally obtained friction forces are several nN and considered to be approximately the same value as inter-layer van der Waals forces [99, 114]. For a planar wall consisting of two planes, the van der Waals force F_{vdw} and van der Waals energy $U(l)$ are expressed as follows [59],

$$F_{vdw} = \frac{dU(l)}{dl}, \quad U(l) = \frac{A\,S(l)}{12\pi D^2} \tag{7.10}$$

where A is a Hamaker constant, $S(l)$ is an overlapping area, and D is a distance between two planes. Applying the eq.1 to a cylindrical layers, we calculate the F_{vdw} for a telescoping nanotube. The overlapping area $S(l)$ is expressed as $S(l) = 2\pi R_i\, l$, where R_i is a radius of the inner core, and l is an overlapping length. Because the contact area of the interlayer is linearly proportional to l, the friction force F_R is expressed as following equation.

$$F_R = \frac{A R_i}{6D^2} \tag{7.11}$$

Using A=1 ×10^{-19} J, R_i=5 nm, and D =0.34 nm, F_R is estimated as ~0.7 nN.

Experimental Results and Discussions

The single MWCNT is partly peeled off its outer layers inside TEM by fixing EBID. The single MWCNT is picked up and its position is set inside a SEM. These experimental processes have been done through hybrid nanorobotic manipulation system inside a SEM and a TEM.

Destructive fabrication process is adapted for the construction of telescoping structure in this work. By applying the mechanical stress on the both-end fixed MWCNTs, its outer layers are peeled off at destructive site, and its inner core is pulled out. In our works, the mechanical stress is applied with the nanomanipulators, and the fixation is done with the deposit of EBID (in Chapter 3.2). Generally, the outer layers will be broken whereas the inner layers will extrude. A perfect MWNT can be broken at just any site in its length. For the as-grown CNTs, it cannot be determined where the destructive site happened, although defects of nanotubes would help the determination of the destructive site. When the tensile stress is reached the fracture stress, the outer layers of the MWCNT are destructed and the inner core is pulled out. We name this process as destructive fabrication process.

We performed *in situ* fabrication of a one-end opened MWCNT inside a TEM. The telescoping structure is produced by peeling off the outer layers of a MWCNT through the destructive fabrication process. The schematic diagram of experimental steps is shown in Figure 7. 32. Single MWCNT is picked up on the tungsten probe, which is etched by FIB, inside a SEM by EBID fixation. The MWCNTs are synthesized by arc discharge method. One end of MWCNT is freed, and the other end is fixed by EBID (Figure 7. 32 (a)). This probe is fixed on the passively driven TEM sample stage. Another tungsten probe etched by FIB is fixed on the TEM manipulator. The MWCNT is set the position using TEM sample stage driven by SEM manipulator within the working area of TEM manipulator. Then, the fee end was also fixed on the tungsten probe fixed TEM manipulator by EBID inside a TEM (Figure 7. 32 (b)). The electron beam was focused to ~100 nm for the EBID. Liquid N_2 is not supplied to the device of TEM in order to make the contamination quickly. The vacuum pressure is 1~2×10^{-4} Pa. After both sides were fixed, the tungsten probe on the TEM manipulator was moved and tensile stress was applied on the MWCNT. When the tensile stress was reached at fracture stress, the outer layers of MWCNT were broken and the inner core was pulled out. We call this process as destructive fabrication process (Figure 7. 32 (c), (d), (e))

Figure 7.33 shows sequential TEM images of the destructive fabrication process. Figure 7. 33 (a) shows a one-end freed carbon nanotube which was picked up inside the SEM on the tungsten probe. The outer diameter of the nanotube is ~28 nm, and the inner diameter is ~2 nm. The other free end is fixed on the tungsten probe by EBID, and then the probe is moved by TEM manipulator. Its outer layers are broken by tensile stress, and the inner core is pulled out (Figure 7. 33 (b)). As shown in Figure 7. 33 (c) and (d), the inner core is smoothly pulled out. Figure 7. 33 (e) shows

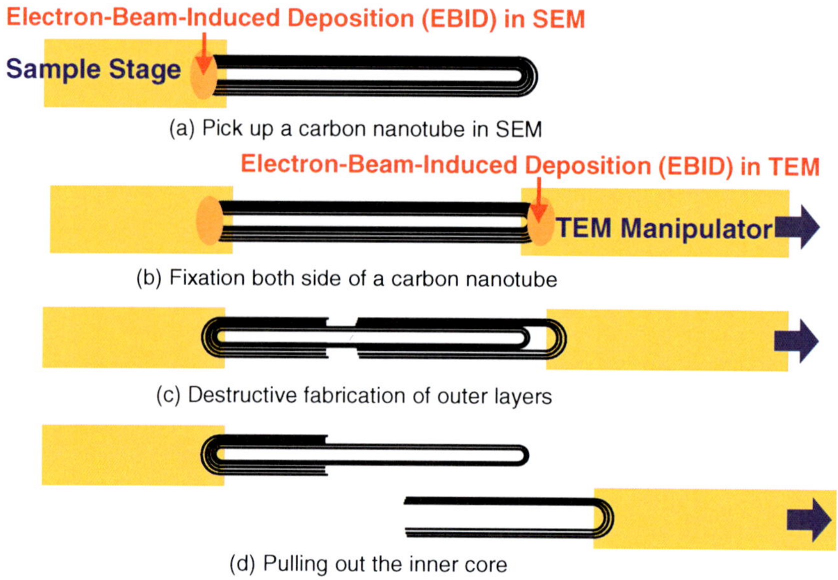

Fig. 7.32 Schematic diagram of the destructive fabrication process

that the inner core was fully pulled out and the sheath was left after the fabrication process. The diameter of the inner core is ~9 nm. The present fabrication process confirms that a carbon nanotube is certainly peeled off its outer layers and its inner core is pulled out by *in situ* experiments inside a TEM.

Here we show the experimental results of electrostatic actuation of a telescoping nanotube. A single MWCNT was also peeled off its outer layers through the destructive fabrication process. The fabricated telescoping nanotube was set against the tungsten probe of TEM manipulator. The probe is used as an anode. Then the DC voltage was applied between the telescoping nanotube and the anode.

Figure 7.34 shows a series of TEM images at various applied voltages. The anode is not seen in these images. The telescoping nanotube has the inner core with diameter of ~6.6 nm, and the outer layers with diameter of ~8.4 nm. The distance between the tip of telescoping nanotube and the tungsten probes G_E is set as ~1.2 μm. It is clearly observed that the length of the inner core l_i is extended and retracted depending on the applied voltage. The length of outer layers l_o seemingly increased and decreased by varying the applied voltage. This may be due to the change of orientation of the nanotube by the applied electric field. On increasing the applied voltage, the telescoping MWCNT is deflected to align to the direction of the electric field. Since the generated electric field is vertical to the substrate, an inclined MWCNT stand vertical to the substrate.

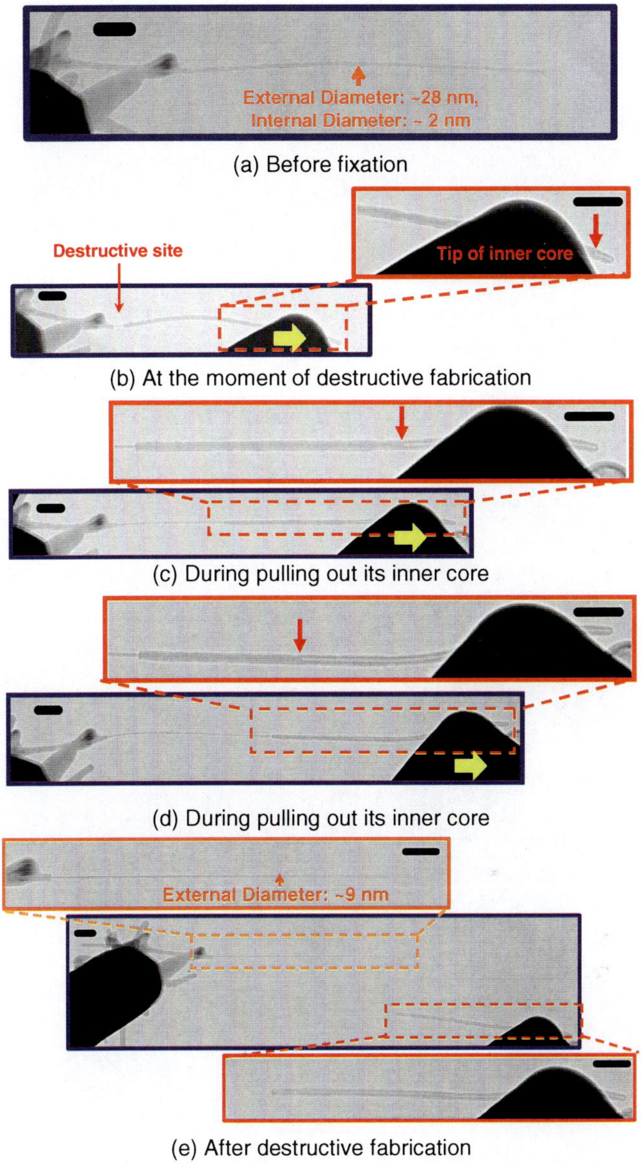

Fig. 7.33 Sequential TEM photographs of destructive fabrication of a multi-walled carbon nanotube (Scale bar: 100 nm)

The ratio of inner core length l_i to outer layers length l_o was calculated at various applied voltage. The increase of the ratio l_i/l_o with the increase of applied voltage means the extraction of the inner core against the outer layers by the electric field. With increasing the applied voltage, the change of the extracted length of inner core becomes insensitive against the voltage higher than ~45 V. The maximum extended inner core length is ~179 nm at ~65 V (initial l_i ~178 nm, maximum l_i ~357 nm). Under the present experimental condition, the electrostatic force F_E is estimated to be ~0.8 to 0.08 nN by the FEM analysis and the friction force F_R is calculated to be ~0.48 nN from eq.2. These forces are approximately the same order of magnitude with each other. Hence, they are considered to be almost balanced with each other. When the applied voltage is decreased, the inner core is not fully retracted; it has approximately ~105 nm hysteresis (final l_i ~283 nm). The hysteresis might be caused by the contamination deposited on the nanotube surface during the observation. These results clearly show that the inner core is extended and retracted with applied voltage. This was the first direct observation of electrostatic actuation of a telescoping nanotube by TEM.

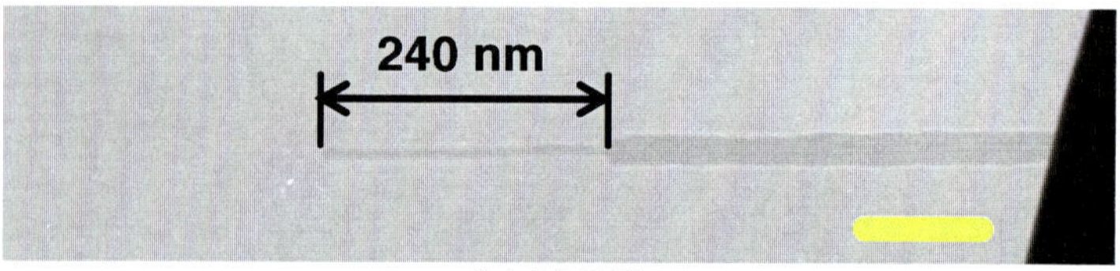

(a) 24.2 V

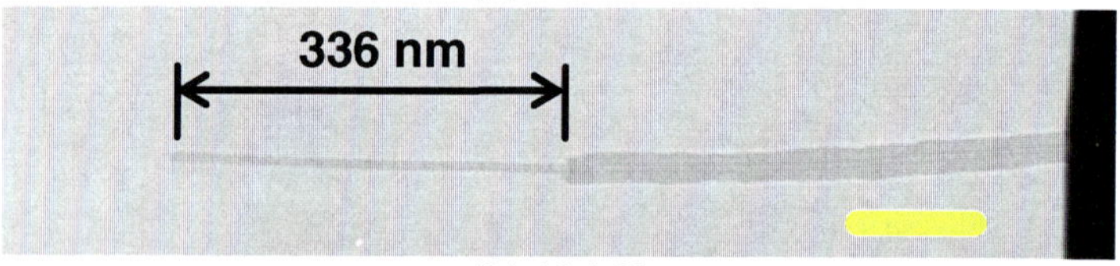

(b) 44.1 V

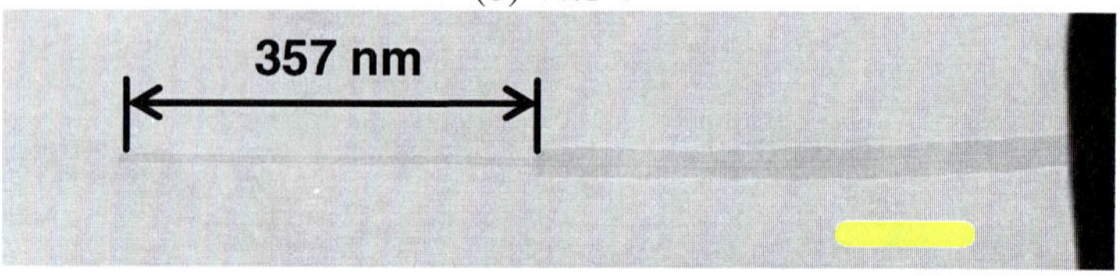

(c) 64.9 V

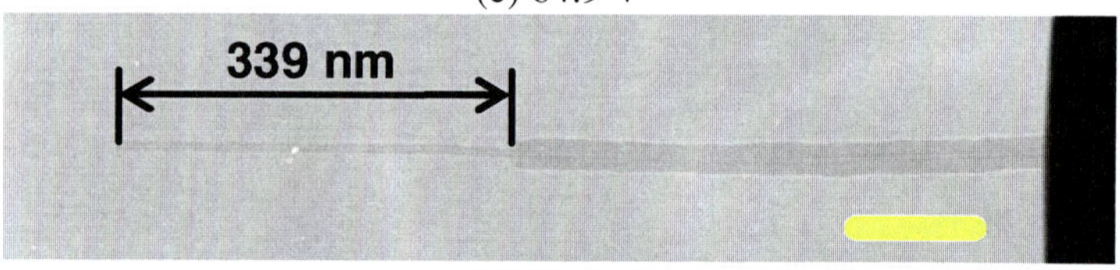

(d) 27.8 V

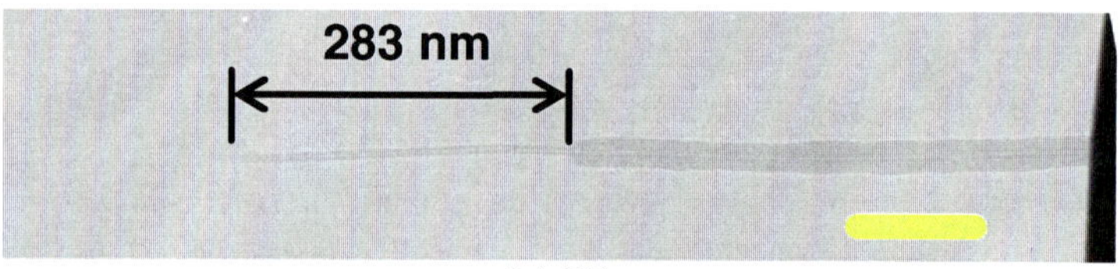

(e) 0V

Fig. 7.34 Actuation of a telescoping MWCNT at various applied voltage (Scale bar: 100 nm)

References

[1] Cumings, J., Zettl, A.: Low-friction nanoscale linear bearing realized from multiwall carbon nanotubes. Science 289, 602–604 (2000)
[2] Fennimore, A.M., Yuzvinsky, T.D., Han, W.Q., Fuhrer, M.S., Cumings, J., Zettl, A.: Rotational actuators based on carbon nanotubes. Nature 424, 408–410 (2003)
[3] Sazonova, V., Yaish, Y., Ustunel, H., Roundy, D., Arias, T.A., McEuen, P.L.: A tunable carbon nanotube electromechanical oscillator. Nature 431, 284–287 (2004)
[4] Rinzler, A.G., Hafner, J.H., Nikolaev, P., Lou, L., Kim, S.G., Tomanek, D., Nordlander, P., Colbert, D.T., Smalley, R.E.: Unraveling nanotubes: field emission from an atomic wire. Science 269, 1550–1553 (1995)
[5] Saito, Y., Uemura, S., Hamaguchi, K.: Cathode ray tube lighting elements with carbon nanotube field emitters. Jpn. J. Appl. Phys. 37, L346–L348 (1998)
[6] Dai, H.J., Hafner, J.H., Rinzler, A.G., Colbert, D.T., Smalley, R.E.: Nanotubes as nanoprobes in scanning probe microscopy. Nature 384, 147–150 (1996)
[7] Kim, P., Lieber, C.M.: Nanotube nanotweezers. Science 286, 2148–2150 (1999)
[8] Liu, P., Dong, L.X., Arai, F., Fukuda, T.: Nanotube multi-functional nanoposition sensors. J. Nanoeng. Nanosyst. 219, 23–27 (2005)
[9] Liu, P., Arai, F., Dong, L.X., Fukuda, T., Noguchi, T., Tatenuma, K.: Field emission of individual carbon nanotubes and its improvement by decoration with ruthenium dioxide. J. Robo. Mech. 17, 475–482 (2005)
[10] Lu, J.P.: Elastic Properties of Carbon Nanotubes and Nanoropes. Phys. Rev. Lett. 79, 1297–1300 (1997)
[11] Hernandez, E., Goze, C., Bernier, P., Rubio, A.: Elastic Properties of C and BxCxNx Composite Nanotubes. Phys. Rev. Lett. 80, 4502–4505 (1998)
[12] Lier, G.V., Alsenoy, C.V., Doren, V.V., Geerlings, P.: Ab Initio Study of the Elastic Properties of Single-Walled Carbon Nanotubes and Graphene. Chem. Phys. Lett. 326, 181–185 (2000)
[13] Poncharal, P., Wang, Z.L., Ugarte, D., de Heer, W.A.: Electrostatic Deflections and Electromechanical Resonaces of Carbon Nanotubes. Science 283, 1513–1516 (1999)
[14] Wong, E.W., Sheehan, P.E., Lieber, C.M.: Nanobeam Mechanics: Elasticity, Strength, and Toughness of Nanorods and Nanotubes. Science 277, 1971–1975 (1997)
[15] Treacy, M.M.J., Ebbesen, T.W., Gibson, J.M.: Exceptionally High Young's Modulus Observed for Individual Carbon Nanotubes. Nature 381, 678–680 (1996)
[16] Yu, M.F., Lourie, O., Dyer, M.J., Moloni, K., Kelly, T.F., Ruoff, R.S.: Strength and Breaking Mechanism of Multiwalled Carbon Nanotubes Under Tensile Load. Science 287, 637–640 (2000)
[17] Salvetat, J.P., Kulik, A.J.: Elastic Modulus of Ordered and Disordered Multiwalled Carbon Nanotubes. Adv. Mater. 11, 161–165 (1999)
[18] Krishnan, A., Dujardin, E., Ebbesen, T.W., Yianilos, P.N., Treacy, M.M.J.: Young's modulus of single-walled nanotubes. Phys. Rev. B 58, 14013–14019 (1998)
[19] Salvetat, J.P., Briggs, G.A.D., Bonard, J.M., Bacsa, R.R., Kulik, A.J.: Elastic and Shear Moduli of Single-Walled Carbon Nanotube Ropes. Phys. Rev. Lett. 82, 944–947 (1999)
[20] Yu, M.F., Files, B.S., Arepalli, S., Ruoff, R.S.: Tensile Loading of Ropes of Single Wall Carbon Nanotubes and their Mechanical Properties. Phys. Rev. Lett. 84, 5552–5555 (2000)

[21] Johnston, B.G.: Guide to Stability Design Criteria for Metal Structures, pp. 64–74. John Wily & Sons (1976)

[22] Koops, H.W.P., Kretz, J., Rudolph, M., Weber, M., Dahm, G., Lee, K.L.: Characterization and application of materials grown by electron-beam-induced deposition. Jpn. J. Appl. Phys. 33, 7099–7107 (1994)

[23] Matsui, S., Kaito, T., Fujita, J., Komuro, M., Kanda, K., Haruyama, Y.: Three-dimensional nanostructure fabrication by focused-ion-beam chemical vapor deposition. J. Vac. Sci. Technol. B 18, 3181–3184 (2000)

[24] Koops, H.W.P., Kaya, A., Weber, M.: Fabrication and characterization of platinum nanocrystalline material grown by electron-beam induced deposition. J. Vac. Sci. Technol. B 13, 2400–2403 (1995)

[25] Weber, M., Rudolph, M., Kretz, J., Koops, H.W.P.: Electron-beam induced deposition for fabrication of vacuum field emitter devices. J. Vac. Sci. Technol. B 13, 461–464 (1995)

[26] Koops, H.W.P., Schössler, C., Kaya, A., Weber, M.: Conductive dots, wires, andsupertips for field electron emitters produced by electron-beam induced deposition on samples having increased temperature. J. Vac. Sci. Technol. B 14, 4105–4109 (1996)

[27] Dong, L.X., Arai, F., Fukuda, T.: Electron-beam-induced deposition with carbon nanotube emitters. Appl. Phys. Lett. 81, 1919–1921 (2002)

[28] Saito, Y., Uemura, S., Hamaguchi, K.: Cathode ray tube lighting elements with carbon nanotube field emitters. Jpn. J. Appl. Phys. 37, L346–L348 (1998)

[29] Lee, N.S., Chung, D.S., Han, I.T., Kang, J.H., et al.: Application of carbon nanotubes to field emission displays. Diamond Relat. Materials 10, 265–270 (2001)

[30] Rosen, R., Simendinger, W., Debbault, C., Simoda, H., Fleming, L., Stoner, B., Zhou, O.: Application of carbon nanotubes as electrodes in gas discharge tubes. Appl. Phys. Lett. 76, 1668–1670 (2000)

[31] Sugie, H., Tanemura, M., Filip, V., Lwata, K., Takahashi, K., Okuyama, F.: Carbon nanotubes as electron source in an X-ray tube. Appl. Phys. Lett. 78, 2578–2580 (2001)

[32] Dong, L.X., Arai, F., Fukuda, T.: Destructive constructions of nanostructures with carbon nanotubes through nanoroboticmanipulations. IEEE/ASME Trans. on Mechatronics 9, 350–357 (2004)

[33] Koops, H.W.P., Kretz, J., Weber, M.: Combined lithographies for the reduction of stitching errors in lithography. J. Vac. Sci. Technol. B 12, 3265–3269 (1994)

[34] Gortz, W., Kempf, B., Kretz, J.: Resolution enhanced scanning force microscopy measurements for characterizing dry etching methods applied to titanium masked InP. J. Vac. Sci. Technol. B 13, 34–39 (1995)

[35] White, V., et al.: Microfabrication of phase shifting zone plates for hard X-rays. Microelectronic Engineering 21, 99–102 (1993)

[36] Rangelow, I.W., Gotszalk, T., Abedinov, N., Grabiec, P., Edinger, K.: Thermal nano-probe. Microelectron. Engng. 57, 737–748 (2001)

[37] Arai, F., Liu, P., Dong, L.X., et al.: Electron-Beam-InducedDeposition of Conductive Nanostructures with Carbon Nanotube Emitters. In: Proc. of the 2003 3rd IEEE Int'l Conf. on Nanotechnology (IEEE-NANO 2003), San Francisco, U.S.A., Aug. 11-14, pp. 811–814 (2003)

[38] Schossler, C., Urban, J., Koops, H.W.P.: Conductive supertips for scanning probe applications. J. Vac. Sci. Technol. B 15, 1535–1538 (1997)

[39] Floreani, F., Koops, H.W.P., Elsaber, W.: Concept of a miniaturized free-electron laser with field emission source. Nuclear Instruments and Methods in Physics Research A 483, 488–492 (2002)

[40] Koops, H.W.P., Kaya, A., Weber, M.: Fabrication and characterization of platinum nanocrystalline material grown by electron-beam induced deposition. J. Vac. Sci. Technol. B 13, 2400–2403 (1995)
[41] Weber, M., Rudolph, M., Kretz, J., Koops, H.W.P.: Electron-beam induced deposition for fabrication of vacuum field emitter devices. J. Vac. Sci. Technol. B 13, 461–464 (1995)
[42] Koops, H.W.P., Weiel, R., Kern, D.P., Baum, T.H.: High-resolution electron-beam induced deposition. J. Vac. Sci. Technol. B 6, 477–481 (1988)
[43] Matsui, S., Mori, K.: New selective deposition technology by electron beam induced surface reaction. J. Vac. Sci. Technol. B 4, 299–304 (1986)
[44] Weber, M., Koops, H.W.P., Rudolph, M., Kretz, J., Schmidt, G.: New compound quantum dot materials produced by electron-beam induced deposition. J. Vac. Sci. Technol. B 13, 1364–1368 (1995)
[45] Matsui, S., Kaito, T., Fujita, J., Komuro, M., Kanda, K., Haruyama, Y.: Three-dimensional nanostructure fabrication by focused-ion-beam chemical vapor deposition. J. Vac. Sci. Technol. B 18, 3181–3184 (2000)
[46] Brodie, I., Westerberg, E.R., Cone, D.R., Muray, J.J., Williams, N., Gasiorek, L.: A Multiple-Electron-Beam Exposure System for High-Throughput, Direct-Write SubmicrometerLithography. IEEE Trans. Electron. Devices 28, 1422–1428 (1981)
[47] Choi, W.B., et al.: Fully sealed, high-brightness carbon-nanotube field-emission display. Appl. Phys. Lett. 75, 3129–3131 (1999)
[48] Fan, S., Chapline, M.G., Franklin, N.R., Tombler, T.W., Cassell, A.M., Dai, H.J.: Self-oriented regular arrays of carbon nanotubes and their field emission properties. Science 283, 512–514 (1999)
[49] Sohn, J.I., Lee, S., Song, Y.-H., Choi, S.-Y., Cho, K.-I., Nam, K.-S.: Patterned Selective Growth of Carbon Nanotubes and Large FieldEmission from Vertically Well-Aligned Carbon Nanotube Field Emitter Arrays. Appl. Phys. Lett. 78, 901–903 (2001)
[50] Dai, H.J., Franklin, N., Han, J.: Exploiting the Properties of CarbonNanotubes for Nanolithography. Appl. Phys. Lett. 73, 1508–1510 (1998)
[51] Okazaki, A., Kishida, T., Akita, S., Nishijima, H., Nakayama, Y.: Direct Nanolithography of Organic Polysilane Films UsingCarbon Nanotube Tips. Jpn. J. Appl. Phys. 39, 7067–7069 (2000)
[52] Good, R.H., Muller, E.W.: Handbuch der Physik, vol. 21, pp. 176–191 (1956)
[53] Gomer, R.: Field emission, field ionization and field desorption. Surface Science 299/300, 129–152 (1994)
[54] Gomer, R.: Field emission and field ionization. Harvard University Press (1961)
[55] Saito, Y., Hamaguchi, K., Hata, K., Tohji, K., Kasuya, A., Nishina, Y., Uchida, K., Tasaka, Y., Ikazaki, F., Yumura, M.: Field emission from carbon nanotubes: purified single-walled and multi-walled tubes. Ultramicroscopy 73, 1–6 (1998)
[56] Saito, Y., et al.: Conical beams from open nanotubes. Nature 389, 554–555 (1997)
[57] Saito, Y.: Field emission patterns from single-walled carbon nanotubes. Jpn. J. Appl. Phys. 36, L1340–L1342 (1997)
[58] Zhu, W., Bower, C., Zhou, O., Kochanski, G., Jin, S.: Large current density from carbon nanotube field emitters. Appl. Phys. Lett. 75, 873–875 (1999)
[59] Saito, Y.: Field emission patterns from single-walled carbon nanotubes. Jpn. J. Appl. Phys. 36, L1340–L1342 (1997)
[60] Dean, K.A., Chalamala, B.R.: Field emission microscopy of carbon nanotube caps. J. Appl. Phys. 85, 3832–3836 (1999)
[61] Lovall, D., Buss, M., Graugnard, E., Anders, R.P., Reifenberger, R.: Electron emission and structural characterization of a rope of single walled carbon nanotubes. Phys. Rev. B 61, 5683–5691 (2000)

[62] Zhong, Z., Wang, D., Cui, Y., Bockrath, M.W., Lieber, C.M.: Nanowire Crossbar arrays as address decoders for integrated nanosystems. Science 302, 1377–1379 (2003)

[63] Iijima, S.: Helical microtubules of graphitic carbon. Nature 354, 56–58 (1991)

[64] Pan, Z.W., Dai, Z.R., Wang, Z.L.: Nanobelts of semiconducting oxides. Science 291, 1947–1949 (2001)

[65] Gao, P.X., Ding, Y., Mai, W., Hughes, W.L., Lao, C., Wang, Z.L.: Conversion of zinc oxide nanobelts into superlattice-structured nanohelices. Science 309, 1700–1704 (2005)

[66] Dickey, E.C., Grimes, C.A., et al.: Visible photoluminescence from ruthenium-doped multiwall carbon nanotubes. Appl. Phys. Lett. 79, 4022–4024 (2001)

[67] Saito, Y., Uemura, S., Hamaguchi, K.: Cathode ray tube lighting elements with carbon nanotube field emitters. Jpn. J. Appl. Phys. 37, L346–348 (1998)

[68] Dong, L.X., Arai, F., Fukuda, T.: Electron-beam-induced deposition with carbon nanotube emitters. Appl. Phys. Lett. 81, 1919–1921 (2002)

[69] Liu, P., Dong, L.X., Arai, F., Fukuda, T.: Nanotube multi-functional nanoposition sensors. J. Nanoeng. Nanosyst. 219, 23–27 (2005)

[70] Arai, F., Liu, P., Dong, L.X., Fukuda, T.: Field emission properties of individual carbon nanotubes in nanorobotic manipulation and electron-beam-induced deposition. J. Robotics and Mechatronics 16, 597–603 (2004)

[71] Thong, J.T.L., Oon, C.H., Yeadon, M., Zhang, W.D.: Field-emission induced growth of nanowires. Appl. Phys. Lett. 81, 4823–4825 (2002)

[72] Wang, Y.G., et al.: In situ growth of nanowire on the tip of a carbon nanotube under strong electric field. Appl. Phys. Lett. 86(133103) (2005)

[73] Tay, A.B.H., Thong, J.T.L.: High-resolution nanowire atomic force microscope probe grown by a field-emission induced process. Appl. Phys. Lett. 84, 5207–5209 (2004)

[74] Tay, A.B.H., Thong, J.T.L.: Fabrication of super-sharp nanowire atomic force microscope probes using a field emission induced growth technique. Rev. Sci. Instrum. 75, 3248–3255 (2004)

[75] Oon, C.H., Thong, J.T.L.: In situ nanowire growth for electrical interconnects. Nanotechnology 15, 687–691 (2004)

[76] Purcell, S.T., Vincent, P., Journet, C., Binh, T.V.: Hot nanotubes: stable heating of individual multiwall carbon nanotubes to 2000 K induced by the field-emission current. Phys. Rev. Lett. 88(105502) (2002)

[77] Regan, B.C., Aloni, S., Ritchie, R.O., Dahmen, U., Zettl, A.: Carbon nanotubes as nanoscale mass conveyors. Nature 428, 924–927 (2004)

[78] Cumings, J., Collins, P.G., Zettl, A.: Peeling and sharpening multiwall nanotubes. Nature 406, 586 (2000)

[79] Li, J., Banhart, F.: The engingeering of hot carbon nanotubes with a focused electron beam. Nano Lett. 4, 1143–1146 (2004)

[80] Banhart, F., Li, J., Terrones, M.: Cutting single-walled carbon nanotubes with an electron beam: evidence for atom migration inside nanotubes. Small 1, 953–956 (2005)

[81] Fukuda, T., Arai, F., Dong, L.X.: Assembly of nanodevices with carbon nanotubes through nanorobotic manipulations. Proc. IEEE 91, 1803–1818 (2003)

[82] Park, J.Y., Yaish, Y., Brink, M., Rosenblatt, S., McEuen, P.L.: Electrical cutting and nicking of carbon nanotubes using an atomic force microscope. Appl. Phys. Lett. 80, 4446–4448 (2002)

[83] Rubio, A., Apell, S.P., Venema, L.C., Dekker, C.: A mechanism for cutting carbon nanotubes with a scanning tunneling microscope. Eur. Phys. J. B 17, 301–308 (2000)

[84] Wei, B.W., D'arcy-Gall, J., Ajayan, P.M., Ramanath, G.: Tailoring structure and electrical properties of carbon nanotubes using kilo-electron-volt ions. Appl. Phys. Lett. 83, 3581–3583 (2003)

[85] Suzuki, S., Kanzaki, K., Homma, Y., Fukuba, S.: Low-acceleration-voltage electron irradiation damage in single-walled carbon nanotubes. Jpn. J. Appl. Phys. 43, L1118–L1120 (2004)

[86] Yuzvinsky, T.D., Fennimore, A.M., Mickelson, W., Esquivias, C., Zettl, A.: Precision cutting of nanotubes with a low-energy electron beam. Appl. Phys. Lett. 86, 053109 (2005)

[87] Smith, B.W., Luzzi, D.E.: Electron irradiation effects in single wall carbon nanotubes. J. Appl. Phys. 90, 3509–3515 (2001)

[88] Falvo, M.R., Clary, G.J., Taylor II, R.M., Chi, V., Brooks Jr, F.P., Washburn, S., Superfine, R.: Bending and buckling of carbon nanotubes under large strain. Nature 389, 582–584 (1997)

[89] Kuzumaki, T., Mitsuda, Y.: Dynamic measurement of electrical conductivity of carbon nanotubes during mechanical deformation by nanoprobe manipulation in transmission electron microscopy. Appl. Phys. Lett. 85, 1250–1252 (2004)

[90] Nakayama, Y., Nagataki, A., Suekane, O., Cai, X., Akita, S.: Current-induced plastic deformation of double-walled carbon nanotubes. Jpn. J. Appl. Phys. 44, L720–L722 (2005)

[91] Knechtel, W.H., Düusberg, G.S., Blau, W.J., Hernandez, E., Rubio, A.: Reversible bending of carbon nanotubes using a transmission electron microscope. Appl. Phys. Lett. 73, 1961–1963 (1998)

[92] Yao, Z., Postma, H.W.C., Balents, L., Dekker, C.: Carbon nanotube intramolecular junctions. Nature 402, 273–276 (1999)

[93] Stevens, R., Nguyen, C., Meyyappan, M.: Nanomanipulation and fabrication by ion beam molding. IEEE Trans. Nanotechnol. 5, 255–257 (2006)

[94] Cumings, J., Zettl, A.: Low-friction nanoscale linear bearing realized from multiwall carbon nanotubes. Science 289, 602–604 (2000)

[95] Fennimore, A.M., Yuzvinsky, T.D., Han, W.Q., Fuhrer, M.S., Cumings, J., Zettl, A.: Rotational actuators based on carbon nanotubes. Nature 424, 408–410 (2003)

[96] Bearden, J.A., Burr, A.F.: Reevaluation of X-Ray Atomic Energy Levels. Rev. Mod.Phys. 39, 125–142 (1967)

[97] Seiler, H.: Secondary electron emission in the scanning electron microscope. J. App. Phys. 54, R1–R18 (1983)

[98] Reimer, L.: Scanning electron microscopy, 2nd edn., p. 118. Springer, Berlin (1998)

[99] Pop, E., Mann, D., Wang, Q., Goodson, K., Dai, H.: Thermal conductance of an individual single-wall carbon nanotube above room temperature. Nano Lett. 6, 96–100 (2006)

[100] Yakobson, B.I., Brabec, C.J., Bernholc, J.: Nanomechanics of carbon tubes: instabilities beyond linear response. Phys. Rev. Lett. 76, 2511–2514 (1996)

[101] Hansson, A., Paulsson, M., Stafström, S.: Effect of bending and vacancies on the conductance of carbon nanotubes. Phys. Rev. B 62, 7639–7644 (2000)

[102] Walkeajärvi, T., Lievonen, J., Ahlskog, M., Åström, J., Koshio, A., Yudasaka, M., Iijima, S.: Bending of multiwalled carbon nanotubes over gold lines. J. Appl. Phys. 98, 104301 (2005)

[103] Wang, X., Yang, H.K.: Bending stability of multiwalled carbon nanotubes. Phys. Rev. B 73, 85409 (2006)

[104] Poncharal, P., Wang, Z.L., Ugarte, D., de Heer, W.A.: Electrostatic deflections and electromechanical resonances of carbon nanotubes. Science 283, 1513–1516 (1999)

[105] Iijima, S., Brabec, C., Maiti, A., Bernholc, J.: Structural flexibility of carbon nanotubes. J. Chem. Phys. 104, 2089–2092 (1996)

[106] Lambin, P., Fonseca, A., Vigneron, J.P., Nagy, J.B., Lucas, A.A.: Structural and electronic properties of bent carbon nanotubes. Chem. Phys. Lett. 245, 85–89 (1995)

[107] Dong, L.X., Nelson, B., Fukuda, T., Arai, F., Nakajima, M.: Towards Linear Nano Servomotors with Integrated Position Sensing. In: Proc. of the IEEE Conf. on Robotics and Automation 2005 (ICRA 2005), pp. 867–872 (2005)

[108] Cumings, J., Zettl, A.: Low-Friction Nanoscale Linear Bearing Realized from Multiwall Carbon Nanotubes. Science 289, 602–604 (2000)

[109] Zheng, Q., Jiang, Q.: Multiwalled Carbon Nanotubes as Gigahertz Oscillators. Phys. Rev. Lett 88, 045503 (2002)

[110] Legoas, S.B., Coluci, V.R., Braga, S.F., Coura, P.Z., Dantas, S.O., Galvao, D.S.: Molecular-Dynamics Simulations of Carbon Nanotubes as Gigahertz Oscillators. Phys. Rev. Lett. 90, 055504 (2003)

[111] Guo, W., Guo, Y., Gao, H., Zheng, Q., Zhoug, W.: Energy Dissipation in Gigahertz Oscillators from Multiwalled Carbon Nanotubes. Phys. Rev. Lett. 91, 125501 (2003)

[112] Hansson, A., Stafstrom, S.: Intershell Conductance in Multiwall Carbon Nanotubes. Phys. Rev. B. 67, 075406 (2003)

[113] Guo, W., Zhoun, W., Dai, Y., Li, S.: Coupled Defect-Size Effects on Interlayer Friction in Multiwalled Carbon Nanotubes. Phys. Rev. B 72, 075409 (2005)

[114] Akita, S., Nakayama, Y.: Interlayer Sliding Force of Individual Multiwall Carbon Nanotubes. Jpn. J. Appl. Phys. 42, 4830–4833 (2003)

Chapter 8
Biological Cell Manipulation/Measurement/ Analysis under E-SEM

8.1 Application Fields of Nano-manipulation System under E-SEM

In this chapter, the historical development of our Environmental-Scanning Electron Microscope (E-SEM) nanomanipulation system has been presented. The working principles which enable the E-SEM system to operate under various pressure levels are explained. The design of the nanomanipulator system is presented and the derivation of its kinematics is showed and confirmed. These two systems, i.e. E-SEM and nanomanipulator, were then integrated for realizing single cells property characterizations.

8.2 Single Cell Nano-surgery System Using Nano-tools

Single cells analysis needs to be investigated through the micro-nano scale techniques based on the local environmental measurements and controls. We developed the Environmental-SEM (E-SEM) nanorobotic manipulation system to manipulate and control the local environments for biological samples in nano scale (Figure 8.1). It realized that direct observation and manipulation of water-containing biological samples under nanometer high resolution imaging.

8.3 Observation of Biological Cells by E-SEM

A model organism is a species that is extensively studied to understand particular biological phenomena, with the expectation that discoveries made in the organism model will provide insight into the workings of other organisms [1]. In particular, model organisms are widely used to explore potential causes and treatments for human disease when human experimentation would be unfeasible or unethical [2].

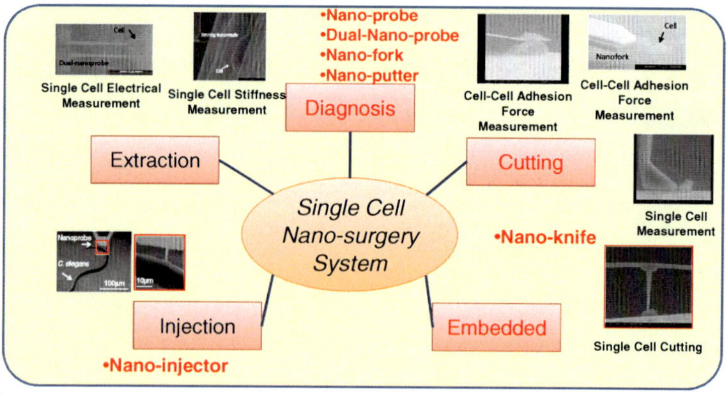

Fig. 8.1 Single Cell Nano-surgery System with various nano-tools

The first and foremost consideration in the selection of any model organisms before conducting any bio-related research is how relevance the selected model organisms to human. If the first consideration is justified, then the second consideration which needs to be addressed is how practical and easy the selected model organisms for experimental endeavors.

Yeast is one of the simplest eukaryotic organisms (organism whose cells contain a clear defining membrane-bound structure of nucleus) but many essential cellular processes are conserved between yeast and humans. There are genes in yeast and mammals that encode very similar proteins [3]. Comparison of the yeast and human genomes, reported in 1997, revealed that 30% of known genes involved in human disease have yeast orthologs (i.e. functional homologs) [4]. Furthermore, hundreds of yeast genes exhibit a link to human disease genes as reported by [5]. Because of these advantageous features, yeast has become the model organism of choice for medicine-related research. For example, studies with yeast have contributed greatly to our knowledge of the regulation of eukaryotic cell division, including the cancer-related disturbances [6]. Up to now, yeast has maintained its role as a useful model system in fundamental studies of disease processes.

8.3.1 Preparing the W303 Wild-Type Yeast Cells for E-SEM Observation

Wild type yeast cells (*W303*) were used for observations and measurements by E-SEM nanomanipulation system. The *W303* cells were cultured on YPD plates (8. 1% yeast extract, 2% peptone, 2% glucose, 2% agar) in a 37 °C incubator for 24 hours and dispersed in pure water. Several micro liters of dispersed cells were placed on an aluminum block on the cooling stage. The HV (high vacuum) mode was operated at 16.7 °C and 3.03 x 10^{-3} Pa pressure. All yeast cells appeared concave and broken under HV mode. The E-SEM mode was operated at 0.0 °C and ~600 Pa pressure (this temperature and pressure setting provides about 98% of humidity level inside the E-SEM's chamber) and acceleration voltage was set at 15 kV. Under the E-SEM mode, decreasing the pressure from ~700 Pa, water gradually evaporates and the samples can be seen underneath.

Their images are shown with HV and E-SEM modes as shown in Figure 8. 2 (a)-(b). The remaining water can be seen at the intercellular spaces of yeast cells as black contrast (Figure 8. 2 (b)). Almost all yeast cells appear sphere shaped by water-contained condition under the E-SEM operation.

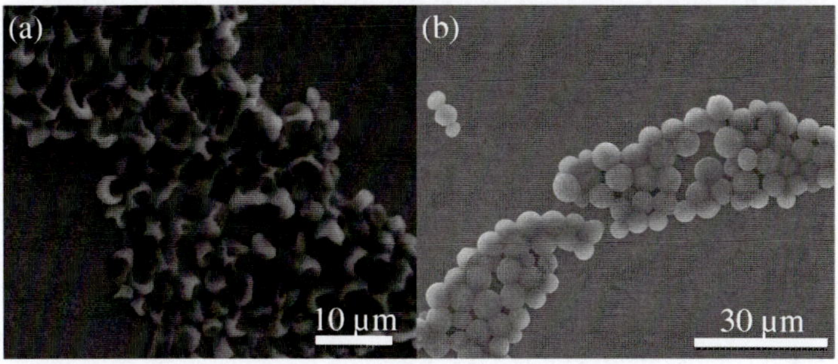

Fig. 8.2 SEM images of the W303 cells under (a) HV and (b) E-SEM modes

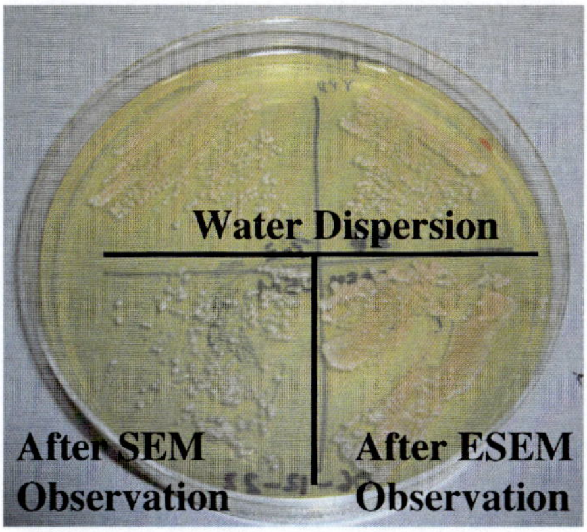

Fig. 8.3 Qualitative evaluation of W303 survivability after SEM (bottom left area) and E-SEM (bottom right area) observations as compared to the control cell growth (top area)

8.3.2 Qualitative Evaluation of Cell Survivability of W303 Cells under E-SEM Observation

To reveal the influence of an electron beam observation under HV and E-SEM modes, yeast cells were cultured once again after observations. First, yeast cells were observed under the E-SEM mode. Then, after the observations, cells were collected from the observation stage (aluminum block) by a sterilized toothpick and inoculated them onto a fresh YPD plate, and finally grow them overnight inside an incubator at 37 °C. The above procedure was repeated for cells that observed under the HV mode. To compare the cell viability with untreated cell, water-dispersed yeast cells were inoculated on the same plate for the control purpose.

The cultured plate is shown in Figure 8. 3. This photo shows the survivability of W303 cells that growth at three different conditions, i.e. after water dispersion (control condition), after SEM observation under HV mode and after SEM observation under E-SEM mode. Here, the cell survivability analysis is based on the number of cell colonies that growth on the plate culture. The number of cell colonies under SEM observation after HV mode is smaller than the cell colonies under the other two conditions, i.e. normal water dispersion and after SEM observation under E-SEM mode. On the other hand, the number of cell colonies under the latter two conditions was almost in the same order. The result indicates that most of the cells were still alive during the SEM observation under E-SEM mode. From this experiment, we can conclude that W303 cells has high survival rate under E-SEM mode as compared to the conventional HV mode under SEM observation. Our findings here agree well with the conclusion made by [7]. They investigated the effects of two E-SEM modes, i.e. low vacuum mode and wet mode (or simply E-SEM mode) on the morphology and defect of the biological cells. Their findings showed that E-SEM modes can provide a necessary environmental condition (by providing a high humidity level) for sustaining biological cells under their native state.

8.4 Mechanical Property Characterization of Single Cell Using Nanotools

8.4.1 Nanoindentation Process

Nanoindentation has recently emerged as a powerful tool for measuring nano- and microscale mechanical properties in tissues, single cells and other biomaterials. Nanoindentation, also known as instrumented or depth-sensing indentation, involves the application of a controlled load to the surface to induce local surface deformation. Load and displacement are monitored during the loading and unloading, and properties such as hardness and reduced modulus are calculated from the loading or unloading curves using well- established equations based on elastic contact theory [8].

With a typical working force range of 1 μN to 500 mN and displacement range of 1 nm to 20 μm, this technique bridges the gap between atomic force microscopy (AFM) and macroscale mechanical testing. Nanoindentation improves upon the spatial, force, and displacement resolutions of these traditional techniques to provide a powerful tool for the characterization of tissues and other biomaterials with submicron resolution. Because of its small probe size, nanoindentation can be used to measure local material properties in small, thin, and heterogeneous samples. Continuing advancements in indentation data analysis will increase the method's utility in the characterization of biomaterials.

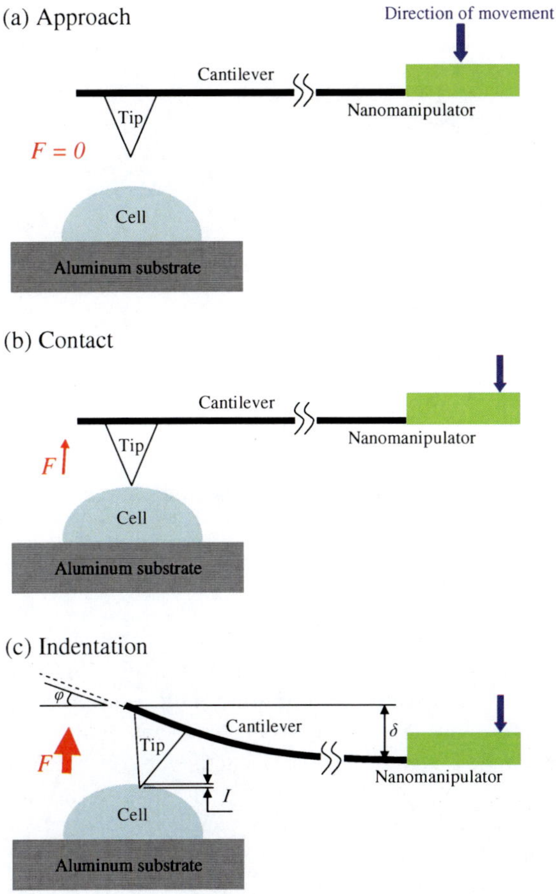

Fig. 8.4 Schematic of nanoindentation process

A schematic of a nanoindentation system is shown in Figure 8.4. An AFM cantilever with a standard or modified tip is mounted on the nanomanipulator stage inside E-SEM. Nanomanipulator is used to approach, contact and indent the cantilever's tip on a single biological cell. Force is calculated from two parameters, i.e. cantilever's deflection angle, θ and cantilever's length, L, which obtained from the SEM images. For accurate force measurement, the AFM cantilever needs to be calibrated before nanoindentation experiment. The nanoindentation experiment is observed in a real-time using E-SEM instrument and an image analysis of the nanoindentation experiment can provide a load-displacement curve as shown in Figure 8. 5. In order to obtain mechanical properties of the biological sample from the load-displacement data, specific model such as the Hertz-Sneddon model or viscoelastic model is fitted to the experimental data.

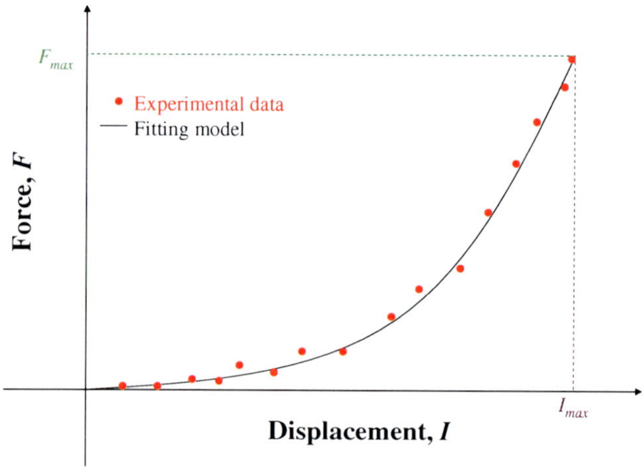

Fig. 8.5 Schematic of typical force-displacement curve

8.4.2 Force Measurement Using AFM Cantilever

The fundamental idea behind the force measurement of the AFM cantilever lies on its ability to deflect when some force is acting on it. Hooke's law made the relationship between the acting force and the deflection of the AFM cantilever becomes more appearance. Once the spring constant of the AFM cantilever is determined, the acting force on the AFM cantilever can be calculated from the famous Hooke's law, i.e. $F=k\delta$, where k is the spring constant of the AFM cantilever and δ is the deflection of the AFM cantilever.

Several techniques are used to measure the deflection of the cantilever, i.e. laser-photodiode, optical interferometry, capacitive sensing, piezoresistive AFM cantilever and visual-based. Among these techniques, we mainly used the visual-based approach to measure the deflection of the AFM cantilever as it provides the most cost-effective and no additional space is needed. When no additional space is needed, the nanomanipulator system has more working space and therefore more flexible.

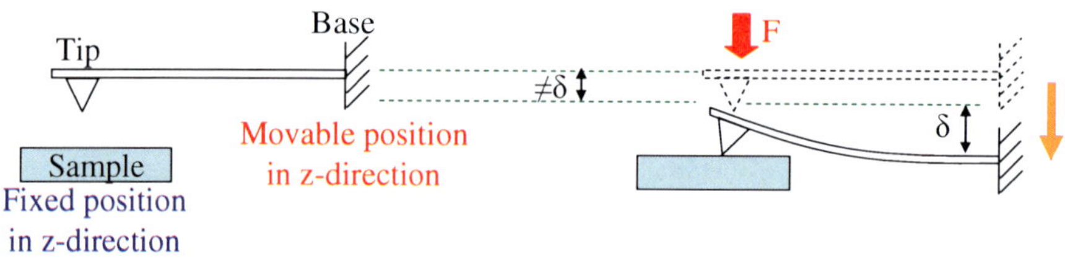

Fig. 8.6 Determination of the deflection of the AFM cantilever

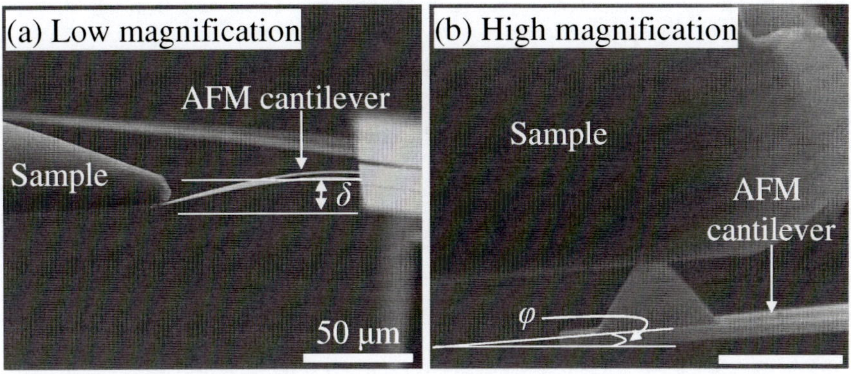

Fig. 8.7 Determination of the AFM cantilever's deflection based on (a) length deflection, δ and (b) angle deflection, φ

In our experimental setup, the AFM cantilever is not only required to navigate in |X|-|Y| axes but also in |Z| axis, therefore the actual deflection of the AFM cantilever must be measured from its base's position and not from its tip's position as shown in Figure 8. 6. This notion seems to give a drawback to the visual-based approach as the observation image need to be zoomed out in order to view the complete structure of the AFM cantilever which therefore reduces the image resolution and eventually reduces the accuracy of the force measurement as shown in Figure 8. 7 (a). In order to overcome the drawback, the standard Hooke's law needs to be modified by changing the parameter of length deflection (δ) with the parameter of angle deflection (φ). By doing so, the observation image can be maintained in its high magnification since the angle deflection is determined from the tip position of the AFM cantilever and not from the AFM cantilever's base as shown in Figure 8. 7 (b). Therefore, the whole structure of the AFM cantilever is not required to be observed. The modification of the Hooke's law is done by relating two equations of beam (cantilever) deflection, i.e. the length and slope of the beam (cantilever) during deflection.

8.4.3 Determination of the Cantilever Deflection via Angular Deflection

Figure 8.8 shows the schematic of the beam deflection. By using the Macaulay's method [9] which is also known as the double integration method, the deflection of

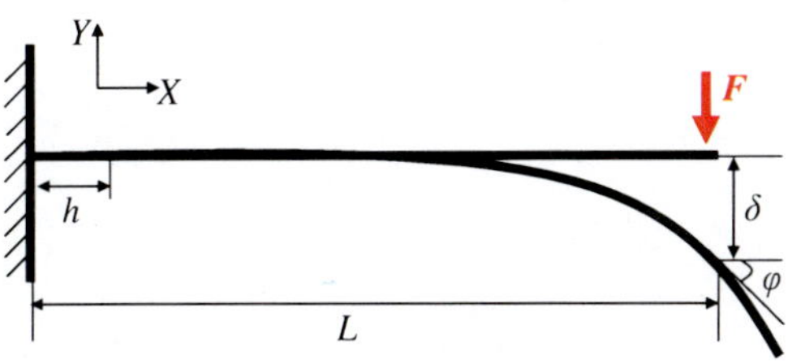

Fig. 8.8 Schematic of the beam deflection

the beam can be represented in two relationships, i.e. the length deflection, δ, and the angular deflection, φ.

The Macaulay's method consists of three steps, i.e. derivation of the bending moment, $M(x)$ that acts internally at every point in the interval between x = 0 and x = L, integration of the $M(x)$ to get the beam slope and second integration to get the elastic curve, $v(x)$. In integrating $M(x)$, two constants integration will emerge, i.e. C_1 and C_2. In order to evaluate these constants, appropriate boundary conditions of the beam will be used to determine the value for the constants. Finally, the value of $x = L$ is substituted into each equation to find the deflection and slope at the tip of the cantilever.

Equilibrium equation for the sum of moments around point h,

$$\sum M_h = -F(L-x) - M = 0. \tag{8.1}$$

Solving the moment equation for M gives,

$$M = Fx - FL. \tag{8.2}$$

The moment-curvature relationship can be expressed as,

$$EI\frac{d^2v}{dx^2} = M = Fx - FL \tag{8.3}$$

where, EI is a constant. Integrating (8.3) to derive the equation for the beam slope dv/dx,

$$EI\frac{dv}{dx} = \int (Fx - FL)dx = \frac{F}{2}x^2 - FLx + C_1. \tag{8.4}$$

Integrating (8.4) to find the equation for the deflection v,

$$EI\,v = \int\left(\frac{F}{2}x^2 - FLx + C_1\right)dx = \frac{F}{6}x^3 - \frac{FL}{2}x^2 + C_1x + C_2. \tag{8.5}$$

In order to evaluate the C_1 and C_2, two boundary conditions for the beam are needed to be determined. It is assumed that the base of the cantilever is fixed. This condition prevents the cantilever to rotate at the base (therefore the slope at the base must be zero) and also prevents the beam to either move up or down at the base (therefore the deflection at the base must be zero). These boundary conditions can be described as,

$$\begin{aligned} &x = 0, \frac{dv}{dx} = 0 \\ &x = 0, v = 0. \end{aligned} \tag{8.6}$$

Substituting these boundary conditions into (8.4) and (8.5), we find the value for C_1 and C_2 are zero. Therefore the general equations for the slope and deflection of the cantilever in Fig. are,

$$EI\frac{dv}{dx} = \int (Fx - FL)dx = \frac{F}{2}x^2 - FLx. \tag{8.7}$$

$$EI\, v = \int\left(\frac{F}{2}x^2 - FLx + C_1\right)dx = \frac{F}{6}x^3 - \frac{FL}{2}x^2. \tag{8.8}$$

Finally by substituting $x = L$ into (8.7) and (8.8), we get,

$$\frac{dv}{dx} = -\frac{FL^2}{2EI} = \varphi\ (angular\ deflection\ in\ radians) \tag{8.9}$$

$$v = -\frac{FL^3}{3EI} = \delta\ (length\ deflection). \tag{8.10}$$

The negative sign indicates the direction of the force. Now, we have the standard equations for the beam deflection under a condition as shown in Figure 8. 8. Back to our discussion regarding the visual-based approach for force measurement of the AFM cantilever, the deflection of the AFM cantilever have to be expressed in angular deflection and not in the length deflection.

By using basic mathematical manipulation from (8. 9) and (8. 10), we arrive to the final equation of the cantilever deflection in respect to the angular deflection,

$$\delta = \frac{2}{3}\varphi L \tag{8.11}$$

where, δ, φ and L are the cantilever deflection, the angular deflection in radians and the total length of the cantilever. The values of φ and L are determined from the image analysis of the SEM images. Finally, by using (8.11), the modification of the Hooke's law can be made as,

$$F = k\delta = k\left(\frac{2}{3}\varphi L\right). \tag{8.12}$$

The evaluation of (8.12) was conducted experimentally by comparing the result of δ which obtained from a direct displacement measurement from the SEM image with the value of δ obtained from (8.11). The experiment was conducted by indenting a tungsten probe on the AFM cantilever tip. By using this setup, since the base of the AFM cantilever was in fix condition, the value of δ can be obtained directly from the length displacement of the cantilever's tip. Therefore, both parameters, δ and φ, can be determined from the high magnification image. Figure 8-9 shows the result of the experiment. It can be seen that the value of δ obtained from both methods, i.e. direct displacement measurement and angular deflection, are in a very good agreement.

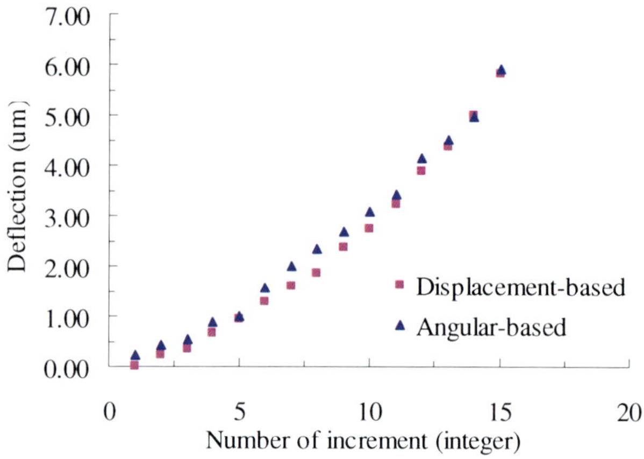

Fig. 8.9 Evaluation of the deflection of the AFM cantilever based on direct length measurement and angular-based approach

Calibration of the Spring Constant of the AFM Cantilever

By using the Hooke's law, the applied force can be determined from two basic parameters, i.e. deflection of the cantilever, δ, and the spring constant of the cantilever, k. In order to obtain more accurate force value from (8.12), the parameter of k needs also to be properly evaluated. Although, the value of k is supplied by the manufacturer, the value is based on a random calibration of the AFM cantilever from the bath unit. It is costly and time consuming for the manufacturer to determine the spring constant for each of the AFM cantilever unit. Therefore, from the user point of view, in order to accurately use the AFM cantilever for force measurement, it is wise action to re-calibrate the AFM cantilever again, so that any deviation of the value of k supplied by the manufacturer from the actual value can be corrected. Two methods are used for the calibration of the AFM cantilever, i.e. nanomanipulation approach and geometrical approach.

8.4.4 Calibration of the Spring Constant of the AFM Cantilever

Geometrical Approach

The calibration of the spring constant, k, of the AFM cantilever using the geometrical approach is based on the Parallel Beam Approximation (PBA) analysis proposed by [10]. By using the PBA analysis, the value of k can be determined if the physical dimensions and the Young's modulus of the cantilever's beam are known as described in the following equation,

$$k = \frac{Et^3 d}{2L^3}\cos\theta\left\{1+\frac{4d^3}{b^3}(3\cos\theta-2)\right\}^{-1} \tag{8.13}$$

where, *E, t, d, L, b* and θ are the Young's modulus of the cantilever, thickness of the cantilever plate, width of the cantilever arm, length of the cantilever, distance between the two cantilever arms and an opening angle of the cantilever arm. The values for *t, d, L, b* and θ are determined from the SEM images while the value for *E* is obtained from the literature. Figure 8. 10 shows the schematic of the AFM cantilever in which the parameters of *t, d, L, b* and θ are being defined.

Although there are several PBA equations proposed by others, (8. 13) has been reported to have 2 % of error as compared to other PBA equations. Table 8. 1 shows some of the available PBA equations and their error percentages.

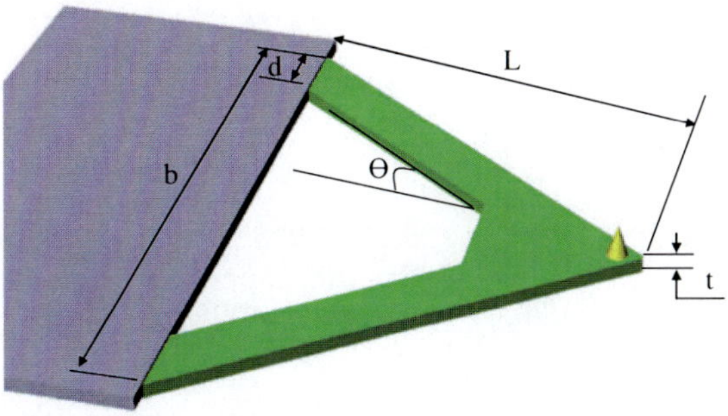

Fig. 8.10 Schematic of the AFM cantilever with its corresponding parameters

Table 8.1 Comparison of the PBA equations from several researchers

Researchers	Formula for *k*	% error
Albrecht *et al.* (1990)	$k = \frac{Et^3d}{2L^3}\left\{1+\frac{b^2}{4L^2}\right\}^{-2}$	25
Butt *et al.* (1993)	$k = \frac{Et^3d}{2L^3}$	16
Sader *et al.* (1993)	$k = \frac{Et^3d}{2L^3}\left\{1+\frac{4d^3}{b^3}\right\}^{-1}$	13
Sader (1995)	$k = \frac{Et^3d}{2L^3}\cos\theta\left\{1+\frac{4d^3}{b^3}(3\cos\theta-2)\right\}^{-1}$	2

The evaluation of (8.13) was conducted by using a commercially available AFM cantilever. The tested AFM cantilever was measured the dimensions of *t, d, L, b* and *θ*. The value of *E* was obtained from [73] which reported 180 GPa for the Si_3N_4 material. Although the AFM cantilever is normally coated with a gold (*Au*) material for conductivity, the effect of *Au* in the value of *E* for Si_3N_4 is small and can be negligible [72]. The obtained values of *t, d, L, b, θ,* and *E* were then plugged into (8.13) and the calibrated value of *k* for the tested AFM cantilever was 0.11 N/m as shown as a straight line in Figure 8. 11.

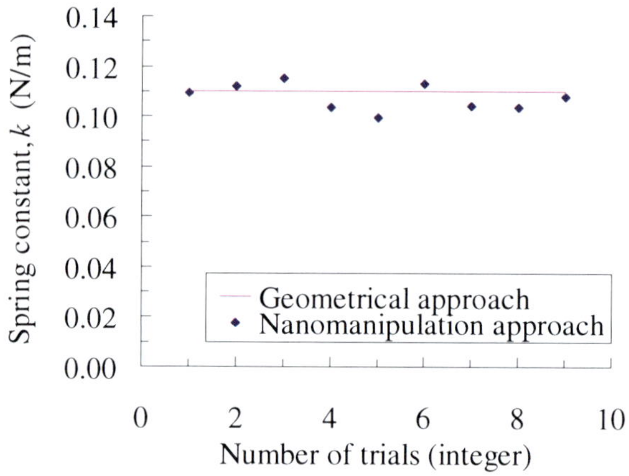

Fig. 8.11 Calibration results of the spring constant of the tested AFM cantilever using geometrical approach (straight line) and nanomanipulation approach (dotted line)

Nanomanipulation Approach

The spring constant of the AFM cantilever can also be calibrated by using the nanomanipulation approach as shown in Figure 8. 12.

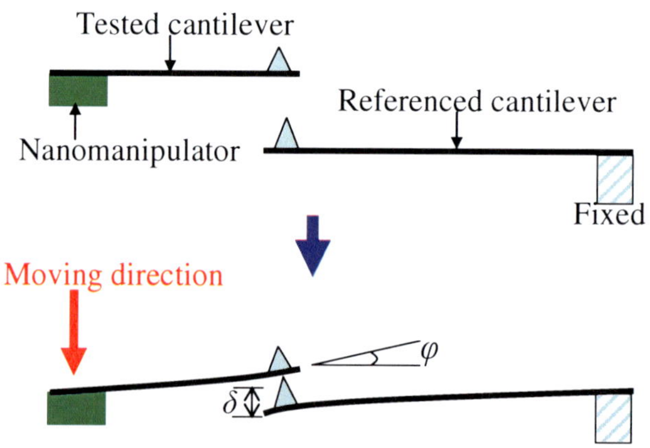

Fig. 8.12 The schematic of the nanomanipulation approach for the calibration of the AFM cantilever

In this approach, the tested cantilever (which its spring constant is unknown) is indenting the referenced cantilever (which its spring constant is known). The standard Hooke's law is used for the referenced cantilever (as its base is not moving) while the modified Hooke's law is used for the tested cantilever as,

$$F_{referenced} = k_{referenced}\delta \tag{8.14}$$

$$F_{tested} = k_{unknown}\left(\frac{2}{3}\varphi L\right). \tag{8.15}$$

Under equilibrium condition, the force acting on both cantilevers must be in the same value. Therefore, by solving (8. 14) and (8. 15) for $k_{unknown}$, the solution for the unknown spring constant for the tested cantilever is,

$$k_{unknown} = \frac{1.5k_{referenced}\delta}{\varphi L} \tag{8.16}$$

The spring constant of the tested cantilever using the nanomanipulation approach was 0.109 ± 0.01 N/m as a dotted line in Figure 8. 11. It is interesting to notice from Figure 8. 11 that both calibration methods, i.e. geometrical and nanomanipulation approaches, arrived at almost the same value of the spring constant of the tested cantilever.

Nevertheless, it must be pointed out that in general any deviation of k of any tested cantilever obtained by using both approaches may be occurred. This is due to the uncertainty of the right spring constant value of the referenced cantilever to be used. If $k_{referenced}$ is too low than $k_{unknown}$, the deflection of the tested cantilever may be so little. On the other hand, if $k_{referenced}$ is too high than $k_{unknown}$, the deflection of the referenced cantilever may be so little. Therefore, the right choice of the spring constant of the referenced cantilever plays an important factor in the accuracy of the nanomanipulation approach for the calibration of the AFM cantilever.

Calibration of Buckling Nanoneedle by Nanomanipulation

The purpose of the calibration of the buckling nanoneedle is to determine its Young modulus and spring constant, i.e. E_{needle} and k_{needle}. Two methods were used in the calibration phase, i.e. nanomanipulation technique and finite element method (FEM).

By nanomanipulation technique, the nanoneedle was slowly indenting another cantilever tip (referenced cantilever). The spring constant of the referenced cantilever was provided by the product's company. It should be noted here, however, that the spring constant obtained from the manufacturer specifications might not be accurate. Possible calibration methods can be performed to determine the exact spring constant of the cantilever prior its application such as a parallel beam approximation approach [17], a scanning vibrometry method [18] and a reference piezolever approach [19]. In order to minimized the error due to the uncertainty of the spring constant's value of the referenced cantilever, a small value of the spring constant was selected, i.e. 0.02 N/m.

The indentation was carried out until the nanoneedle was begun to buckle. Then, the buckled nanoneedle was slowly retracted until the nanoneedle was returned to a straight condition as shown in Figure 8.13.

From the nanomanipulation experiment, force-buckling length curve was plotted as dotted points as shown in Figure 8.14. This experimental data was used to determine the E_{needle} and k_{needle}. The value of E_{needle} was determined from the Euler buckling equation as shown in (8.18) where ℓ_{needle} is the length of the buckling nanoneedle. Equation (8.18) was obtained by substituting $I = (wb^3)/12$ into (8.17) in the derivation:

$$E = \frac{F_n (K\ell_{needle})^2}{\pi^2 I} \tag{8.17}$$

$$E_{needle} = \frac{7.68 F_n \ell_{needle}^2}{\pi^2 wb^3} \tag{8.18}$$

where, F_n is the buckling force applied to the nanoneedle, E_{needle} is the Young's modulus of the nanoneedle, I is the second moment of area, K is the nanoneedle effective length factor, whose value depends on the conditions of the end support of the nanoneedle. The value of the length of the buckling nanoneedle was corrected with $K = 0.8$ for the structure which has one end fixed and the other end pinned [59]. I is a property of a shape that is used to predict its resistance to buckling. The buckling nanoneedle which has a rectangular cross section has the value of $I = (wb^3)/12$ where w and b are the width and height of the rectangular cross section, respectively. The values for w and b which was determined from SEM images are 190 nm and 210 nm respectively.

For calculating the applied indentation force, F_n in (8.2), Hooke's law was used as shown in (8.19) where k_{cant} and δ_{cant} are the spring constant and deflection of the referenced cantilever.

$$F_n = k_{cant}\delta_{cant}. \tag{8.19}$$

From this experiment, the value for E_{needle} of the buckling nanoneedle is 97.8 ± 2.1 GPa which obtained from (8. 18). The statistical analysis was based on the Student's *t* test (GraphPad software). The Student *t* test is normally used when the mean population of the sample was unknown and the number of tested sample was small. The confident level of 95 % was used. The value of $E_{needlee}$ is in agreement with the bulk modulus of the Silicon material which is about 100 GPa [11]. The value for k_{needle} of the buckling nanoneedle, i.e. 0.45 N/m, was obtained from the second slope of the linear portion of the force-buckling length which occurring just after the nanoneedle started to buckle as shown in Figure 8-37. The value of k_{needle} is reasonable as it must be higher than the spring constant of the referenced cantilever (0.02 N/m).

The calibration was re-done again after the usage of more than 20 times of indentation and the calibration result shows a good similarity of force-buckling length curve as initially obtained. The buckling nanoneedle also strongly maintains its straight shape after its usage.

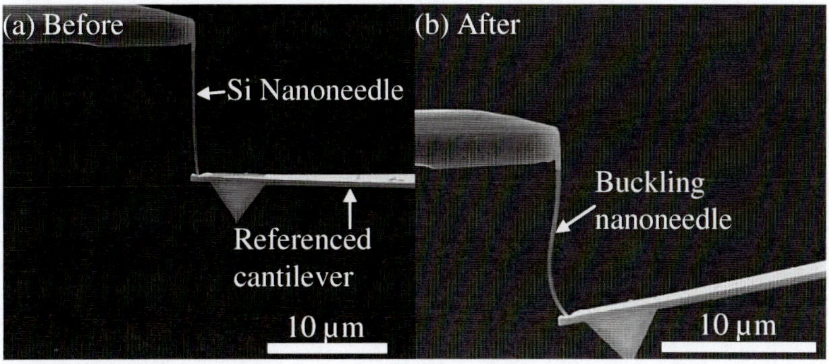

Fig. 8.13 Calibration of the buckling nanoneedle (a) before buckling and (b) after buckling by using nanomanipulation technique

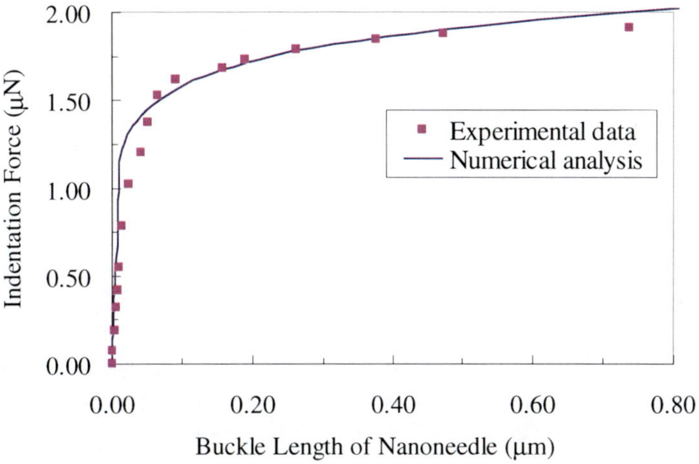

Fig. 8.14 Force-buckle length of buckling nanoneedle obtained from nanomanipulation experiment and numerical analysis

8.4.5 Modeling of Single Cell Stiffness

Hertz-Sneddon Model for Cell Stiffness Estimation

The nanoindentation experiment usually provides a force-displacement data of the sample as shown in Figure 8. 5. Useful information of the sample mechanical properties can be extracted from the force-displacement curve by fitting the curve with some specific mechanical models.

For time-independent mechanical properties such as Young's modulus, Hertz-Sneddon model has been widely used by many researchers. For time-dependent mechanical property such as viscoelasticity, the Hertz-Sneddon model has been used as the foundation of the development of the viscoelastic equations.

Brief Definition of the Hertz-Sneddon Model

Figure 8.15 shows the schematic of the indenter tip-sample contact for three different indenter tips, i.e. cylindrical, conical and spherical. The parameters of *I, h, r,* θ and *R* are indentation depth, sample height, radius of the cylindrical tip and radius of the spherical tip, respectively. The Hertz model is used for the indentation based on a spherical tip while the Sneddon model is used for the indentation based on the cylindrical and conical tips. Therefore, the selection between Hertz and Sneddon models depends on the shape of the indenter tip. In 1882, Hertz published a classic paper which aroused considerable interest [12]. The load-displacement relation based on spherical tip from the Hertzian analysis is,

$$F_{spherical} = \frac{4}{3}\frac{E}{(1-v^2)}R^{1/2}I^{3/2} \tag{8.20}$$

where, v is the Poisson's ratio of the sample. In 1965, Sneddon published a paper on indentation of linear half spaces by rigid punches of arbitrary profile [32]. Although, in his paper, he showed the load-displacement relationship of several tip shapes, only the cylindrical and conical shapes are applied in our case. The load-displacement relations based on cylindrical and conical tips are,

$$F_{cylindrical} = 2\frac{E}{(1-v^2)}aI \tag{8.21}$$

$$F_{cone} = \frac{2}{\pi}\tan\theta\frac{E}{(1-v^2)}I^2. \tag{8.22}$$

The indenter is described by an arbitrary function, $z=f(\rho)$ which is rotated about z axis so as to produce a solid of revolution. Nevertheless, Oliver and Phar have shown that (8.20), (8.21), (8.22) are also valid for non-revolute indenter as well such as pyramidal tip [13].

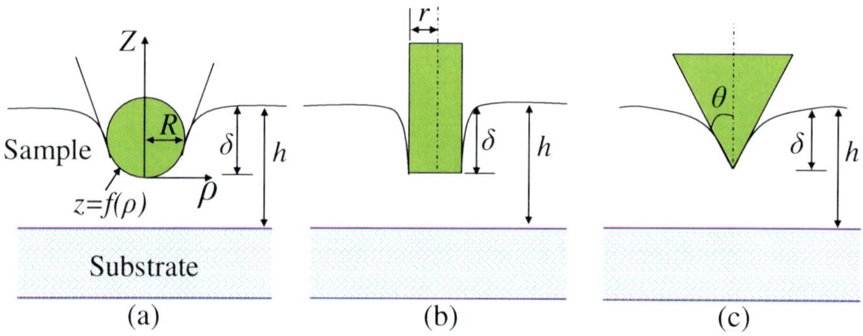

Fig. 8.15 The schematics of the tip-sample contact for (a) spherical, (b) cylindrical and (c) conical tips

Validity of the Hertz-Sneddon Model for Biological Applications

The application of the Hertz model to biological cells requires further justification. The justification can be weighted by analyzing the assumptions made in the Hertz model to our system. In general, the Hertz model makes at least three assumptions for its feasibility [14], i.e. the sample as an isotropic and linear elastic solid occupying an infinitely extending half space, the indenter is not deformable, and there is no additional interaction force such as van der Waals, capillary or electrostatic force between the indenter and the sample. By analyzing our sample condition and the assumptions made in the Hertz model, we could analyze the feasibility of the Hertz model in our sample.

In order to establish a good contact between an indenter and a cell, the W303 cells were adhered on the silicon substrate before the indentation experiment by evaporating water around the cells. The water evaporation was done by setting the chamber pressure and temperature of the sample to 600 Pa and 0 °C. This setting provides a high humidity environment (up to 95 %) to sustain the condition of the cells. The W303 cells have strong cell wall that protects the membrane layer, cytoplasm and organelles of the cells [15]. Due to this rigid cell wall, the W303 cells can be modeled as an isotropic or homogeneous property. Several studies also reported that the cell wall of the yeast cells is fully elastic [15, 16].

By considering the above points on the W303 cells, i.e. adherent cells, isotropic and homogeneous cells, and elastic cells, we believe that the application of the Hertz model in our analysis is justifiable. Further evaluation of the feasibility of the Hertz model for biological cells can also be done from numerical analysis of cell modeling which would be interesting and useful.

Determination of Cell Stiffness and Operation of Cell Surgery Based on Nanomanipulation

We propose several approaches for achieving two biomechanic tasks, i.e. determining the stiffness of a single cell and performing single cell surgery as shown in Figure 8. 16.

In the cell stiffness measurement, the measurement of cell stiffness is based on the deformation of the cell from an indentation by a $W_{0.09}$ nanoneedle. By detecting the applied compression force and the deformation of the cell, the stiffness of the cell can be measured (Figure 8. 16 (a)). In order to fabricate the hard nanoneedle, strong material which is hard to bend is needed. Here, Tungsten (W) material was used. To prevent excessive indentation force on the cell by the hard nanoneedle, this process has to be performed by avoiding any sudden pressure which can affect on the activities of the cell, e.g. mechanotransduction effect. The other preventive step is to use a lower cantilever spring constant. We used 0.09 N/m cantilever spring constants.

The hard nanoneedle can also be used for single cell surgery. In theory, all hard nanoneedles must be able to penetrate the cell, however, in practice, for the yeast cells, only some nanoneedles (Si-Ti and W_2 nanoneedles) can penetrate the cells while others (e.g. $W_{0.09}$ nanoneedle) are only limited to cell stiffness measurement only. This is because to penetrate the cell, more compression force is needed. This excessive force may exceed the force limits of the nanoneedle and cause the failure of the nanoneedle. This so-called "cell surgery" operation is shown schematically in Figure 8. 16 (b).

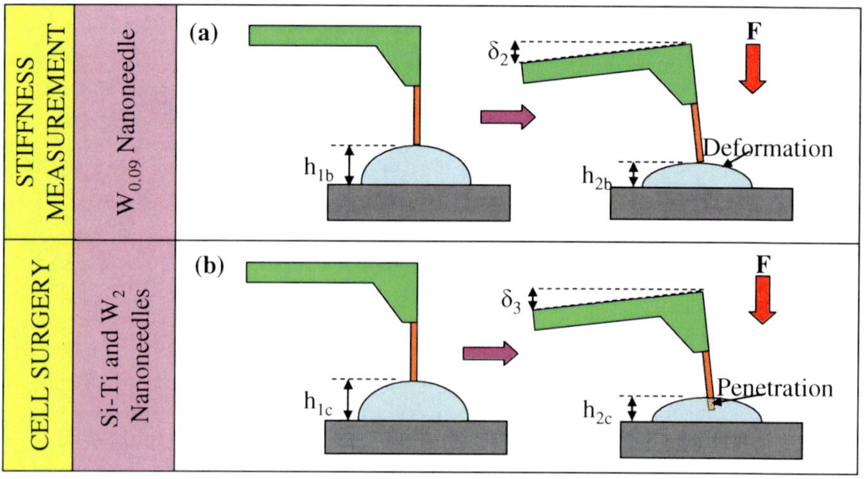

Fig. 8.16 The schematics diagrams for the nanoneedles indentation experiments indicate local single cell stiffness measurement using hard nanoneedle which is based on Tungsten deposition on the cantilever (spring constant = 0.09 N/m) ($W_{0.09}$) as shown in (a). Single cell surgery is shown in (b) by using hard Si-Ti and W_2 nanoneedles.

Estimation of Cell Stiffness Using Hard Nanoneedle

The applied force, $F_{cylindrical}$ in (8. 18), was calculated by using (8. 12), where k, φ and L are the spring constant, the displacement angle in radians and the length of the cantilever, respectively. Values of φ, L and I were determined from direct measurements of E-SEM images using image analysis software, while k was obtained from the manufacturer. It should be noted here, however, that the spring constant obtained from the manufacturer specifications might not be accurate. Possible calibration methods can be performed to determine the exact spring constant of the cantilever prior the application of the cantilever on the cell stiffness measurement such as a parallel beam approximation approach [17], a scanning vibrometry method [18] and a reference piezolever approach [19]. Equation (8. 12) was derived from Hooke's law, i.e. $F=k\delta$, where δ is the displacement of the cantilever which was obtained by using $\delta = \varphi(8.\ 2/3)L$. The final equation of the Young modulus of the cell obtained using a hard nanoneedle is expressed in (8. 23).

$$E_{cell} = \frac{(2.930\times10^{6})F_{cylindrical}}{\delta} \tag{8.23}$$

Determination of Cell Stiffness Using Buckling Nanoneedle

Under the nanomanipulation technique, the nanoneedle and the cell can be modeled as two springs in series as shown in Figure 8.17. In order to model the nanoneedle as a spring, the nanoneedle must be able to buckle linearly. The results from the calibration analyses confirmed the linearity of the buckling phase of the nanoneedle during the indentation as can be seen from the slant slope of Figure 8.37.

The spring constant of a single cell, k_{cell} can be calculated from the relationship of two springs in series as described in (8. 24),

$$k_{cell} = k_{needle}\left(\frac{\Delta_{total} - \Delta_{cell}}{\Delta_{cell}}\right) \tag{8.24}$$

where k_{needle} is the spring constant of the buckling nanoneedle, Δ_{total} is the total displacement of the nanoneedle and the cell, and Δ_{cell} is the deformation of the cell.

The Young's modulus of the cell, E_{cell} is obtained from the Hertz-Sneddon mechanics model as described in (8. 28),

$$F = \frac{2}{\pi}\tan\alpha\frac{E}{(1-v^2)}\Delta_{cell}{}^2 \tag{8.25}$$

where α is half opening angle of conical tip and v is the Poisson's ratio ((0.5 for soft biological materials) of the elastic half space (cell's surface). The Hertz-Sneddon models predict that the load depends on the indentation according to a power law related to the tip geometry. In order to choose the correct tip geometry an equation of the form $F = aI^b$ was fitted to force versus indentation (I) curves using commercial fitting software (DataFit), where the exponent b depends on the tip shape. For buckling nanoneedle experiments, we obtained a value of b close to 2, characteristic of the conical tip as formulated by (8. 5). The equation is derived from the classic Hertz's mechanics model for linear elastic material [32]. Equation (8. 28) is based on the assumption that the indenter has a revolute shape whereas for the buckling nanoneedle, it has a rectangular which is not the true body revolution. However, it has been shown that the error by using (8. 28) for non body of revolution is very small as was explained by [13].

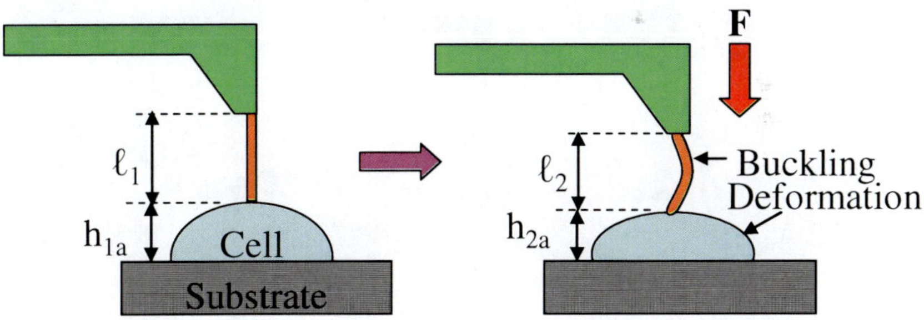

Fig. 8.17 The schematics diagrams for the nanoneedles indentation's experiments indicate local single cell stiffness measurement using buckling nanoneedle

The calculation of an applied force, F in (8. 28), was based on Hooke's law as shown in (8. 26),

$$F = k_{cell}\Delta_{cell}. \tag{8.26}$$

The final equation of the Young modulus of the cell obtained using a buckling nanoneedle is expressed in (8. 27).

$$E_{cell} = \frac{(3.237)F_{cone}}{\delta^2} \tag{8.27}$$

8.4.6 Fabrication of Nanoprobe

Three kinds of nanoneedles were fabricated. Figure 8.26 shows the general fabrication processes for each of the nanoneedles. The standard Platinum coated tetrahedral cantilever tips having a spring constant of 2 N/m were used in the fabrication of Si-Ti and W_2 nanoneedles. Whereas for the $W_{0.09}$ nanoneedle, the standard sharp pyramidal cantilever tip which has a 0.09 N/m spring constant was used.

The first type of hard nanoneedle ($W_{0.09}$) was fabricated by first flattening the apex of the sharp pyramidal cantilever tip by using FIB etching. Then W material was deposited on the flat surface by using a small area of FIB deposition, i.e. 100 nm2 as shown in Figure 8. 18 (a). The ion dose parameter for deposition is 2.672×10^{20}. The time for the deposition is about 13 min. The typical diameter and the length of the W0.09 nanoneedles are about 250 nm and 3 μm respectively.

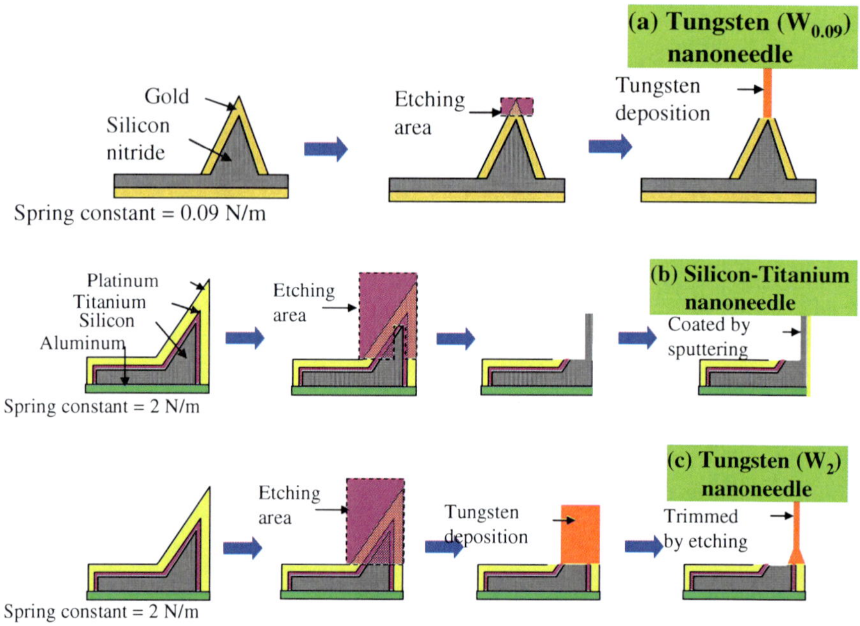

Fig. 8.18 Fabrication procedures of the nanoneedles: (a) Tungsten ($W_{0.09}$) nanonedle, (b) Silicon-Titanium (Si-Ti) nanoneedle and (c) Tungsten (W_2) nanoneedle, respectively

The second type of hard nanoneedle, i.e. Silicon-Titanium (Si-Ti) nanoneedle, was fabricated by coating the former Si nanoneedle with Ti material by sputtering as shown in Figure 8. 18 (b). The time for the sputtering operation is about 3 min to obtain a coating of approximately 20 nm of Ti material. The coated area of the Ti material is only on the one side of the nanoneedle facing the Ti source during the sputtering operation. This coating increases the strength of the basic Si nanoneedle from easily buckling to unbuckling. The typical diameter and length of the Si-Ti nanoneedles are about 185-200 nm and 4 μm respectively.

Finally, the third type of hard nanoneedle (W_2), was fabricated by first flattening the apex of the sharp tetrahedral cantilever tip by using FIB etching, then followed by W deposition of large area of FIB deposition, i.e. 160,000 nm^2, and finally trimmed to produce nanoneedle structure by using FIB etching as shown in Figure 8. 18 (c). The diameter and length of the W_2 nanoneedle are around 150-170 nm and 3 μm. The actual images of each nanoneedle are shown in the Figure 8. 19. All of the fabricated nanoneedles have flat ended tips.

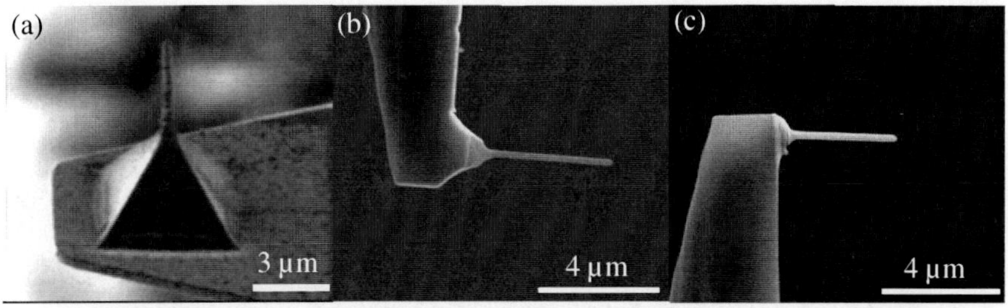

Fig. 8.19 SEM images of (a) $W_{0.09}$, (b) Si-Ti, and (c) W_2 nanoneedles

The buckling nanoneedle was fabricated by using only the FIB etching of a commercially available AFM cantilever. The process is almost same with the fabrication procedure as shown in Figure 8. 18 (b). The etching process consists of two steps, i.e. side and front etchings. The side etching of a cantilever tip produced a very thin layer, i.e. 150-200 nm thick, while the front etching step trimmed the width of the thin layer to make it a needle-like structure. The cross section of the nanoneedle has a rectangle shape. The diameter and length of the nanoneedle are about 190-210 nm and 10 μm respectively. The quantitative analysis of the detailed material of the nanoneedle was done by using an energy dispersive X-ray analysis (EDS) method. From the EDS analysis, the main element of the nanoneedle is Silicon (83.03 %) while the rest is Oxygen.

The material characterization of the nanoneedle was performed in order to have an expected Young's modulus from the material property. This expected Young's modulus was then used to compare the Young's modulus of the nanoneedle that determined from the experimental and numerical calibrations. The calibrations of the nanoneedle were done experimentally and numerically.

8.4.7 Measurement of Single Cell Stiffness by E-SEM Nanomanipulation

Effect of Cell Sizes, Environmental Conditions and Cell Growth Phases on the Strength of the Single Cells

The conventional approach to characterize cellular biology is called biochemistry. This developed science is used for studying physiological aspects, mainly genetics, by characterizing protein and other biomaterials. Since single cells are difficult to study, a collection of cells are used for characterizing cellular physiology and inturn used to describe behavior of single cell [20]. However, in addition to this advance understanding of cellular genetics, information about mechanical properties of cells is also needed. The molecular structure of the cell-wall is only partially understood, and its mechanical properties are an area of "near-total darkness" [21]. Moreover, the approximation of single cell behavior from a group used in conventional approach also requires further justification whether it can be applied to all cell types [22, 23]. This knowledge of cell mechanics could be valuable in future for biomedical applications, for example, variations in cell mechanics of healthy and unhealthy cells can be linked to a specific disease. In this paper, we show three kinds of cell mechanic characterization of single yeast cell based on cell size, environmental conditions and growth phases.

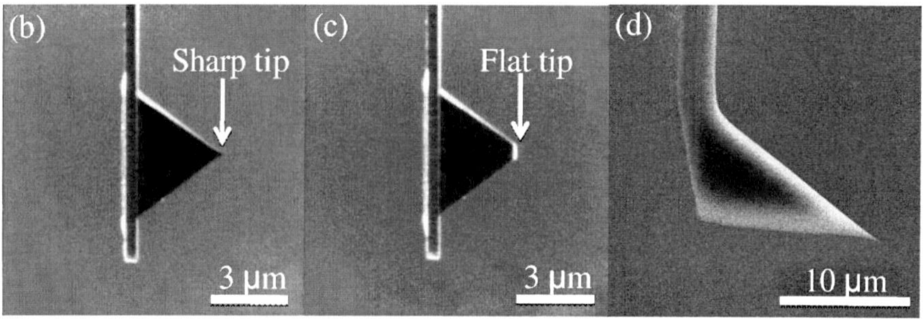

Fig. 8.20 Schematic diagram of (a) an indentation experiment by using (b) sharp and (c) flat AFM pyramidal cantilever tips under the E-SEM mode. For the HV mode, (d) a tetrahedral cantilever tip was used.

Numerous methods such as micropipette aspiration, osmotic swelling/shrinking, cell poking, and cell compression [24] have enabled the response of the whole *Saccharomyces cerevisiae* cells to applied forces to be measured [25], however the mechanical properties of different cell sizes remain unknown. Recent advances in atomic force microscopy (AFM) have shown extensive research progress in the biomechanics of the living cells [16, 26, 27], however the effect of environmental conditions on these properties are still unknown. The mechanical characterization of yeast cells at log and saturation phases have been reported [25, 28], but detailed mechanics at early, mid and late log have not been investigated.

We proposed to measure the single cell property by our constructed E-SEM-nanomanipulator system. Unlike the conventional SEM, E-SEM enables soft, moist, and/or electrically insulating materials to be imaged without prior specimen preparation.

The first advantage is that AFM system is difficult to provide a real-time sample observation and the image is mainly constructed after the scanning of the AFM tip on the sample surface is finished. Therefore, it is difficult, if not impossible, to directly observe the response of the object during the sample manipulation. On the other hand, E-SEM system provides real-time sample observation which can be obtained before, during and after the manipulation. This real-time observation capability of the E-SEM system has many advantages such as in analyzing the dynamic response of the sample and sample selection/sorting for certain properties like cell size, become much easier. The mechanical characterization and analysis capabilities which can be done on the sample can also be enhanced.

The second advantage is related to the manipulation aspect. The AFM system can only provide two-dimensional (8. 2D) manipulation, due to the difficulty of the real-time observation. Although 2D manipulation can measure the stiffness property of the sample, it lacks in the flexibility of manipulation thus limiting the area of sample which can be analyzed. Unlike AFM system, E-SEM-nanomanipulation system can provide both 2D and 3D manipulations on the sample, thus increasing the flexibility of the measurement such as the stiffness characterizations in different area of the sample.

The characterization of mechanical properties of the cell requires no interference from other external pressures or forces on the cell surface. These external forces such as the osmotic pressure may prevent investigation of the true mechanical properties of cell experimentally. E-SEM system can provide environmental conditions for the cells in their native state without any effect of the osmotic pressure. Cells can be sustained by releasing low pressure of H_2O molecule gas inside the E-SEM chamber to provide high relative humidity (up to 100%). Therefore, the stiffness data obtained from this experimental setting shows better representation of the mechanical properties of the cell. Our objective is to show the capability and effectiveness of the E-SEM system for reliable mechanical stiffness measurement of biological samples. In the present study we discuss an in-situ characterization of the mechanical properties of the W303 yeast cells using E-SEM-nanomanipulator under three different cases, i.e. cell sizes, environmental conditions and cellular growth phases.

Determination of the Penetration Force and the Cell Stiffness

The indentation experiments were conducted by using the integrated E-SEM-nanomanipulation system. The cells were fixed on the cooling stage in perpendicular direction with an AFM cantilever (0.02 or 0.7 N/m spring constant). The AFM cantilever is controlled by the nanomanipulator and used to apply the compression force on each cell.

The penetration of a single cell by the AFM cantilever in this experiment was based on a direct observation of the cell when it was bursted by the AFM cantilever. Due to the shape and size of the indenters, i.e. pyramidal or tetrahedral tips (sharp or flat), the cell bursted phenomenon cannot be avoided as soon as the cell wall is penetrated [29].

The E-SEM can provide real-time observation, i.e. a cell being approached, touched, indented, and finally penetrated/bursted by the AFM cantilever can be seen clearly. Therefore, as far as this paper is concerned, the penetration of a single cell by the AFM cantilever was based on the occurrence of the cell bursting via real-time observation.

Force (*F*)-cantilever indentation (*I*) curves were determined from the experiment. A typical *F-I* curve using a sharp cantilever tip shows a response of a cell before and after the penetration. The value of *F* was calculated by using (8. 12), where *k*, φ and *L* are the spring constant, the displacement angle in radians and the length of the cantilever, respectively. Values of φ, *L* and *I* were determined from direct measurements of E-SEM images, while *k* was obtained from the manufacturer.

The Hertz-Sneddon models which based on the shape of the tips, i.e. spherical, cylindrical and conical were used to estimate the Young's modulus of the cells as expressed in (8.17)-(8.19). Parameters of *E*, v, θ, *R* and *a* are the Young's modulus, the Poisson's ratio (v = 0.5 for soft biological materials [90]) of the elastic half space (cell's surface), the half opening angle of a conical tip, the radius of curvature of a spherical tip and the radius of a cylindrical tip, respectively. Values for θ, *R* and *a* were obtained from E-SEM images.

The models predict that the load depends on the indentation according to a power law related to the tip geometry [90]. In order to choose the correct tip geometry, an equation of the form $F=mI^b$ was fitted to force versus indentation curves using commercial fitting software (DataFit), where the exponent *b* depends on the tip shape.

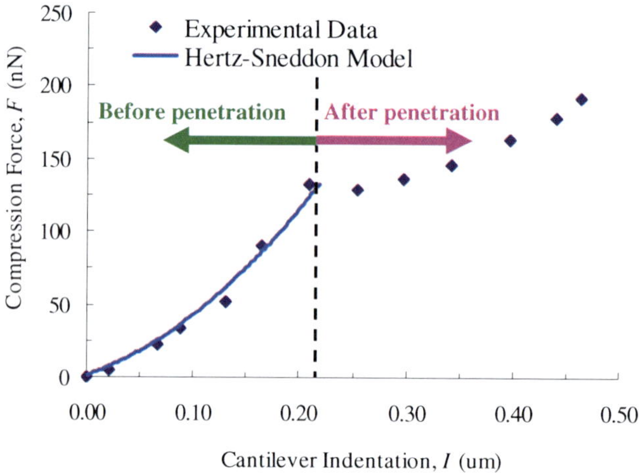

Fig. 8.21 Typical force – indentation (F-I) curve under E-SEM mode using a sharp tip. The curve shows points before penetration (green arrow) and after penetration (pink arrow) of the cell. The blue line shows the best fit of the Hertz-Sneddon model of the experimental data until the penetration point (dotted line).

For the sharp pyramidal tip under E-SEM mode, we obtained a value of *b* close to 3/2, characteristic of the spherical tip. The curves were then fitted by using the spherical model. Interestingly, under HV mode, the value of *b* close to 2 was obtained. Therefore, the canonical model was applied under HV mode. Finally, for the flat pyramidal tip, the *b* value close to 1 was obtained which is for the cylindrical model. The fitting model for the sharp tip using the spherical tip was reported by Lanero et al. [30] while Touhami et al. [31] used the conical tip as their fitting equation. Obataya et al. [29] used the cylindrical equation to model their flat nanoneedles.

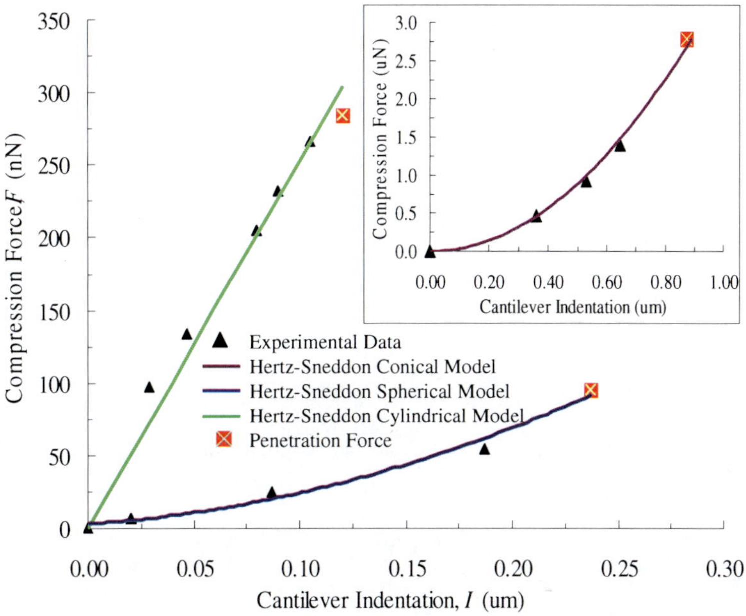

Fig. 8.22 Typical force – indentation (F-I) curves under E-SEM mode using spherical and cylindrical models for sharp and flat tips, respectively. Inset shows the F-I curve under HV mode using conical model.

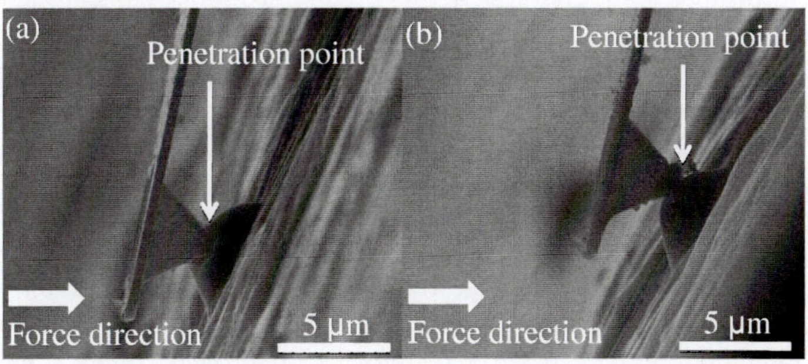

Fig. 8.23 Penetrations of yeast cells by different cantilever tips, i.e. (a) sharp cantilever tip and (b) flat cantilever tip

The Hertz-Sneddon models were derived from the classic Hertz's mechanics model for linear elastic material [32]. For flat cylindrical punch relating to an applied force, F, with an indentation depth, I, yields the following expression for the initial unloading (load is released after indentation) response, $dF/dI = cA^{1/2}(8.1\text{-}v^2)/E$, where c is coefficient ($c=2/(\pi)^{1/2}$) and A is the contact area. The equation was based on the flat-ended punch, but holds true for any punch that can be described by a smooth solid revolution (spherical, conical, elliptical, etc.) [13]. Berkovich and Vickers indenters (8. 3 and 4 sided pyramidal cones, respectively) commonly applied in instrumented indentation techniques cannot be described as bodies of revolution. However, it has been found experimentally and by means of finite element simulations that the deviation from the above equation of pyramidal and other geometrical shapes is negligible [13, 33, 34]. The constant c=1.142 for the pyramidal indenter and c=1.167 for tetrahedral indenter differ little from $c=2/(\pi)^{1/2}$=1.1284 from the above equation. In other words, the derived equation for the sharp conical tip can be used without large error, even the sharp indenter is not a true body of revolution.

The final equations of the Young's modulus based on the Hertz-Sneddon models, and the parameter's values for three different tip shapes, i.e. conical (HV mode), spherical (E-SEM mode) and cylindrical (E-SEM mode and HV mode), are expressed in (8. 28)-(8. 30).

$$E_{cell} = (6.68)\frac{F_{cone}}{I^2} \tag{8.28}$$

$$E_{cell} = (3.98\times10^3)\frac{F_{spherical}}{I^{3/2}} \tag{8.29}$$

$$\begin{cases} E_{cell} = (1.19\times10^6)\dfrac{F_{cylindrical}}{I} & \text{(ESEM mode)} \\ E_{cell} = (1.83\times10^6)\dfrac{F_{cylindrical}}{I} & \text{(HV mode)} \end{cases} \tag{8.30}$$

Figure 8. 22 shows typical F-I curves using the spherical and cylindrical models under E-SEM mode while the inset shows a curve using the conical model under HV mode as obtained in our experiments. The curves are shown until the bursting point of the cells which is the maximum force which the cell can sustain before the cell is bursted. After the bursting point, the force will drop for a while due to the reversed force and will increase again when the tip traverses inside the cell or touch the bottom of the cell wall.

The curve leading up to the failure is expected to give a combined elastic property of indenting the whole yeast cell and the localized deformation of the surface ruled by the properties of the cell-wall surface. Therefore, the reported E values in this paper indicate the maximum strength of the cell before bursting from a local indentation point. The E values must be higher than the local elastic cell property as the strength of the cell is dominated by the whole cell stiffness [85]. Nevertheless, examining the cell strength based on the Young's modulus parameter per se is not adequate since its value may remains constant at different situations, e.g. cell growth phases [85]. Therefore, the penetration force was used to determine the cell strength under several factors, i.e. cell sizes, environmental conditions and cellular growth phases.

The Effect of the Environmental Conditions on the Strength of the Cells

The required force was measured to penetrate a single yeast cell under HV mode using sharp and flat tips for different cell sizes. Higher spring constant of the cantilever, i.e. 0.7 N/m was used for the compression experiment as lower spring constant values failed to penetrate the cell. Figure 8. 20 (c) shows the image of a tetrahedral cantilever tip used in this experiment. The penetration force to burst a cell wall increases about twenty times from E-SEM mode to high vacuum (HV) condition (cells which 5 μm diameter were used for comparison and analysis). Since this data under high vacuum condition is reported for the first time, it is difficult to compare with other reports. Nevertheless, the increase in penetration force from E-SEM to HV modes which was obtained in our experiment can be logically analyzed.

Table 8.2 Young's Modulus of the Single Cells at Different Environmental Conditions

Environmental Conditions (Pa)	Young Modulus (MPa)
ESEM mode (~ 600)	3.31 ± 0.11 (n=4)
HV mode (~ 0.00314)	26.02 ± 3.66 (n=6)

Cell wall construction is tightly controlled since polysaccharide composition, structure and thickness of the cell wall vary considerably, depending on the environmental conditions [78]. Therefore, we can infer that under extreme condition of high vacuum, yeast cell was increasing its wall strength as suggested by the increased compression force needed to penetrate its cell wall. Under high vacuum condition, cells may become too dry which eventually increases their cell hardness. These findings reveal that besides the shape of a cantilever tip, other external factors such as environmental conditions also influence the penetration force of a single cell. Like the E-SEM mode, cell size remains the internal factor that influences the penetration force under high vacuum condition. By using a sharp tip, an average of 3 μN penetration force was required for a 5 μm diameter cell. As expected, this force increased (up to 14 μN) when a flat cantilever tip was applied, for cells with an average diameter of 4 to 6 μm.

For a quantitative evaluation of the elasticity of the cell, the Hertz-Sneddon mechanics models for cylindrical and conical tips as expressed in (8. 17) and (8. 19) are used for sharp and flat tips, respectively. Average elastic properties of yeast cells under the influence of two environmental conditions are shown in Table 8. 2, where the cell stiffness increases about ten times under HV mode. The cell surface becomes rigid when it is placed in the HV mode. It is believed that in the absence of water, cell surface becomes harder than in the presence of water. This leads to a significant increase in the Young's modulus (8. 26.02 ± 3.66 MPa) obtained from (8. 20) under HV mode as compared to the E-SEM mode (8. 3.31 ± 011 MPa) as expected.

The Effect of Growth Phases on the Strength of the Cells

All cells have a unique growth phase curve during their life cycle which is normally divided into four phases; lag, log, saturation and death phases. In the present study we did not investigate the death phase. These phases can be easily identified from the optical density (OD) absorbance-time curve. Lag phase located at the initial lower horizontal line where the cells adapt themselves to growth conditions and the cells mature and divide slowly. Log phase, can be seen from an exponential line where the cells have started to divide and grow exponentially. The actual growth rate depends upon the growth conditions. Saturation phase resembles a final upper horizontal line where the growth rate slows and ceases mainly due to lack of nutrients in the medium. During the death phase, all of nutrients are exhausted and the cells die [35]. It is believed that the mechanics of the normal cells have different properties at each of the phases. The mechanical properties of *W303* yeast cells at four different growth phases (early, mid, late log and saturation) are discussed. Since each cell has a unique growth phase, we determined the growth phase of *W303* yeast cells by measuring the absorbance at 660 nm using a Spectrophotometer at one hour intervals.

The compression experiments were done from early log phase to saturation phase. Penetration force was analyzed for each phase. As expected, the force needed to penetrate a single cell increased from early log to saturation phase (8. 161 ± 25, 216 ± 15, 255 ± 21 and 408 ± 41 nN), whereas, the elastic properties of the cells seem constant for all the phases obtained from (6), i.e. 3.28 ± 0.17, 3.34 ± 0.14, 3.24 ± 0.11 and 3.38 ± 0.11 MPa . From this analysis, it can be inferred that the strength of the *W303* cell wall is increased during the growth phase since the penetration force increases at each phase, and at the same time, maintaining their elastic properties during the growth phase as verified by the constant elastic

Table 8.3 Young's Modulus of the Single Cells at Different Cell Growth phases

Cell Growth Phase	Young Modulus (MPa)
Early Log	3.28 ± 0.17 (n=8)
Mid Log	3.34 ± 0.14 (n=9)
Late Log	3.24 ± 0.11 (n=8)
Saturation	3.38 ± 0.11 (n=8)

modulus values shown in Table 8. 3. The increase in the strength of the cell wall does not affect the elastic properties of the cell. We believe that the increase in the strength of cell wall is mainly due to an increment of β-glucans molecules [94].

Stiffness Measurement of Single Cell Using Nanoprobe

Available experimental techniques to probe single cells include micropipette aspiration, optical tweezers, magnetic tweezers [36], atomic/molecular force probes [37], nanoindenters, microplate manipulators and optical stretchers [38, 39]. The functionality of nanoneedles is not limited only to the stiffness measurements but it can also be used for single cell surgery [40] which can be further applied to a novel single cell drug delivery system [41, 42] or as a delivery tool for nanoparticles [43]. Conventional drug therapy suffers from the problems of inefficacy or nonspecific effects; hence, nanosystems are being developed for targeted drug delivery [44, 45].

In order to successfully deliver materials; e.g. bioactive peptide, proteins, nucleic acids or drugs inside the cell, carriers must be able to penetrate the cell wall or cell membrane without causing death or create any mechanotransduction to the cell [46], i.e. the process of converting physical forces to biochemical signals and integrating them into cellular responses [47]. Therefore, the knowledge of biomechanics of the cell is crucial in providing *prior-estimation* of required insertion force to deliver drug material inside the cell. Without this information, the insertion process may be unsuccessful due to inadequate applied force or the cell may be seriously damaged due to the excessive applied force.

Based on this information we have developed a system called E-SEM nanomanipulator which unlike the conventional SEM, enables soft, moist, and/or electrically insulating materials to be imaged without prior specimen preparation. A low pressure (about 1333 Pa) of gas can be accommodated around the sample. When the gas is vapor, hydrated samples can be maintained in their native state [28]. The temperature of the sample can be controlled by using the cooling stage assembled inside the chamber of E-SEM, while E-SEM itself can control the pressure of the chamber. This integrated system can perform real-time local observations of biological samples and nano-scale manipulation at the same time. This is different from a nonlocal direct observations [37, 48] and nonbiological sample observations [49, 50]. In this report, we present the use of several types of nanoneedles inside E-SEM-nanomanipulator system which have several advantages, e.g. can be used in local stiffness measurements and penetration without cell bursting as it is very important in the cell surgery application.

Single Cell Local Stiffness Characterization by Hard Nanoneedles ($W_{0.09}$ Nanoneedles)

The local stiffness property, i.e. the deformation of the surface of the cell upon the applied load, on several points on mother cell and daughter cell, to the best of our knowledge, has never been reported. In order to achieve multi-point indentations on a single cell, we have used the standard indentation technique together with the image analysis to measure the cell deformation and compression force. The compression force was calculated using (8. 12) and the value of φ was obtained from the bending of the cantilever. These values (cell deformation and compression force) were then fitted in (8. 23) (Hertz-Sneddon cylindrical tip model) to obtain the local stiffness of the cell. $W_{0.09}$ nanoneedle which has small value of cantilever's spring constant (0.09 N/m) was used in this experiment as shown in Figure 8. 24.

Three random indentation points on each mother and daughter cell were performed as shown in Figure 8. 24 (a) and (b) respectively. The cells can be recognized as mother and daughter cells from a direct image observation or from the cell volume comparison. The shape of the cells was assumed to be a prolate spheroid, and the volumes were calculated by using the formula $V=(\pi/6)lw^2$, where V is the volume, l is the height, and w is the width at maximum circumference (diameter) [51]. Thomas et al. [52] reported about 38% difference in cell volume between fully separated mother and daughter cells of *Candida utilis* type of yeast cell. In our experiment, the volume difference of the mother and daughter cells is about 45% and this is reasonable since the daughter was not fully grown as it was still intact with its mother at the time of measurement.

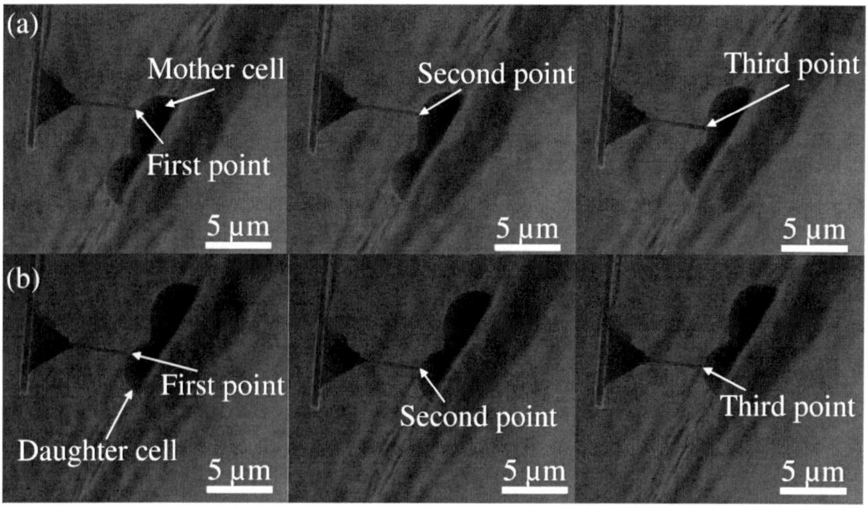

Fig. 8.24 Local stiffness measurement using the $W_{0.09}$ nanoneedle on three points of the (a) mother cell and (b) daughter cell

The results from the indentations were summarized in Table 8. 4. From the data, all the three points on the mother and daughter cells show very good agreement to each other in local stiffness profile, i.e. for mother cell (1.47 MPa, 1.43 MPa and 1.47 MPa) and for daughter cell (1.11 MPa, 1.09 MPa and 1.09 MPa). On the other hand, the data revealed the heterogeneous local stiffness property between mother and daughter cell. This can be comprehended since the daughter cell was still in its early growth phase and the insertion of new cell wall macromolecules, as required for its strength, into its existing polymer network [15] was still carried out at the time of measurement and therefore its Young's modulus exhibited slightly lower value than the mother cell.

Table 8.4 Local Stiffness Characteristics of Mother Cell and Daughter Cell using $W_{0.09}$ Nano needle

		Cell Height (μm)	Cell Diameter (μm)	Young Modulus (MPa)
Mother cell	Point 1	2.98	5.57	1.47
	Point 2			1.43
	Point 3			1.47
Daughter cell	Point 1	2.49	4.47	1.11
	Point 2			1.09
	Point 3			1.09

The local stiffness characteristic of the mother cell as reported here falls in the same stiffness range as the data obtained from the literature. Peeling et al. [16] reported 0.72 ± 0.06 MPa of the local stiffness property of the yeast cell. Lanero et al. [30] reported 1.79 ± 0.08 MPa, 1.12 ± 0.02 and 2.0 ± 0.2 MPa for coated cell, uncoated cell and coated cell-bud scar of the *Saccharomyces cerevisiae* respectively. As compared to the *Saccharomyces cerevisiae* strain, baker-type yeast cells which has been widely used for the cell stiffness analysis, we have shown, for the first time, the mechanical property of the *W303* strain, wild-type yeast cell.

Bucking Nanoneedle for Characterizing Single Cells Mechanics Inside E-SEM

The birth of nanotechnology has given new ways for characterizing a single cell mechanics. Numerous methods such as micropipette aspiration, osmotic swelling/shrinking, cell poking, and cell compression [24] have enabled the response of the whole cells, i.e. global stiffness, to an applied force to be measured, but it is unknown what is the mechanical property of the cell at a single point indentation, i.e. local stiffness. The ability to perform single point indentation on a single cell is a crucial starting step for obtaining a more complete stiffness mapping on the whole cell surface. Single point indentation is also useful in estimating the required force needed to penetrate a single cell in the injection task. Recently, Tan et al. [53] proposed a new mechanical model based on membrane theory for investigating the relationship between the injection force and the deformation of biological cells.

Realizing this, several researchers have performed a local stiffness characterization on several points on a cell surface using a nanoindentation method using either a standard atomic force microscope (AFM) cantilever [30] or a rigid nanoneedle [29]. The application of a standard AFM cantilever which has a tapered tip in various shapes, e.g. vickers, berkovich, knoop, conical, spherical and rockwell

[31], is however limited when the sample is placed in a confined area such as a trench with has a smaller width that can obstruct a wider area of the cantilever tip to pass through. Even if the sample is placed on a flat surface substrate as implemented in this work, it is difficult for the cantilever tip to reach the sample when encounters several situations, e.g. the position of the cantilever tip is slanted away from the sample or the cantilever tip is blocked by a higher obstacle or even other higher samples which placed next to a lower height targeted sample. The above problems faced by a standard AFM cantilever tip can be overcome when a rigid nanoneedle tip is used. A nanoneedle which has a high aspect ratio, e.g. 10 μm long and 100-200 nm uniform width, can easily reach to a deeper and narrow area where samples may be located. However, a too rigid nanoneedle may damage the cell if an excessive force is applied. Controlling the indentation force not to exert an excessive force is not an easy task since the cell stiffness is varied among cells. For an instance, previously we have reported that the stiffness of the W303 yeast cells differ significantly at difference cell sizes, environmental conditions and growth phases.

By acknowledging this notion, we overcome the limitation of a standard AFM cantilever and a rigid nanoneedle by using a buckling nanoneedle. A nanoneedle which can buckle under a compression force by the nanomanipulation approach, at the best of our knowledge, is a novel technique in determining the stiffness of the cells. In addition to a high aspect ratio characteristic, the buckling nanoneedle can also compensate an excessive force through buckling. In order to ensure the nanoneedle will buckle during an indentation on single cells, the spring constant of the nanoneedle, k_{needle}, and the tested single cells, k_{cell}, must be about the same range. If $k_{needle} >> k_{cell}$, the nanoneedle will behave like a rigid nanoneedle. On the other hand, if $k_{needle} << k_{cell}$, the nanoneedle will buckle too easily and will not provide an adequate indentation force to deform the cells. Since the fabrication process of the buckling nanoneedle is straightforward as it involves only a focused ion beam (FIB) etching on a tip of a commercial available AFM cantilever, it is easy to control the spring constant of the buckling nanoneedle by adjusting the diameter and/or length of the buckling nanoneedle via etching/trimming during the fabrication stage.

Furthermore, in term of the implementation of a buckling nanoneedle, at least two requirements are needed, i.e. a high image resolution and a direct observation of the buckling nanoneedle. An optical microscopy (OM) system has superior capability in a real-time/direct observation of micro-scale samples [54]. However, it has relatively low image resolution when dealing with nano-scale objects (smaller than the wavelength of visible lights) due to the diffraction limits. On the other hand, an AFM system can provide a high image resolution for nano-scale objects [55]. However, the AFM system is difficult to provide a 'true' real-time observation of a sample when fast response rate (less than a second) is concerned. The AFM image is mainly constructed after the scanning of the AFM tip on the sample surface is finished. Therefore, it is difficult, if not impossible, to directly observe the mechanical response of an object during the manipulation of the object. Most of the real-time observations by the AFM dealt with a slow response rate (in several seconds to minutes intervals) [56, 57, 58]. Another issue with the AFM observation is that the cells being scanned suffer from injuries to various degrees, depending on the cell type and viability, and the injured cells may undergo remarkable changes in morphology [59]. In order to realize the above two requirements, we have developed a system called an integrated environmental scanning electron microscope (E-SEM) – nanomanipulator.

In addition to the advantages of a high aspect ratio and the compensation of an excessive indentation force by a buckling nanoneedle, its implementation inside the E-SEM-nanomanipulator system can provide fast visual detections on apparent stiffness variations of different samples. These fast visual detections can be realized from one of two ways, i.e. from an apparent dented marks appeared on the sample surface after the indentation experiment and/or from the buckling length of the buckling nanoneedle during the indentation experiment. The capability to quickly detect an apparent variation of sample stiffness is very important as it can reduce a testing/checking/diagnosis time when involving many samples. I want to apply these techniques in a novel fast disease detection based on cell mechanics analysis.

Implementation of the Buckling Nanoneedle in the Mechanical Properties Characterizations of Single *W303* Cells

The whole stiffness response of a single cell, i.e. the deformation of the entire cell upon the applied load from a single point indentation, to the best of our knowledge, has not been previously reported. The difficulty to achieve such data is due to the stiff indenter which may penetrate or burst the cell. Our buckling nanoneedle can be used for preventing the cell penetration by its ability to buckle during indentation. The mechanism of the cell deformation is stated as the following; upon indentation of the soft nanoneedle on the cell, local deformation occurred below the tip of the nanoneedle. Further indentation has induced the whole cell to deform in addition to the local cell deformation. The ability to produce a large local point indentation could give more information regarding the mechanical property of the organelles inside the cell. The global stiffness measurement of single cells from single point indentation was performed using the buckling nanoneedle as shown in Figure 8. 25.

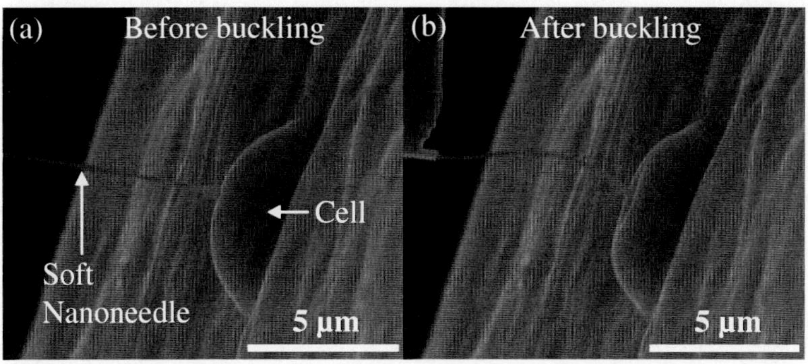

Fig. 8.25 Single cell global stiffness measurement from a single point using the buckling nanoneedle. Two images show the Si nanoneedle at (a) straight condition and (b) buckling condition.

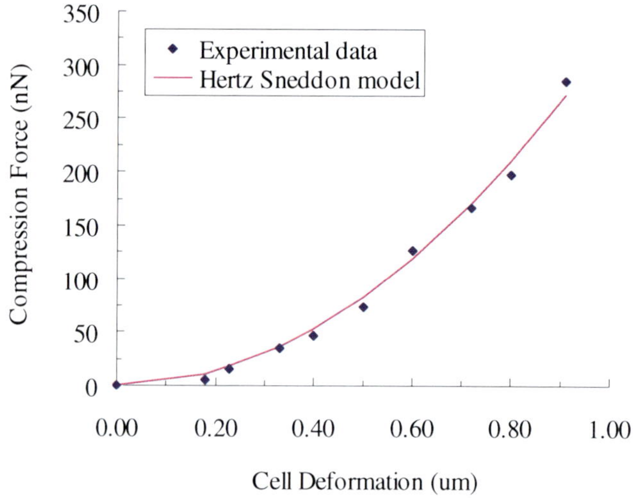

Fig. 8.26 A typical force-deformation curve of a single cell using the buckling

The measurement was performed using a standard indentation procedure. The buckling nanoneedle started to buckle after the cell deformed for about 0.5 μm. From this point, the buckling rate of the buckling nanoneedle increased with the increase of the indentation depth. Figure 8. 26 shows the force-cell deformation response of a single cell using the buckling nanoneedles. By using (8. 27) (based on Hertz-Sneddon conical tip model) as described earlier, the spring constant and Young modulus of single *W303* yeast cells were estimated as presented in TABLE 8-8. The values of k_{cell} for about the same physical parameters of two yeast cells, i.e. 0.92 ± 0.12 N/m and 0.95 ± 0.36 N/m show strong mechanical property similarity. The values which represent the whole cell spring constants are reasonable as compared to the reported local spring constant of the *Saccharomyces cerevisiae* yeast cell (0.06 ± 0.025 N/m) [79]. The values of whole E_{cell}, i.e. 3.64 MPa and 3.92 MPa are also rational since they represent the whole cell stiffness property as compared to the reported local Young modulus of the yeast cell, i.e. 0.72 ± 0.06 MPa [16]. Our method shows improved data sensitivity as compared to the other global stiffness measurement method that relies totally in a compression of a single cell between one big indenter and a substrate. Data from this method may not represent the actual global cell stiffness as extra dissipation force may be included in the measurement as reported by [85] for the global cell spring constant and Young modulus of yeast cell as 11.1 N/m and 112 MPa.

Finer Local Stiffness Characterization of Single *W303* Cells at Different Environmental Pressures

We described a global stiffness property of *W303* yeast cells [60]. The global stiffness and local stiffness properties are referring to the stiffness of a whole cell and a cell wall of the cell, respectively. The value of the Young's modulus for the whole cell body is expected to be higher than the Young's modulus for a local cell wall. It is suspected that these Young's moduli which obtained globally and locally on cells are referring to a difference aspect of the cell elasticity [90]. The former modulus is taking the influence of not only the cell wall of the cell but also the cytoplasm and all of the cell's organelles. On the other hand, the latter modulus is suspected to be influenced by the cell wall only. For an instance, Smith et al. [85] reported 100 MPa for the Young's modulus of the *Saccharomyces cerevisiae*. They performed the micromanipulation test by pressing a wide flat indenter to a cell (the diameter of the indenter was larger than the diameter of the cell) until the cell started to burst.

On the other hand, our indentation test was performed by using a buckling nanoneedle which has a small tip diameter, i.e. about 170 – 200 nm. By using this nanoneedle, several local indentations can be performed on the surface of the cell. Therefore, local stiffness property on various areas of the cell could be estimated.

The experiments were conducted at two different environmental pressure levels, i.e. higher pressure (600 Pa) and lower pressure (8. 1.04×10^{-3} Pa). Higher pressure is obtained by conducting the experiments under E-SEM mode which can provide up to 100 % humidity level, while lower pressure is obtained by conducting the experiment under HV mode (~ 0 % humidity level). The aim of these experiments was to emulate the changes of the cell stiffness due to external factors which increase its normal stiffness. Suresh et al. [34] reported a dramatically increase of the red blood cells (RBCs) stiffness after the intrusion of the bacteria, *P. falciparum* inside the cells. By using the buckling nanoneedle, this significant increase can be quickly detected by observing the dented mark on the cell surface which remained after the indentation of the buckling nanoneedle.

Here, we show that once the stiffness of the cell increases dramatically, the dented mark on the cell after the indentation is not clearly visible even observed under high image resolution. However, this dented mark can be seen clearly when the indentation was performed on the cells which have normal stiffness. *W303* cells were placed inside the lower pressure condition (HV mode). The dented mark on the cells was appeared due to a prolonged hold on the cell surface by the nanoneedle tip like a standard plastic deformation test [61]. However, for a short hold period, the deformation of the cell was fully recovered. Even for a long hold period, it is unknown whether the dented mark will be remained after some duration of time. The dented mark was appeared due to the strain has reached its yielding point. The yield strength or yield point of a material is defined in engineering and materials science as the stress at which a material begins to deform plastically.

The measurements of the finer stiffness property of single cells under their native state were done at 600 Pa and 0 ºC. On the other hand, the stiffness measurement of single cells under tough condition which can increase the stiffness of the cells was performed at high vacuum (8. 1.04×10^{-3} Pa) condition. The results from the experiment are tabulated in Table 8.5 where the values of E_{cell} and k_{cell} were calculated from (8. 28) and (8. 26), respectively. From the experimental data, the average k_{cell} and E_{cell} taking from three *W303* cells under E-SEM mode are 0.35 ± 0.03 N/m and 1.24 ± 0.11 MPa, respectively. Interestingly, when the *W303* cells were placed under HV mode, their Young's modulus were increased dramatically, i.e. 16.26 ± 0.89 MPa. Under HV mode, higher force (~ 900 nN) was needed to produce a relatively small cell deformation (~ 200 nm). The average spring constant of the cells under HV mode is 4.55 ± 0.25 N/m. The statistical analysis was based on the Student's *t* test. The confident level of 95 % was used. If this significant stiffness variation needs to be detected quickly, the qualitative approach can be used. This qualitative approach is based on a direct observation of a dented mark on the cell surface after the indentation.

Table 8.5 Finer Local Stiffness Characteristics of Single Cells at Native Condition

		Cell Physical Parameters		Cell Stiffness Characteristics	
		Height (μm)	Diameter (μm)	Spring Constant (N/m)	Young Modulus (MPa)
Cell 1	Point 1	3.689	5.513	0.35	1.25
	Point 2			0.35	1.24
	Point 3			0.34	1.23
Cell 2		3.710	5.905	0.36	1.29
Cell 3		2.197	3.777	0.34	1.20

8.5 Viscoelastic Measurement of Single Cell Using Sharp, Flat and Bucking Tips Inside E-SEM

The conventional tensile test for the measurement of Young's relaxation modulus is difficult to be performed for very small structures such as biological cells. Nanoindentation endows an effective way to measure the local mechanical properties at the microscale and nanoscale [62]. Techniques to measure time-independent elastic-plastic properties, such as Young's modulus and hardness, have been well established [13]. In addition to the measurement of the time-independent mechanical properties, nanoindentation has also being used for the measurement of some time-dependent viscoelastic properties, e.g. Young's relaxation modulus.

Most viscoelastic indentation solutions are originated from the method of functional equations developed for linear viscoelastic contact problems by Radok [63] and completed by Lee and Radok [64]. The method of functional equations is actually an extension of the Laplace transform method, as formulated by Lee [65]. Most of the works in the viscoelastic characterizations are based on a single approach. Rationally, realizing on only one single approach for any measurement or characterization tasks will not give high confidence satisfactory. Here, we show three methods for viscoelastic characterizations which based on sharp tip, flat tip and buckling tip. By having more than one test, i.e. sharp tip test, flat tip test and buckling tip test, we may use one test as properties measurements and others as a correctness check, or we could use all results from the tests to arrive at the best possible estimation of the sample properties. In this way, more accurate comparison could be made and stronger conclusion could be achieved.

In addition to provide an additional approach for the viscoelastic characterization, buckling nanoneedle has another advantage. For a quite some time, the viscoelastic solution proposed by Cheng and Cheng [66] seems difficult to be applied experimentally. They derived an elegant closed form solution of an unloading stiffness for a linear viscoelastic solid. From the slope of the unloading curve, an instantaneous Young's modulus could be determined easily. However, the solution is only valid if the unloading rate is sufficiently fast. Due to this strict requirement, the solution has only been implemented by using numerical analyses. Furthermore, an execution of the unloading task by a standard AFM cantilever will further indent the sample which make the analysis difficult, if not impossible, to be done. Here, we propose a buckling nanoneedle, which can prevent the negative force effect during the unloading phase. By incorporating the buckling nanoneedle with the nanomanipulation system inside an environmental scanning electron microscope (E-SEM), for the first time, relatively fast unloading test, i.e. 2 μm/s unloading rate, could be performed experimentally for the determination of the instantaneous modulus of the sample.

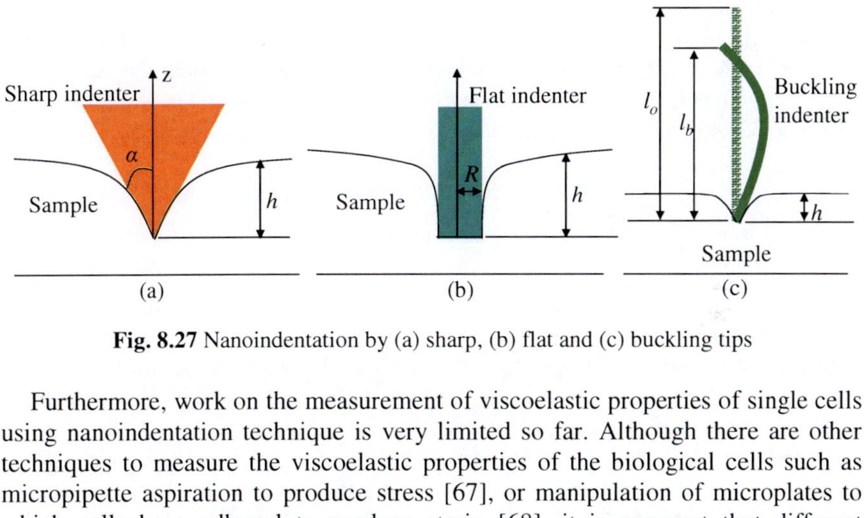

Fig. 8.27 Nanoindentation by (a) sharp, (b) flat and (c) buckling tips

Furthermore, work on the measurement of viscoelastic properties of single cells using nanoindentation technique is very limited so far. Although there are other techniques to measure the viscoelastic properties of the biological cells such as micropipette aspiration to produce stress [67], or manipulation of microplates to which cells have adhered to produce strain [68], it is apparent that different experimental techniques and theoretical models yield very different viscoelastic parameters even for the same cell type under similar conditions. It would be nice if the measurement data could be comparable with different techniques within the same scope of the measurement. Here, we propose nanoindentation approaches based on three indenters, i.e. sharp tip, flat tip, and buckling tip, for the viscoelastic characterizations of single cells.

The application of the correspondence principle and functional equation to the viscoelastic characterizations on the biological cells is not without issues. Here, we are examining the applicability of these principles for the yeast cells viscoelastic characterizations. The elastic-viscoelastic correspondence principle and functional equation are probably the most useful tools in viscoelasticity. The application of these principles demands the tested materials to be isotropic (homogeneous) and incompressible. Most of the biological cells, e.g. red blood cells (RBCs), on the other hand, are anisotropic due to the lack of proper structure of cell wall although there are incompressible (Poisson's ratio = 0.5). However, unlike the RBCs, yeast cells have proper and strong cell wall. The existence of this rigid cell wall has made the yeast cells to be modelled as isotropic, incompressible material. For instance, Smith et al. [85] and Stenson et al. [69] reported the homogeneous and isotropic characteristic of the yeast cells. The other issue related to the correspondence principle and functional equation to the biological cells is the finite size of the cells relative to the indentation contact diameter. Here, this issue can be justified provided the deformation is sufficiently small, i.e. 500 – 900 nm, that the specimen cross section does not change appreciably.

8.5.1 Modeling for Viscoelastic Measurement of Single Cell

Nanoindentation by a Sharp Tip Indenter

Indentation of a sharp indenter to a material, as shown in Figure 8.27 (a), will produce a non-constant contact area between the indenter and the material. For this particular problem, the method of functional equations can be applied to find the solution for the indentation of a sharp indenter on a time-dependent material [70].

For the indentation of a sharp indenter on a linearly elastic and isotropic material, Sneddon [32] has derived the indentation load-displacement relation as expressed by

$$F = \frac{2Eh^2 \tan\alpha}{\pi\left(1-v^2\right)} \tag{8.31}$$

where h, α, v and E are the indentation depth, tip half opening angle, Poisson's ratio and Young's modulus, respectively.

Using Lee-Radok [64] approach for a linearly viscoelastic, isotropic, and homogenous materials with a constant Poisson's ratio, (8. 31) leads to the following time-dependent load-displacement relation

$$F(t) = \frac{2\tan\alpha}{\pi\left(1-v^2\right)} \int_0^t E(t-\tau)\left[\frac{dh^2(\tau)}{d\tau}\right] d\tau. \tag{8.32}$$

Under a constant-rate displacement history, $h(t)=V_o t$, where V_0 is a constant indentation velocity, (8. 32) becomes

$$F(t) = \frac{4V_0^2 \tan\alpha}{\pi(1-v^2)} \int_0^t (\tau)E(t-\tau)d\tau. \tag{8.33}$$

Rewrite (8. 33) and differentiate it with respect to time t, we get

$$\int_0^t E(\tau)d\tau = \frac{\pi\left(1-v^2\right)}{4V_0^2 \tan\alpha}\frac{dF(t)}{dt} \tag{8.34}$$

and applying another time derivative to the both sides of (8. 34), yields

$$E(t) = \frac{\pi\left(1-v^2\right)}{4V_0^2 \tan\alpha}\frac{d^2F(t)}{dt^2} \tag{8.35}$$

where $E(t)$ is the Young's relaxation modulus of the time-dependent material which can be calculated from the experimental data of the indentation force function $F(t)$.

Equation (8. 35) appears simple to use. However, due to the fact that the load-displacement data from nanoindentation measurements are scattered, the use of raw nanoindentation data to determine the derivative can induce errors. The errors, however, can be reduced with the used of appropriate curve fitting technique to cast the nanoindentation data into smooth functions such as power/polynomial/exponential function.

Another approach is to consider the correlation between nanoindentation data and the load-time relation described by (8. 33) with the use of appropriate viscoelastic model for the material. For the linearly viscoelastic material, we use the standard linear solid model (SLS) (i.e. a parallel spring dashpot in series with another spring) for the Young's relaxation modulus

$$E(t) = E_\infty + E_i \exp^{\left(-t/\tau\right)} \tag{8.36}$$

where E_∞, E_i are relaxation coefficients, and τ is the relaxation time.

Substitute (8. 36) into (8. 33), we obtain

$$F(t) = \frac{4V_0^2 \tan\alpha}{\pi\left(1-v^2\right)}\left(\frac{E_\infty t^2}{2} - E_i\tau^2\left(1-\exp\left(-t/\tau\right)-\frac{t}{\tau}\right)\right). \tag{8.37}$$

After fitting (8. 37) with the load-time curve from nanoindentation tests, the parameters such as relaxation numbers and the relaxation times can be determined. These parameters can be used in (8. 36) to determine the Young's relaxation modulus.

The force data was calculated by using the Hooke's law (8. 12), where k, φ and L are the spring constant of the cantilever, the angle displacement of the cantilever and the length of the cantilever. The parameters φ and L were obtained from E-SEM images while the value for k was obtained from the calibrations of the cantilever using a geometrical approach and nanomanipulation approach.

The value for the Poisson's ratio was assumed to be 0.5 for soft biological materials [90] and the value for the tip half opening angle for the sharp tip was obtained from an E-SEM image as 22.5°. Therefore, (8. 37) can be simplified as

$$F(t) = 0.7V_0^2\left(\frac{E_\infty t^2}{2} - E_i\tau^2\left(1 - \exp\left(-t/_\tau\right) - \frac{t}{\tau}\right)\right). \tag{8.38}$$

Nanoindentation by a Flat Tip Indenter

The correspondence principle can be applied for a flat tip indentation on a material since the contact area is constant. Figure 8. 27 (b) shows a schematic diagram of a flat tip indentation. For the indentation of a flat indenter on a linearly elastic and isotropic material, Sneddon [32] has derived the following load-displacement relation

$$F(t) = \frac{2ERh}{1-\nu^2} \tag{8.39}$$

where R is the tip radius of the indenter. By using the standard linear viscoelastic solid (SLS) model and assuming that the material is incompressible, i.e. Poisson's ratio $\nu_1 = \nu_2 = 0.5$, Cheng et al. [71] showed that the corresponding viscoelastic solution of (8. 39) can be obtained in terms of the Laplace variable s as

$$\overline{F}(s) = 2R\overline{h}\frac{q_0 + q_1 s}{1 + p_1 s}. \tag{8.40}$$

The parameters p_1, q_0 and q_1 are given by

$$p_1 = \frac{\eta}{G_1 + G_2}, q_0 = \frac{2G_1G_2}{G_1 + G_2}, q_1 = \frac{2G_1\eta}{G_1 + G_2} \tag{8.41}$$

where G_1 and G_2 are shear moduli of the first and second spring elements in the SLS model, while η is the viscosity. The relationship of the shear moduli to their Young's moduli E_1 and E_2, and Poisson's ratio is

$$G_1 = \frac{E_1}{2(1+\nu_1)}, G_2 = \frac{E_2}{2(1+\nu_2)}. \tag{8.42}$$

Under a step displacement, i.e. a sample is subjected to a suddenly applied constant displacement, $h(t)=h_0H(t)$, where h_0 is a constant and $H(t)$ is a Heaviside unit step function, the indentation force, $F(t)$ can be obtained by taking the Laplace transform of $h(t)$ and substitute it into (8. 40) and then inverting the Laplace transformation. By doing so, the following equation, i.e. the indentation force-relaxation time, is obtained

$$F(t) = \frac{8Rh_0E_1}{3}\left[\left(\frac{E_1}{E_1 + E_2}\right)e^{-\left[(E_1+E_2)t/3\eta\right]} + \left(\frac{E_2}{E_1 + E_2}\right)\right] \tag{8.43}$$

where E_l is the instantaneous Young's modulus, E_0. The equilibrium Young' modulus can be obtained by subtracting E_2 from E_l. The force data was calculated from image analysis software (ImageJ, National Institute of Health) using (8. 12). The value for the flat tip radius, R, was obtained from an E-SEM image as about 195 nm. Therefore, (8. 43) can be rewritten as

$$F(t) = (5.2\times10^{-7})h_0 E_0\left[\left(\frac{E_0}{E_0+E_2}\right)e^{-\left[(E_0+E_2)t/3\eta\right]}+\left(\frac{E_2}{E_0+E_2}\right)\right]. \tag{8.44}$$

The value for the instantaneous indentation, h_0, was obtained from an E-SEM image. Experimental data of force-relaxation time were then fitted with (8. 44) to determine the values for E_0, E_2 and η using non-linear regression analysis (GraphPad Prism software).

In actual indentation tests, an ideal step loading or displacement history is difficult to be achieved. Instead we apply a ramp displacement with a very short rise time (usually t_0 = 1 s in this study) first and maintain a constant displacement thereafter. In order to minimized the error in (8. 44) due to the existence of the rise time in the step displacement, the period of which the relaxation phase was measured need to be at least ten times of the rise time [72]. Here, patience of the experimenter plays an important role. For our experiments, the duration of the whole experiment was set to 120 seconds.

Nanoindentation by a Buckling Nanoneedle Indenter

The indentation of the buckling nanoneedle, as shown in Figure 8. 27 (c), was analyzed based on the closed form solutions proposed by Cheng et al. [73]. For a rigid, axisymmetric indenter of arbitrary shape, $f(x)$, indenting an elastic solid, the solution, in term of force (F) and displacement (h) relationship, has been derived by Sneddon [32] as

$$F = \frac{4Ga}{1-\nu}\int_0^1 \frac{x^2 f'(x)}{\sqrt{1-x^2}}dx \tag{8.45}$$

where a is the radius of the circle of contact and $f'(x)=df(x)/dx$.

Applying the theory developed by Lee and Radok [64] to the problem of indentation in viscoelastic solids and assuming time-independent Poisson's ratio, (8. 45) becomes

$$F(t) = \frac{4}{1-\nu}\int_0^t G(t-\tau)\frac{d}{d\tau}\left[a(\tau)\int_0^1 \frac{x^2 f'(x)}{\sqrt{1-x^2}}dx\right]d\tau. \tag{8. 46}$$

Cheng et al. [73] has shown that by differentiating (8. 46) with respect of time and by using the fundamental theorem of calculus, the following relationship, i.e. initial unloading slope immediately after $t=t_m$ with a constant unloading rate, $-v_h$, can be obtained as

$$\frac{dF(h)}{dh}=\frac{dF/dt}{dh/dt}=\frac{4}{1-v}\left[G(0)a(t_m)-\frac{1}{v_h}\int_0^{t_m}\frac{dG}{d\eta}\bigg|_{\eta=t_m-t}a(\tau)\frac{dh(\tau)}{d\tau}d\tau\right]. \quad (8.47)$$

When the unloading rate, v_h, is sufficiently fast, the second term on the right side of (8. 47) approaches zero, independent of the loading history

$$\frac{dF}{dh}=\frac{4}{1-v}G(0)a(t_m)=\frac{2E(0)}{\left(1-v^2\right)}(Rh)^{1/2} \quad (8.48)$$

where *E(0)* is the instantaneous Young's relaxation modulus of time-dependent material.

At the best of our knowledge, (8. 48) has only been verified using a numerical approach. The absentee of an experimental validation is believed due to the difficulty to obtain fast unloading data using the conventional AFM cantilever tip. The main advantage of the buckling nanoneedle over the standard AFM cantilever is the prevention of the negative force on the sample during the unloading phase. The negative force will make the tip further indenting the sample during the unloading phase. For the buckling nanoneedle, no negative force is existed. The main factor behind the occurrence of the negative force is due to the spring inertia. From our observation, the simple way to identify the existence of the spring inertia is by examining the difference between the axis of the applied force and the axis of the spring source, i.e. cantilever beam or the buckling nanoneedle. If the axes are different, then the spring inertia is existed during the unloading phase. For the standard AFM cantilever, the axis of the applied force is in |Z|-direction while the axis for the spring source, i.e. cantilever beam, is in |Y|-direction. Due to this, during the unloading phase, tip tries to balance the movement of the base of the cantilever by further moving downward. Consequently, the sample is further indented by the tip of the AFM cantilever.

Interestingly, for the buckling nanoneedle, since the axis of the applied force and the spring source, i.e. buckling nanoneedle, is in the same direction, i.e. |Z|-axis, there is no negative force during the unloading phase. Therefore, the sample is not further indented during the unloading phase of the buckling nanoneedle. In this work, (8. 48) was analyzed, for the first time, using experimental data based on the buckling nanoneedle tip, where the instantaneous Young's relaxation modulus has been obtained.

The force data in (8. 48) was calculated by using the Euler buckling equation as follows

$$F=\frac{E_{needle}\pi^2 wb^3}{7.68\ell_{buckle}^2} \quad (8.49)$$

where E_{needle} is the Young's modulus of the buckling nanoneedle, w and b are the lengths of the cross section of the buckling nanoneedle and ℓ_{buckle} is the buckling length of the buckling nanoneedle. The value for E_{needle} was obtained from the calibration of the buckling nanoneedle using a nanomanipulation approach, while the values for w, b and ℓ_{buckle} were obtained from E-SEM images.

8.5.2 Fabrication and Calibration of Nanoprobe for Viscoelastic Measurement of Single Cell

Fabrications of the Flat Tip and the Buckling Tip

The sharp tip indenter, as shown in Figure 8. 28 (a) was based on the standard AFM cantilever. The half opening angle of the sharp tip is 22.5° as measured from the SEM image. The flat tip indenter, as shown in Figure 8.28 (b), was fabricated by using Focused Ion Beam (FIB) deposition and etching on the standard AFM cantilever. The acceleration voltage and ion doses for FIB deposition and etching were 30 kV, 8.56×10^{18} dose and 2.67×10^{20} dose. The diameter and height of the flat tip are 390 nm and 470 nm. The buckling nanoneedle, as shown in Figure 8. 28 (c), was fabricated by etching/trimming the standard AFM cantilever using FIB etching. The parameters for FIB etching were 30 kV. The diameter and height of the buckling nanoneedle are 150 nm and 9.95 μm, respectively.

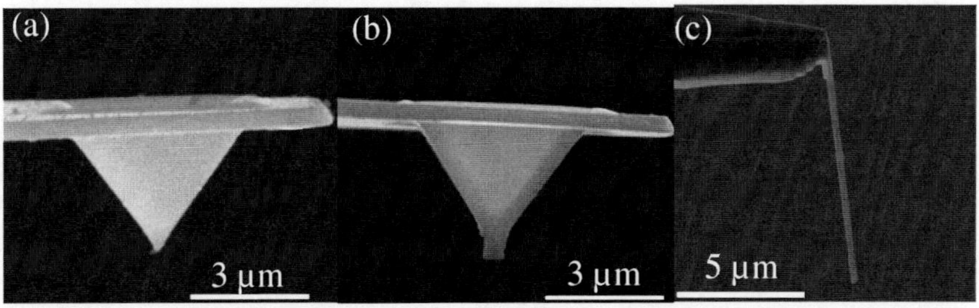

Fig. 8.28 SEM images of (a) sharp, (b) flat and (c) buckling tips

Calibration of the Sharp Tip, Flat Tip and Buckling Nanoneedle Tip

The spring constants of the sharp tip indenter and the flat tip indenter were calibrated using the geometrical approach and the nanomanipulation approach. The geometrical calibration of the cantilever's spring constant was based on the Parallel Beam Approximation (PBA) method [72] as described in the following equation,

$$k = \frac{Ec^3 d}{2L^3} \cos\theta \left[1 + \frac{4d^3}{m} (3\cos\theta - 2) \right]^{-1} \tag{8.50}$$

where E, c, d, m, L and Θ are Young's modulus of the cantilever's material, thickness of the cantilever's beam, width of the cantilever's arm, distance between two cantilever's arms, length of the cantilever and half angle of the opening cantilever's arms. The material for the sharp tip and flat tips is silicon nitride (Si_3N_4). The values for c, d, m, L and Θ for the sharp tip (i.e. 398 nm, 20.17 μm,

107.82 μm, 100.20 μm and 27.80°) and the flat tip (i.e. 390 nm, 21.47 μm, 110.66 μm, 102.54 μm and 25.96°) were obtained from the SEM images, while the value for E was obtained from [73] which is 180 ± 20 GPa. Figure 8-54 shows the physical parameters of the cantilever in the PBA method for estimating the cantilever's spring constant.

The nanomanipulation approach was based on the work by [74]. This approach amounts the deflecting of an 'unknown' cantilever with a 'referenced' cantilever, where the spring constant of the 'referenced' cantilever has been precalibrated. The 'referenced' cantilever has a spring constant of 0.09 N/m. The calibration was done inside E-SEM instrument under high vacuum condition. By using the nanomanipulator unit assembled inside the E-SEM chamber, the indentation of the 'unknown' cantilever on the 'referenced' cantilevers could be performed.

The Young's modulus and the spring constant of the buckling nanoneedle, E_{needle} and k_{needle}, were calibrated using a nanomanipulation approach. Under this technique, the buckling nanoneedle was slowly pressed against another referenced cantilever (0.110 N/m). The indentation was carried out until the nanoneedle begun to buckle. Then, the buckled nanoneedle was slowly retracted until the nanoneedle returned to the straight condition.

The value of Eneedle was determined from the Euler buckling equation as shown in (8. 52) where ℓ_{buckle} is the length of the soft nanoneedle during the buckling condition. Equation (8. 52) was obtained by substituting $I = (wb^3)/12$ into (8. 51) in the derivation,

$$E = \frac{F_n (KL)^2}{\pi^2 I} \tag{8.51}$$

$$E_{needle} = \frac{7.68 F_n \ell^2_{buckle}}{\pi^2 wb^3} \tag{8. 52}$$

where F_n is the buckling force applied to the nanoneedle, E_{needle} is the Young's modulus of the nanoneedle, I is the second moment of area, K is the nanoneedle effective length factor, whose value depends on the conditions of the end support of the nanoneedle and ℓ is the length of the nanoneedle. The value of the length of the soft nanoneedle was corrected with $K = 0.8$ for the structure which has one fixed end and the other pinned end. I is a property of a shape that is used to predict its resistance to buckling. The soft nanoneedle which has a rectangular cross section has the value of $I = (wb^3)/12$ where w and b are the width and height of the rectangular cross section. The values for w and b, which obtained from an image analysis, are 140 nm and 145 nm, respectively.

In order to calculate the applied indentation force, F_n in (8. 52), Hooke's law was used as shown in (8. 53) where k_{cant} and δ_{cant} are the spring constant and deflection of the referenced cantilever.

$$F_n = k_{cant} \delta_{cant}. \tag{8.53}$$

From the experiment, the value for E_{needle} for the buckling nanoneedle which was obtained from (8. 52) is 91.3 ± 13.4 GPa. The value of E_{needle} is in agreement with the bulk modulus of the Silicon material which is about 100 GPa [11]. The value for k_{needle} of the soft nanoneedle is 0.12 N/m, which obtained from the second slope of the linear portion of the force-buckle length occurring just after the soft nanoneedle started to buckle.

8.5.3 Viscoelastic Measurement of Single Cell by E-SEM Nanomanipulation

Viscoelastic Characterization of Single Cell Using a Sharp Tip Indenter

The indentation experiments using a sharp tip, as shown in Figure 8. 29 (a), were performed inside E-SEM chamber at 600 Pa and 0 °C. The sharp tip was attached on the nanomanipulator unit by using a conductive tape. By using a direct control of the nanomanipulator, the sharp tip was then slowly approaching and indenting a single cell. Only loading part of the indentation was executed as unloading part was not necessary in the analysis. The indentation duration was about 2 minutes. Five *W303* yeast cells were tested. Figure 8. 30 (a) shows the force-indentation responses for five single cells. The loading rates were in the range of 2.1 nm/s to 3.7 nm/s as can be seen in Figure 8. 30 (b). The indentation activities were recorded using E-SEM's real-time movie recorder with a frame rate of 40 frames per second. The real-time movies were then analyzed as series of stilled images.

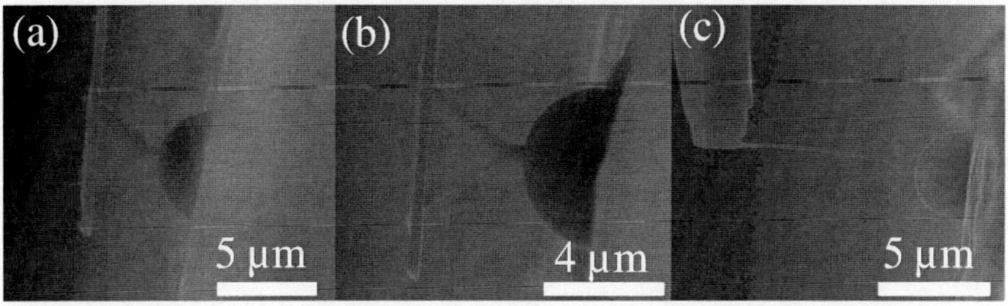

Fig. 8.29 SEM images from the nanoindentation experiment using (a) sharp tip, (b) flat tip and (c) buckling nanoneedle tips

In order to calculate the Young's relaxation modulus of the single cells from (8. 36), experimental data of force-time responses are required. Since the fact that the force-indentation data from the nanoindentation measurements are scattered, use of the raw nanoindentation force-time data in (8. 38) can induce errors. The errors, however, can be reduced with the use of an appropriate curve fitting technique to cast the nanoindentation data into a smooth function. After several trials, power series function has been chosen for fitting the force-time data. Figure 8. 30 (c) shows one example of the fitting result for Cell 1 using the Lavenberg-Marquardt method (GraphPad Prism software). The goodness-fit value which is 0.9974 indicates a good fitting result. The fitting power series function for Cell 1 is

$$F(t) = 0.1062t^{1.586}. \quad (8.54)$$

The power series functions for force-time data of five single cells were then fitted in (8. 38) to extract the relaxation numbers and the relaxation time which later on these parameters were used in (8. 36) to determine the Young's relaxation modulus of the cells. For an instance, the Young's relaxation modulus for Cell 1 is

$$E(t) = 2.794 + 7.413\exp\left(-t/_{6.748}\right). \quad \text{MPa} \quad (8.55)$$

The results of the Young's relaxation modulus for five cells are shown in Figure 8. 30 (d). The instantaneous Young's modulus, *E(0)*, taken at *t*=0.1s, for Cell 1 is about 10.10 MPa, while for Cell 2 to Cell 5, their *E(0)* are about at the same range which are 3.32 MPa, 3.86 MPa, 3.61 MPa, and 3.89 MPa, respectively. The equilibrium Young's modulus, *E(∞)*, taken at t=120s, for Cell 1 is about 2.79 MPa, while for Cell 2 to Cell 5, their *E(∞)* are about at the same range which are 0.90 MPa, 1.05 MPa, 1.29 MPa and 1.38 MPa, respectively.

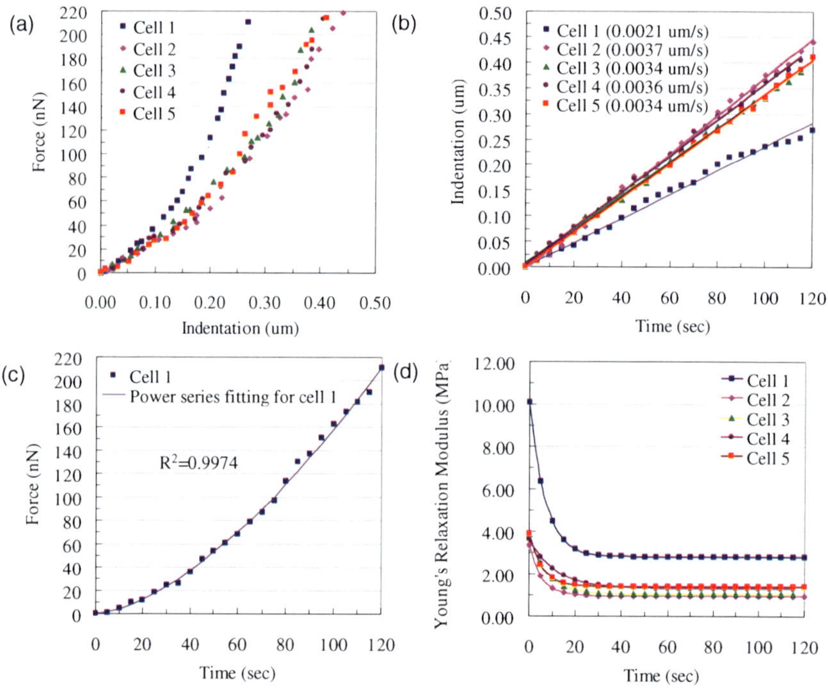

Fig. 8.30 Experimental results obtained by using the sharp tip indenter: (a) force-indentation responses, (b) indentation-time responses, (c) force-time response and (d) Young's relaxation modulus-time responses.

The results of the Young's relaxation modulus for the five cells clearly show the influence of the indentation rate toward the cell dynamics response during the indentation process. The slower the indentation rate of the indentation is, the slower the relaxation response of the cell will be. This can be seen from the Cell 1's relaxation response which takes longer time to settle down to the equilibrium Young's modulus as compared to the relaxation responses of Cell 2 to Cell 5. The data for the Young's relaxation moduli obtained in this experiment are reasonable as the elastic Young's modulus for the yeast cells are in the range of a several MPa.

Several authors reported viscoelastic behaviour of yeast suspension. Mancini and Moresi [75] measured rheology of baker's yeast using dynamic stress rheometer in the concentration range of 25 to 200 g.dm^{-3}. Their result revealed definitive Newtonian liquid behaviour. Lo Presti and Moresi [76] showed that at 30 °C, both cell suspensions of baker's yeast cells and *Yarrowia lipolytica* cells were found to exhibit Newtonian behaviour with quite apparent dynamic viscosities, at least up to yeast concentrations of 150 kg.dm^{-3}. Speer et al. [77] used controlled shear-rate rheometer with plate-cone system to measure viscosity of suspensions of flocculating and nonflocculating strains of *S. cerevisiae* and *S. uvarum*.They concluded that flocculating strains had higher both apparent viscosity and yield stress than nonflocculating ones.

Viscoelastic Characterization of Single Cell Using a Flat Tip

The indentation experiments using a flat tip, as shown in Figure 8-57(b), were performed inside E-SEM chamber at 600Pa and 0 °C. The flat tip was attached on the nanomanipulator unit by a conductive tape. The flat tip was firstly approaching a single cell for getting a smooth contact. Then, by using a step indentation (i.e. indentation rate of 2 μm/s), the flat tip was rapidly indenting the cells and remained constant hereafter for about 2 minutes. The whole activities were recorded using the E-SEM's real time movie recorder with the frame rate of 40 frames per second. Seven yeast cells were tested in the experiment. Figure 8-59 shows the force relaxation responses for seven cells. The dotted points are the experimental data which then fitted in (8. 44) to extract the values for E_1, E_2 and η.

In general, the values for the instantaneous Young's modulus, $E(0)$, the equilibrium Young's modulus, $E(\infty)$, and the viscosity, η, for seven cells are in very good agreement to each other. The average values for $E(0)$, $E(\infty)$ and η are 2.62 ± 0.21 MPa, 1.13 ± 0.64 MPa and 29.7 ± 11.6 MPa.s, respectively. As for the sharp tip indenter experiment, the results from the flat tip indenter experiment also reasonable as the values for the moduli should be in the range of a several MPa. Interestingly, the equilibrium Young's moduli obtained from the flat tip indenter experiment, i.e. 1.13 ± 0.64 MPa, shows good agreement with the results for the equilibrium Young's moduli obtained from the sharp tip indenter, i.e. 1.16 ± 0.22 MPa. This indicates that the sharp and the flat tip indenters can be used for validating each other measurements of the equilibrium Young's modulus.

Despite of the good agreement in the values of $E(\infty)$ obtained from the both sharp and flat tip indenters, their values for the instantaneous Young's modulus seems differs from each other. As for the sharp tip indenter, it was 3.67 ± 0.27 MPa, while for the flat tip indenter, it was 2.62 ± 0.21 MPa. The difference in the *E(0)* for the sharp and flat tip indenters can be comprehended by acknowledging the differences in the tip shape used for the both indenters. Under the same amount of indentation force, the sharp tip is expected to extract more viscoelastic property of single cell as compared to the flat tip. The sharp tip requires less force to indent further inside a cell surface as compared to the flat tip. Due to this, the measured *E(0)* of single cells from the sharp tip was little bit higher from the *E(0)* measurement by the flat tip. Other explanation maybe due to the diversity of the viscoelastic properties of the cells as some biological factors such as cell growth phases may influence the viscoelastic properties of the cells [78].

Viscoelastic Characterization of Single Cells Using a Buckling Nanoneedle Tip

The indentation experiments using a buckling nanoneedle tip were performed inside E-SEM chamber at 600 Pa and 0 °C. The buckling nanoneedle tip was attached on the nanomanipulator unit by using a conductive tape. In order to show the inadequate of slow unloading data to satisfy the (8. 48) and the necessity to provide fast unloading data for calculating the instantaneous Young's modulus, two types of loading profiles were performed, i.e. slow loading-slow unloading and slow loading-fast unloading. The slow loading/unloading rate was about 20 nm/s while the fast unloading rate was about 2 μm/s. During the slow loading -slow unloading profile, the nanoneedle was slowly indenting a cell which also resulting its nanoneedle to start buckle. The buckling length of the nanoneedle is proportional to the amount of the force being applied. After the nanoneedle has buckled for several μm (about 1 to 2 μm), the unloading part was taking place by slowly moving away the buckling nanoneedle from the cell. On the other hand, during the fast unloading profile, the retraction of the buckling nanoneedle away from a cell was performed in a sudden movement. The total time for the indentation test was last within a few seconds (less than 3 seconds). The experiments were recorded using the E-SEM movie recorder with the frame rate of 40 frames per second. Five cells were tested in the experiment.

The fast unloading experiment was performed at 2 μm/s rate. This motion rate is capable to remove the indenter (buckling nanoneedle in this case) away from the single cell within 1 second or less. Cheng et al. [79] reported an unloading time of 0.01s or 0.1s as the best unloading time to reach the instantaneous limit. Furthermore, their numerical results also showed that the unloading speed of 1 s has closely approached the convergence limit. Due to this, we assumed that the 2 μm/s rate is adequate for the analysis.

Furthermore, an attempt to increase the unloading rate (although it is highly possible by using the nanomanipulation system) in our experiment resulted a distortion in the movie due to the limitation of the resolution of the E-SEM movie recorder (8. 40 frames per second).

Careful observation and calculation of the initial unloading slope from the slow unloading curve revealed that it has a negative slope, as expected. On the other hand, the slope from the fast unloading curve has a positive value. Therefore, the slope from the fast unloading curve can be fitted into (8. 48) to calculate for the instantaneous Young's modulus, *E(0)*, of the cells. The values of the *E(0)* for each of the cells calculated by fitting their slope's values into (8. 48) are 3.43 MPa, 4.07 MPa, 3.01 MPa, 2.72 MPa and 4.12 MPa, respectively.

Interestingly, the values of the instantaneous Young's moduli obtained from the buckling nanoneedle experiment, i.e. 3.47 ± 0.62 MPa, are in a very good agreement which the same moduli obtained from the sharp tip experiment, i.e. 3.67 ± 0.27 MPa. This finding has shown that our methods for viscoelastic properties characterization by using several types of indenters, i.e. sharp tip, flat tip and buckling tip, can be very useful in term of the ability to compare and to validate the viscoelastic parameters obtained from each experiment under the same experimental setting.

8.6 Single Cell Adhesion Force Measurement Using Nanofork

Cell-substrate adhesion and cell-cell adhesion are very important aspects in tissue engineering [80]. The quality of the artificial scaffolds depends on its ability to promote various kinds of cell activities, e.g. cell adhesion, cell growth, cell proliferation and cell differentiation [81]. Therefore, the analysis of cell activities on the scaffold is very important in order to evaluate the functionality of the scaffold. As compared to the other cell activities, i.e. cell growth, cell proliferation and cell differentiation, the investigation of cell adhesion is challenging as the analysis is difficult, if not impossible, to be done qualitatively by means of observation per se, but rather quantitatively by means of force measurement.

Although there are several force measurement techniques available, e.g. atomic force microscopy (AFM) cantilever, optical tweezer and micropipette, these techniques are difficult to be implemented directly when dealing with the cell adhesion measurement. AFM cantilever is normally implemented in the nanoindentation experiment for cell stiffness characterization where the moving direction of the cantilever is towards the cell. However, in order to measure cell adhesion force, the AFM cantilever must move perpendicularly away from the cell. Therefore the normal settings of the nanoindentation experiment cannot be used for cell adhesion force measurement. The optical tweezer method is almost impossible to measure cell adhesion force (which expected to be high) as the current force resolution is relatively low in the range of several hundred picoNewton. The micropipette approach is very useful for measuring the suction force on the cell surface but seems useless for measuring the cell adhesion force. Although the micropipette is expected to be able to pick up a cell by applying a suction force on the cell, it does not have a capability to measure the releasing force/adhesion force of the cell from the substrate. In term of cell pick-up point of view, microgripper seems to be the best approach. It would be very interesting if the microgripper could pick-up a cell and at the same time be able to measure the releasing/adhesion force between the cell and the substrate. However, the current design of the microgripper does not provide this interesting capability, i.e. measuring the releasing force while pick-up the sample.

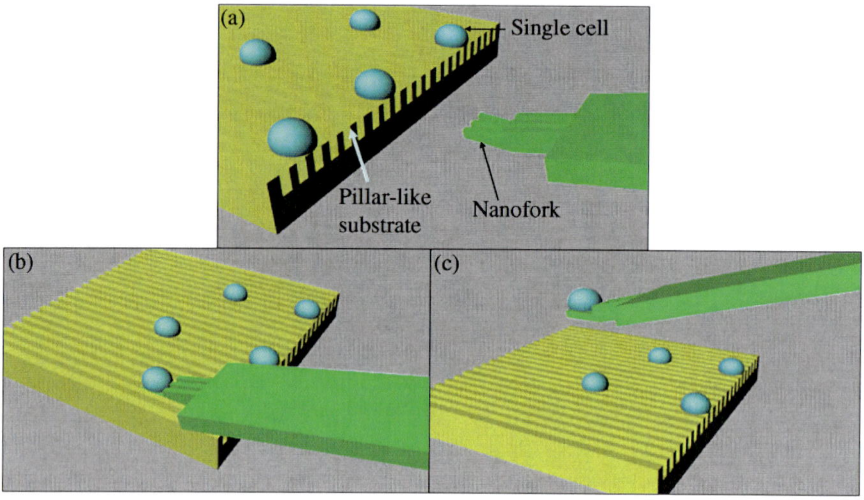

Fig. 8.31 Overview of the (a) cell pick-up and cell adhesion force measurement strategy using nanofork and pillar-based substrate. The pillar-based substrate enables the nanofork (b) to be inserted between the cell and the base of the substrate and (c) to pick-up the cell.

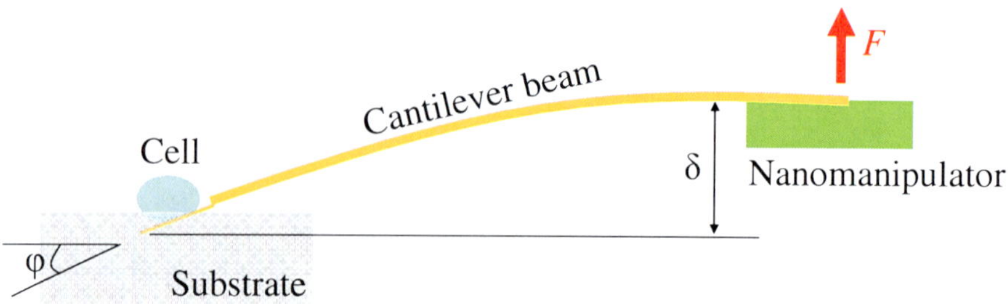

Fig. 8.32 Schematics of the parameters determination for the force measurement

Previously, we have modified the current setting of the nanoindentation technique which makes it possible to measure force data when the cantilever is moving way from the cell. In this modified approach, we coated the tip of the AFM cantilever which a bio-adhesion protein (from mussel). By touching the bio-adhesive coated tip on the cell surface, adhesion force between the tip and the cell was created. Then, by moving the cantilever away from the cell, the cell elongated and eventually released from the substrate. We have demonstrated the local and whole cell elongations by using the bio-adhesive protein coated on a nanoneedle and flat tip AFM cantilever, respectively. However, since the cell elongated before released from the substrate, the analysis of the cell adhesion force becomes complicated. It is difficult to give a clear separation between cell stiffness and cell adhesion from the whole cell elongation and cell release process.

We proposed a new strategy for measuring the cell adhesion force. We believe that in order to pick-up a whole cell and at the same time measuring the cell adhesion force, a special device is required in E-SEM. We called this special device a nanofork since it has three nanoneedles structure which resembles the appearance of a normal fork. Due to this, the nanofork not only can be used for picking up a whole cell but also can measure a force data during the pick-up process. In order to effectively implement this nanofork, a special design of substrate is required. The substrate must be able to provide gaps between cell and the substrate. Therefore, the nanofork can be inserted through these gaps in order to successfully pick up the cell. In order to realize this idea, we have fabricated a line-patterned structure on the surface of the substrate.

Figure 8. 31 shows the schematics of the cell pick-up and cell adhesion measurement strategy using nanofork and line-patterned substrate. Single cells were placed on the line-patterned substrate by dropping several microliters sample using micropipette. Nanofork was attached on the nanorobotics manipulation system by using an electrically conductive tape. The nanorobotics manipulator has 3 D.O.F. (three translational directions, i.e. X, Y, and Z axes) for maneuvering the nanofork inside the E-SEM chamber. The task is initiated by moving the nanofork to the targeted cell's area. The alignment matching between the cell and the nanofork was achieved by adjusting the Z-axis of the nanofork until the focused images between the cell and the nanofork are equal. The nanofork was then maneuver through the gaps between the cell and the base substrate. The nanofork was then moved upward to release the cell from the substrate. During this endeavor, the cantilever beam of the nanofork gradually deformed as shown in Figure 8. 32. The deformation of the cantilever beam can be used for force calculation which is equal to the adhesion force between the cell and the substrate. The force calculation was based on the Hooke's law as shown in (8. 12). When the force exerted by the nanofork is exceeded the adhesion force between the cell and the substrate, the cell will be released, the so called cell pick-up activities. The maximum force needed to release the cell from the substrate is defined as the cell adhesion force.

8.6.1 Fabrication and Calibration of Nanoprobe and Line-Patterned Substrate for Viscoelastic Measurement of Single Cell

Fabrication of the Nanofork

The structure of the nanofork can be divided into three parts, i.e. handle, support of the tines and three times. The handle of the nanofork functions as a force sensing mechanism. From the deflection of the cantilever, i.e. handle, the applied force can be determined. The support of the tines provides apertures between the nanofork and the line-patterned substrate. The apertures will enable the nanofork to slide along the line-patterned substrate. The three tines of the nanofork function as a holder for the cell. The nanofork was fabricated by FIB etching. A commercially available AFM cantilever was used to fabricate the nanofork. The spring constant of the nanofork is about 2 N/m which obtained from the manufacturer's product specifications.

The fabrication of the nanofork consists of six steps. First, the tip of the AFM cantilever was removed by using FIB etching. Second, FIB etching was performed on the top area of the cantilever's tip to get fork-like structure which dimensions for the width and gap of the nanofork are 1 μm each. Third, refine top etching was performed on the nanofork to get the width of less than 1 μm. Forth, the cantilever was tilted 15 degrees angle. Fifth, FIB etching was performed on the side area of the cantilever to obtain 1 μm thickness. Finally, refine side etching was performed to obtain less than 1 μm thickness of the nanofork. The width and thickness of the nanofork have to be less than 1 μm for smooth insertion through the line-patterned substrate.

Figure 8. 33 shows the SEM images of the fabricated nanofork. The structure of the nanofork can be divided into three parts, i.e. handle, support of the tines and three times. The handle of the nanofork functions as a force sensing mechanism. From the deflection of the cantilever, i.e. handle, the applied force can be determined. The support of the tines provides apertures between the nanofork and the line-patterned substrate. The apertures will enable the nanofork to slide along the line-patterned substrate. The three tines of the nanofork function as a holder for the cell. The nanofork was fabricated by FIB etching. A commercially available AFM cantilever was used to fabricate the nanofork. The spring constant of the nanofork is about 2 N/m which obtained from the manufacturer's product specifications

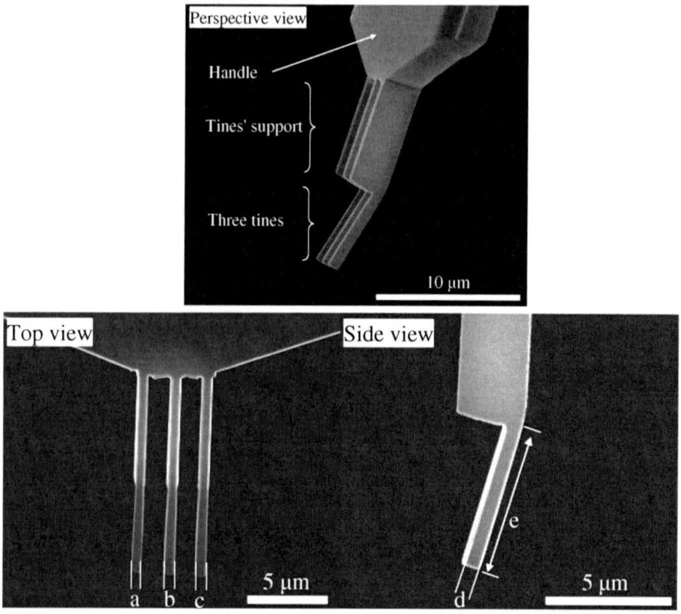

Fig. 8.33 SEM images of the fabricated nanofork. (a, b, c, d: 0.6 μm, e: 5.6 μm)

Calibration of the Nanofork

Since the tip of the cantilever was modified in order to obtain the nanofork structure, it is expected that the spring constant of the cantilever will also change. Therefore, the calibration of the spring constant of the nanofork is required before it can be used as a force sensor in the single cell adhesion measurement. Since it is difficult to place the referenced cantilever (*rc*) on the tines area of the nanofork, the contact performed at the handle part of the nanofork. In this configuration, different equation for estimating the spring constant of the nanofork is used as shown below,

$$k_{nanofork} = \frac{k_{rc} * 2 * L_{rc} * \varphi_{rc}}{\varphi_{nanofork} * (3 * L_{nanofork} - a)} \tag{8.56}$$

where L, φ and a are length of the cantilever, angle deflection of the cantilever and distance between the base of the nanofork and contact point with the referenced cantilever. Table 8. 6 shows the values for the parameters L, φ and a which obtained from the SEM images.From the experimental result of the calibration of the nanofork, the new spring constant of the nanofork is 2.59 ± 1.29 N/m with 95 % confidence level.

Table 8.6 Values for the parameters for the calibration of the nanofork

Parameter	Value (as obtained from SEM images)
L_{cant}	98.79 μm
φ_{cant}	Variable
L_{sample}	240.15 μm
φ_{sample}	Variable
a	220.37 μm

AFM Surface Profile Analysis of the Line-Patterned Substrate

The main purpose of the AFM surface analysis is to obtain height information of the line pattern structure on the substrate. Figure 8. 34 shows an AFM image of the line-patterned substrate which obtained under room temperature. This figure shows an AFM quantitative analysis of the line structure of the substrate. The height of the line structure is about 0.94 μm.

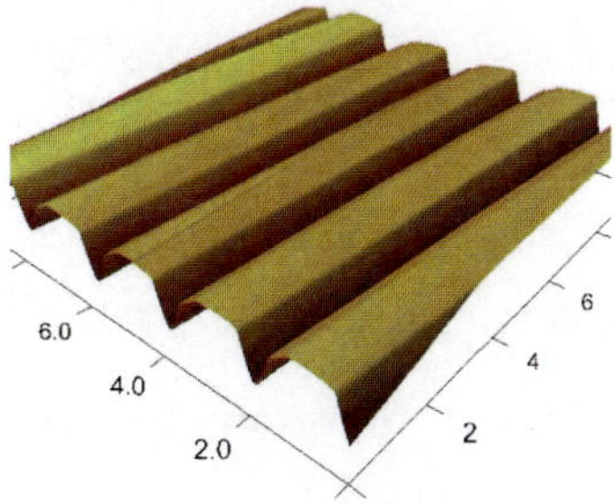

Fig. 8.34 An AFM image of the line-patterned substrate

In addition to the height information of the line-patterned substrate, the width of the line pattern and the gap between two line structures can also be determined from the AFM analysis. In order to correctly determine the values for the width and the gap, two considerations are needed. Data which marked with the red dotted lines were obtained from the contact between the body of the cantilever's tip and the edge of the line structure. Due to this, false slopes were detected. Therefore, although the side line of the line structure is almost straight, a false slope was detected as the tip of the cantilever was unable to sense the side area of the line structure. Another consideration is the transition of the tip of the cantilever from the gap area to the top surface of the line structure.

After considering these two issues, i.e. the inability of the tip of the cantilever to make a contact with the side area of the line structure and the transition of the tip of the cantilever from the gap area to the top surface of the line structure, correct determinations of the width of the line structure and gap between two line structures, i.e. 1.03 μm and 0.88 μm, respectively.

Verification of the Cells Attachment on the Line-Patterned Substrate

In order to successfully perform cell adhesion force measurement using the nanofork, cells must be able to adhere on the line-patterned substrate. Figure 8. 35 shows images of adherent cells on the line-patterned substrate. The *W303* cells were dropped several microliters on the line-patterned substrate by using micropipette. The water solution that covers cells was evaporated by controlling the pressure inside the E-SEM chamber (600 Pa) and temperature of the cells (0 °C). From this result, we can conclude that the cell attachment of the pillar-based substrate is possible.

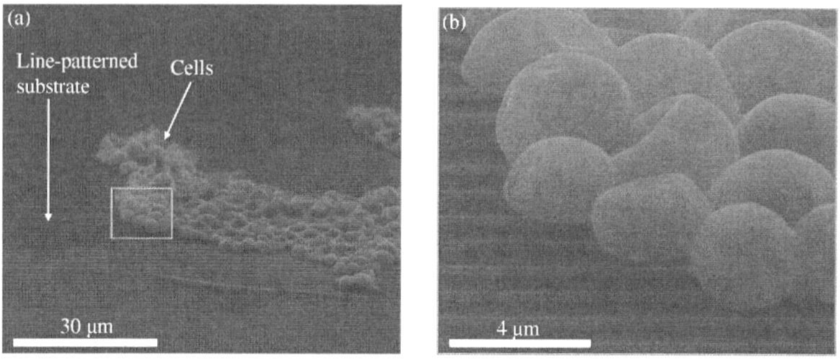

Fig. 8.35 Attachment of W303 cells on the line-patterned substrate (a). (b) a magnification image of some area of (a) which marked by a green rectangle line.

8.6.2 *Experimental Result of Adhesion Measurement of Single Cells by Nanofork*

Needs for the Line-Patterned Substrate on the Single Cell Adhesion Measurement

In order to appreciate the important of the line-patterned substrate in the cell adhesion force measurement and cell pickup task, we have tested the nanofork on the flat surface. This experiment was conducted to emphasize that the flat surface is not adequate for protecting cells during the cell adhesion force measurement and cell pick-up task by nanofork. The flat surface was based on the aluminum material. Several microliters of sample were dropped on this aluminum substrate. Then, the pressure inside the E-SEM chamber and temperature of the sample were set to 600 Pa and 0 °C for smoothly evaporating the water that covers the cells. An attempt to pick-up the cells using nanofork was followed soon after that.

Figure 8. 36 shows the result of an attempt to pick-up a cell by using the nanofork. The nanofork was driven by the nanorobotics manipulator system. The speed of the nanorobotics system was adjusted from coarse to fine modes when the nanofork was near to the cell. Then, the speed of the nanorobotics manipulator system was reduced further before executing the cell pick-up task. Clearly, due to the tight space between the cell and the substrate, the nanofork did not have enough space to pick up the cell appropriately. As a consequence, the cell was seriously damaged due to excessive hit by the nanofork. This result confirms that new surface structure on the substrate is required for the successful pick-up by the nanofork. This issue, i.e. inadequate gap between cell and substrate, can be overcome by using different pattern structure on the surface of the substrate. From this point of view, the line-patterned substrate can be used as a support for the cells which can provide enough space between the cells and the base of the substrate for successful cell pick-up and cell adhesion measurement tasks.

Application of Nanofork and Line-Patterned Substrate in Single Cells Adhesion Measurement

The nanofork and the line-patterned substrate were used in the measurement of an adhesion force of single cells. Wild-type yeast cells (*W303*) were used in the experiment. The parameters of the environment inside E-SEM chamber are 600 Pa and 1 °C which give about 98 - 100 % humidity. The experiment was conducted by performing the alignment test between the nanofork and the line-patterned substrate. Once the alignment between the nanofork and the line-patterned substrate was achieved, a sample (about 2 μl of sample solution) was dropped on the line-patterned substrate using a micropipette.

It was confirmed that the nanofork was successfully slid beneath the cell. The cell was then lifted by the nanofork. During lifting the cell, the cantilever of the nanofork was deformed and this deformation can be used to estimate the adhesion force between the cell and the line-patterned substrate. The maximum adhesion force was acknowledged when the cell was completely released from the substrate. The calculation of the force was based on the Hooke's law as shown in (8. 12). The result of the experiment is shown in Table 8. 7.

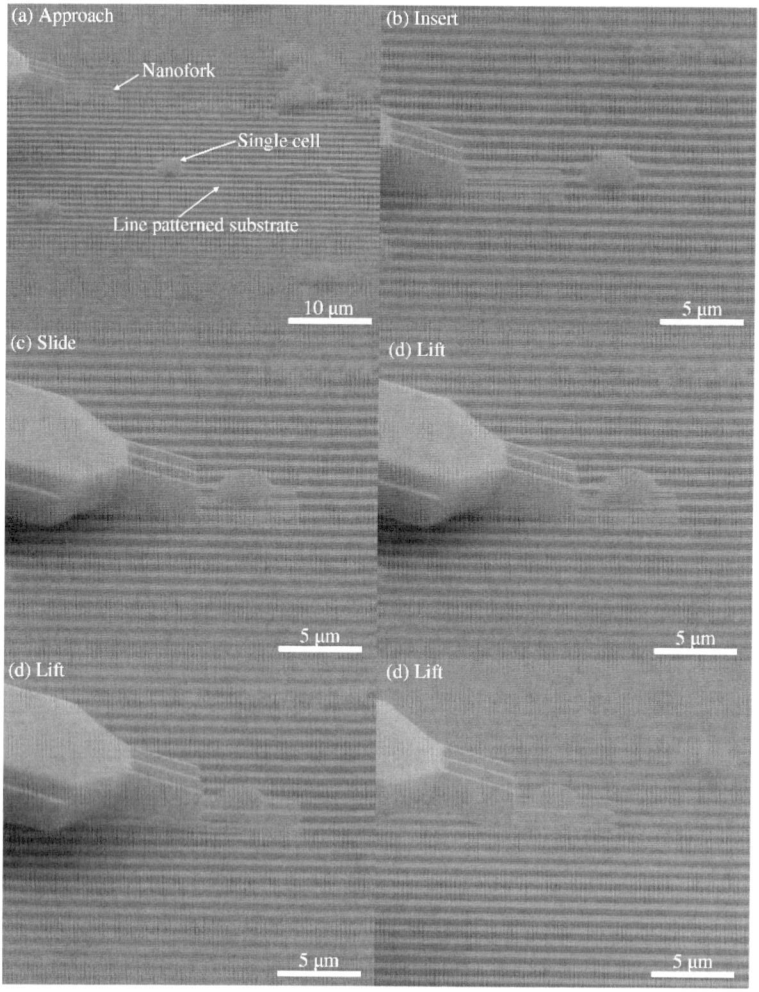

Fig. 8.36 Application of the nanofork in a single cell adhesion measurement

Table 8.7 Result of Single Cell Adhesion Measurement using Nanofork

Parameters for Nanofork			Release Force (nN)
Spring constant (N/m)	Length (μm)	Deflection Angle (Radian)	
2.59	191.9	0.002	555

8.7 Dynamic Single Cell-Cell Adhesion Force Measurement Using Nanopicker

Cell's activities, such as embryogenesis, mitosis, morphogenesis, cell orientation, cell motility, and survival depend on the attachment to the extracellular matrix and neighboring cells. We have developed a micro putter for the study the cell adhesion to substrate.

Currently, there are mainly two types of techniques for cell-cell adhesion study, i.e. dura aspiration pipette and AFM manipulation, Dual aspiration pipette assay is widely for cell-cell adhesion force measurement task. Quantitative analysis result can be obtained by using this method. However, part of the cell where contact with the pipette was always sucked into the internal pipette due to the sub pressure during the experiment [82]. The large deformation may cause physical damage to cell. It is also difficult to calculate the adhesion force accurately due to the large cell deformation.

AFM system has been widely used for single cell adhesion force measurement as well. In this technique, one cell was immobilized to a modified AFM cantilever tip by chemical glue. The adhesion force between this cell and the other cell on substrate can be obtained by using the AFM system [83]. Quantitative and exact measurement results can be obtained owing to the precious AFM manipulation system. However, it is difficult to change the cell on the cantilever tip once it was immobilized by the glue. Some researchers also worried about the cell activity and cell surface properties after the immobilization [84]. Furthermore, AFM image is constructed based on the scanning of AFM tip on sample surface. Therefore, it is difficult, if not impossible, to manipulate the sample during the imaging process. In order to observe the position of the cantilever and cell during the experiment, additional optical microscope (OM) has to be employed. Considering the resolution of OM is usually limited to approximately 200 nm, it is still a challenge to realize the observation of bio sample at nano scale during the manipulation procedure.

We proposed a novel method for cell-cell adhesion force measurement based on the nanorobotic manipulation system inside E-SEM. A nano picker is fabricated from AFM cantilever by using FIB etching technique. The adhesion force of cell-cell is measured based on the deflection of the nano picker. The effect of the length of contact time on cell adhesion was also evaluated.

Experimental Setup

Figure 8. 37 shows the experimental setup of single cell adhesion force measurement inside E-SEM. Several drops of solution contained yeast cells were put on the substrate inside E-SEM by micropipette. By gradually decreasing the chamber pressure, the water would evaporate. Therefore, yeast cells become visible and attach to the substrate. There were two nano robotic manipulators inside the E-SEM chamber, which were called as NM1 and NM2. The nano-picker and micro probe were assembled to NM1 and NM2 respectively Figure 8. 37 A). In order to measure the adhesion force between two cells, the first step was to make two cells adhere together. Here, the micro probe was used to transfer one cell onto the top of the target cell

(Figure 8. 37 B-C). The adhesion force between these two cells was measured using the nano-picker. First, the nano-picker tip inserted to the contact area of the two cells. Next, it moved upward to lift the upper cell and finally separated the two cells. The adhesion force F can be calculated based on the bounce height of the nano-picker by Hooke's law (Figure 8. 37 D-E). The time effect on the strength of the cell-cell adhesion force was studied by changing the contact time of the two single cells.

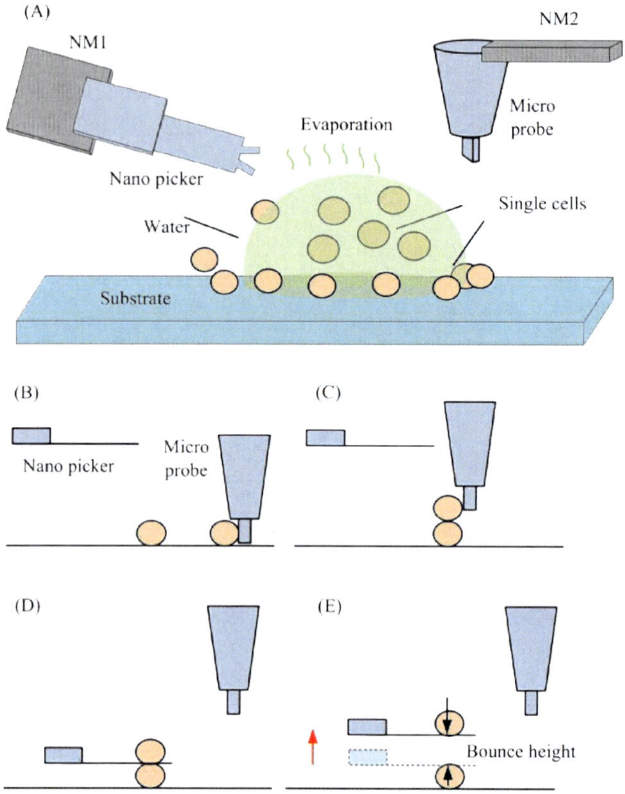

Fig. 8.37 Scheme drawing of the single cell adhesion force measurement using nano picker inside E-SEM. (A) Water solution containing yeast cells were pipetted and put on the substrate surface by micro pipette. The nano picker with V-shape tip was assembled to the nano manipupator 1 (NM1) and the micro probe was assembled to the nano manipulator 2 (NM2). (B) and (C) One single cell was transferred onto the top of another cell by using the micro probe. (D) and (E) The adhesion force is measured by the nano picker.

A schematic drawing of the single cell adhesion force measurement is shown in Figure 8. 38 . Under the driving of the nanorobotic manipulation system, the nano picker moved towards, and inserted to the contact area of the two cells. Then the nano picker moved upward until the two cells separated to each other. The dynamic movement of the nano picker and yeast cell can be observed from the real time image directly. Since the nano picker was fabricated from AFM cantilever, the value of adhesion force F can be calculated by Hooke's law: $F=k\delta$ where k is the spring constant of the nano picker and δ is the rebounded height of the nano picker, which can be obtained from direct measurements of E-SEM images. In this experiment, all of the dimensions were determined from the E-SEM images using image analysis software. The relative humidity around the cell was fixed to 70 % by setting the chamber pressure and temperature to 444 Pa and 0.5 °C respectively.

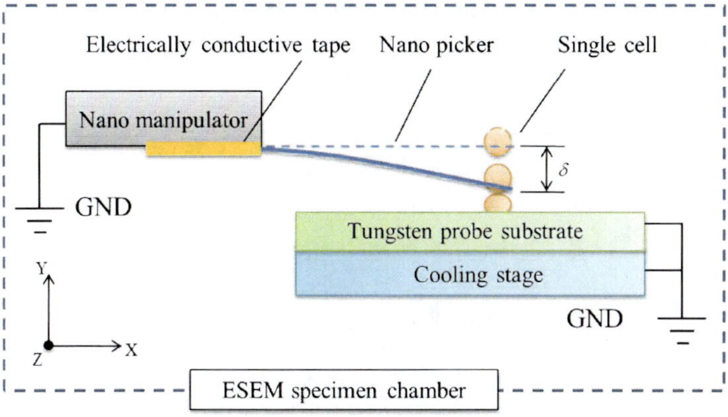

Fig. 8.38 Schematic drawing of cell-cell adhesion force calculation method

Single Cell Manipulation

8.7.1 *Preparation of a Micro Probe*

In micro and nano scale, surface tension forces, electrostatic forces, and Vander Waals forces, are the dominant forces [85]. Therefore, the end effector with a larger contract area can provide a larger manipulation force. Here, we fabricated micro probe from a tungsten probe to transfer the single cell using the adhesion force.

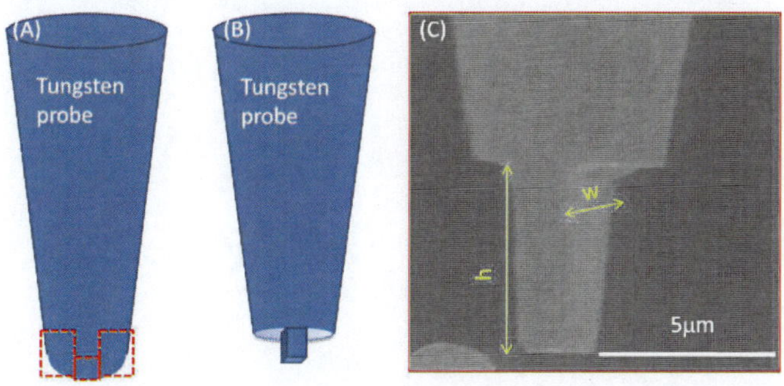

Fig. 8.39 Fabrication process of the end effector from tungsten probe

The micro probe was fabricated from a tungsten probe with a 3 μm round tip. A flat area was milled on the tungsten probe tip by focused ion beam (FIB) etching. The fabrication process was shown in Figure 8.39. Figure 8.39 (A) shows the schematic section drawing of the tungsten probe and the etching areas. Figure 8.39 (B) shows the schematic drawing of the micro probe. Figure 8. 39 (C) shows the final structure of the end effector under E-SEM observation. During the experiment, the flat area of the end effector will touch and push the cell. The height and width of the flat area were approximately 5 μm and 4 μm respectively. The dimensions are show in Table 8.8.

Table 8.8 Designed dimensions of the micro probe and nano picker

Parameters	Micro probe		Nano picker					
	h	w	a	b	c	d	θ	e
Values (μm)	5.0	4.4	7.0	4.2	0.7	4.1	20.0°	0.4

8.7.2 *Cell Transfer by the Micro Probe*

Cell transfer was achieved based on the adhesion force between the micro probe and cell. Figure 8. 40 shows the manipulation process. Figure 8. 40 (A) shows the initial position of cell and the micro probe. Under the gently pushing of the micro probe, the cell was released from the substrate and attached to the tip of probe (Figure 8. 40 (B)). Then, the single cell was raised to a height around 10 μm and moved toward the target cell (Figure 8. 40 (C)). Because the adhesion force between cell to cell is larger than the force to probe tip, the single cell can be released from the probe and attach to the target cell surface successfully (Figure 8. 40 (D)).

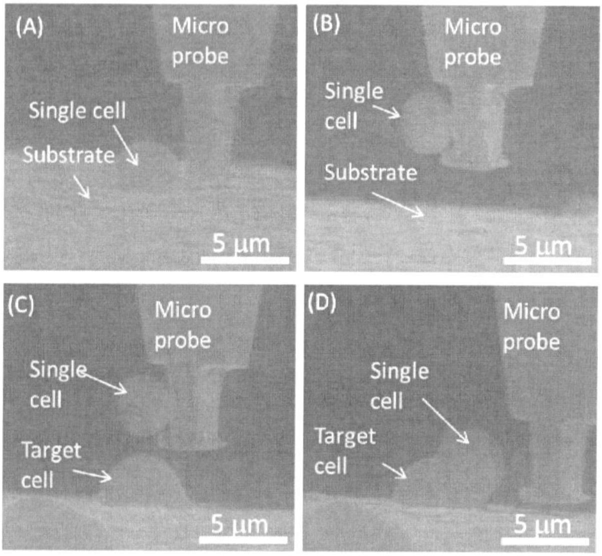

Fig. 8.40 Cell transfer by the micro probe

Cell-Cell Adhesion Force Measurement

Preparation of the Nano Picker

The nano picker was fabricated from a commercial AFM cantilever. This type of cantilever has two cantilever beams with different spring constant values, i.e., 0.02 N/m and 0.09 N/m. The cantilever with a spring constant 0.09 N/m was removed and the remained cantilever with a spring constant 0.02 N/m was put inside FIB chamber.

The V-shape nano-picker was fabricated by FIB etching technique. The distance between the two branches gradually decreased from 4 μm (open-end) to 1 μm (root-end). Therefore, it can insert into the contact area between two cells and to pick up the top cell more easily. Figure 8. 41 shows the detail fabrication process of the nano picker from the AFM cantilever. Figure 8. 41 (A) shows the schematic drawing of the commercial AFM cantilever from the top view. First, the tip of the cantilever was cut by FIB etching, as shown in Figure 8. 41 (B). The areas encircled by red dash line indicate the etching areas. And then, the cantilever beam was fabricated to a two branch structure as shown in Figure 8. 41 (C). The fabricated nano picker was observed under FIB and the image is shown in Figure 8. 42 Figure 8. 42 (A) shows the top view of the nano picker, and Figure 8. 42 (B) shows the 3D view of the nano picker. The designed dimensions of the nano-picker are also shown in Table 8. 8.

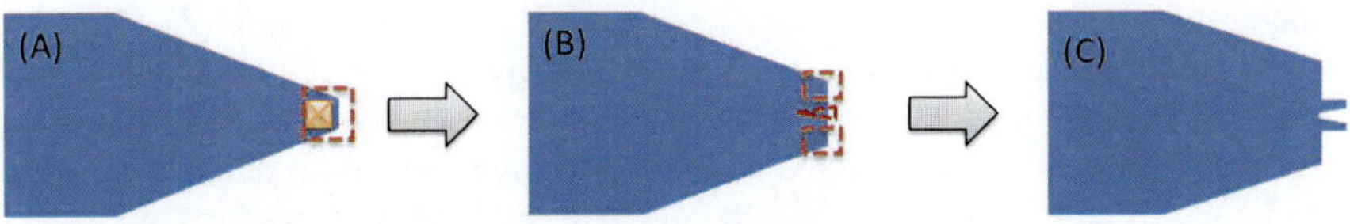

Fig. 8.41 Fabrication process of the nano picker. (A) Schematic drawing of the commercial AFM cantilever, and the areas encircled by red dash line indicate the etching areas. (B) Fabrication of the two branches. (C) Final structure of the nano picker.

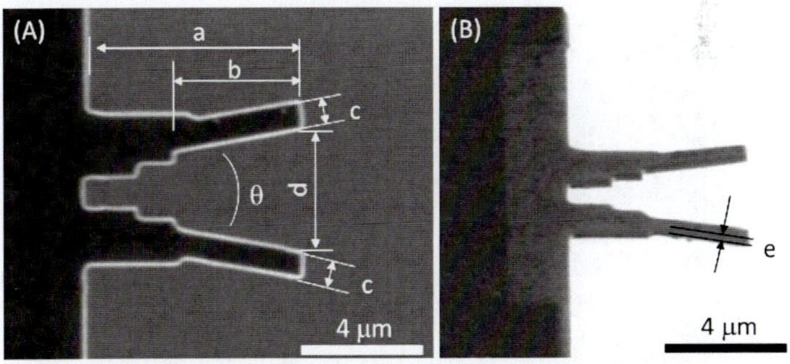

Fig. 8.42 Images of the nano picker. (A) Top view. (B) 3D view.

The AFM cantilever tip was etched during the fabrication and thereby the length of the AFM beam became shorter. Hence, the spring constant of the nano picker should become larger than the initial value. Considering the spring constant is the key parameter for the adhesion force measurement, the value of it was calibrated by using the nanomanipulation approach. The nano picker pressed a referenced cantilever (0.02 N/m) slowly using the nanomanipulator inside the E-SEM chamber. The deflections of the nano-picker and the 'referenced' cantilever could be obtained from the real time E-SEM image. Under equilibrium condition, the force acting on the cantilever and nano-picker must be in the same value. Therefore, the spring constant of the nano-picker can be calculated based on Hooke's law. The calibrated spring constant value is 0.025 ± 0.006 N/m for this nano picker.

Cell-Cell Adhesion Force Measurement via Nano Picker

The processes of single cell adhesion force measurement using the nano-picker is shown in Figure 8. 43. The body of the nano picker was assembled to NM1 using double side electro conductive tape. Therefore, the electrostatic force during the measurement can be eliminated. First, the nano picker inserted into the join area between the two cells driven by NM1 (Figure 8. 43 (A)). Once the V-shape branches were underneath the upper cell, the nano-picker moved up slowly to lift the single cell. With the increasing of the lift force, the upper cell was detached from the underneath cell finally (Figure 8. 43 (B)). The cell-cell adhesion force can be calculated based on the bounce height of the nano-picker tip by Hooke's law.

One single cell was picked up by the micro probe and put on the top of the target cell. The time at this initial condition is defined as time zero. After incubation for a certain time, the top cell was picked up using the nano picker to measure the adhesion force. Next, the single cell hold by the nano picker was moved and touched the cell on the substrate again. Finally, the adhesion force of cell-cell at different length of contact time can be measured by this procedure. The adhesion force versus different contact time period are shown in Figure 8. 44.The experiment was done with contact time period from 0 to 15 minutes. More data are added at the first few minutes, during which time the adhesion force has a sharp increase. It shows clearly that the adhesion forces are related to the length of contact time. At the first five minutes, the adhesion force between cell and cell has a sharp increase to 25 nN. After that, the fluctuation of adhesion force became stable.

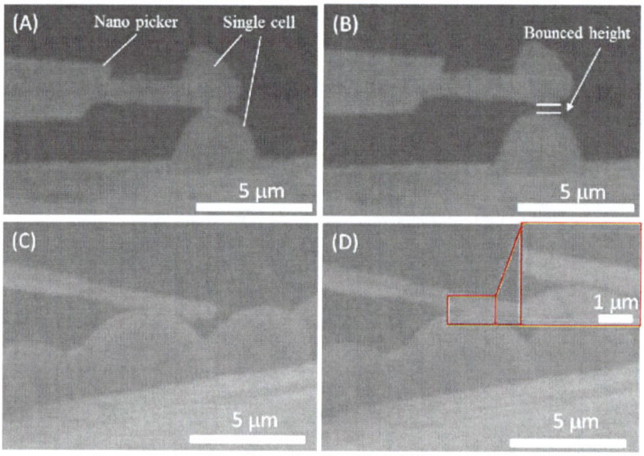

Fig. 8.43 Single cell adhesion force measurement experiment. (A) and (B) cell-cell adhesion force measurement. (C) and (D) cell-probe adhesion force measurement.

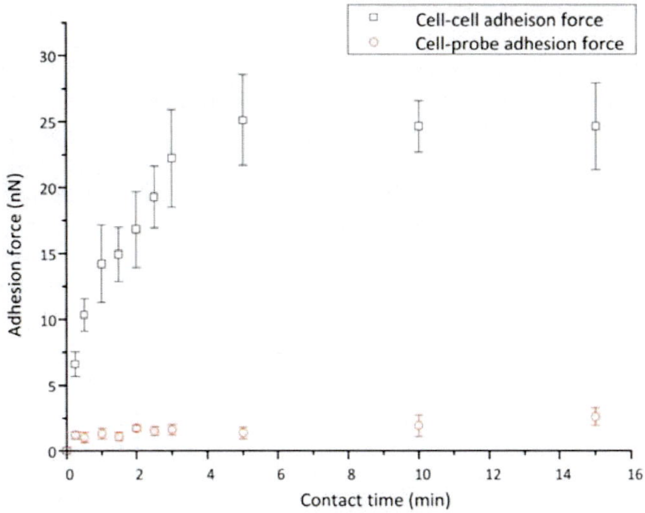

Fig. 8.44 Adhesion forces of cell-cell and cell-probe versus different contact time period

Evaluation of the Adhesion Force between the Nano Picker and a Single Cell Considering the adhesion between the nano picker and cell may influence the accuracy of the cell-cell adhesion force measurement. The adhesion force between the nano picker and the single cell was measured to evaluate such effect. As shown in Figure 8. 43, the nano picker touched the single cell for a certain time, and then it moved up until bounced back from the cell surface driven by NM1. Thereby, the adhesion force between the nano picker and the single cell can be obtained based on the bounce height. The adhesion forces between the nano-picker and the single cell at different contact time are also shown in Figure 8. 44.

The results demonstrate that the adhesion force is around 3 nN, though it has a slight rise with the increasing of contact time. It indicates that the adhesion force of cell-probe is much smaller than cell-cell at each contact time.

8.7.3 Results and Discussion

It has reported that the adhesion force of a melanoma cell and an endothelial cell is around 0.5 nN for approximately one second contact [84]. The adhesion force of two S180 cells is ranged from several nN to hundreds nN depending on the culture medium and the contact time [83]. In this paper, the adhesion force of single yeast cell-cell (*W303*) rises to approximately 25 nN with increasing of contact time. Because the cell types studied by each group were different to each other, the exact measured adhesion forces were also different. However, it demonstrated that the adhesion force of single cell-cell should be at nN scale. Consequently, it means that our result is reasonable and this methodology is suitable for single cell adhesion force measurement.

The interaction of cell to cell is usually regulated by specific chemical bonds, which are treated as springs in the cell adhesion kinetics model [86]. The totally adhesion force is the sum of the each bond's destructive force. It has reported that the number of active connection bond is related to the contact time, i.e., the number of bonds would increase with time until to a balance condition. Initial interaction events of cell-cell may occur from sub second to minutes timescale involve cytoskeletal redistribution, reinforcement of linkages and changes in cell motility. This interaction can be demonstrated on the way of the changing of the cell adhesion force. Our result shows the adhesion force was enhanced with the increasing of contact time at the first few minutes. After that, the force became constant. The force increasing time period demonstrates the initial interaction between cells. On the other hand, the constant force indicates that the interaction between the cells reached the balance condition. This result would benefit the understanding of how the cell interaction happened with time.

The influence of the adhesion between the nano picker and single cell can be ignored during the cell-cell adhesion study. There are two major reasons for this statement. First, the nano picker has a V-shape tip, which has two branches with a width approximately 0.7 μm and a height 0.4 μm each. The nano picker was parallel to the substrate during the experiment. Therefore, the contact area between the nano picker tip and the single cell, which has a spherical shape, was very small. Second, the attachment of cell-picker was smaller comparing to cell-cell at the same contact area, since there were more chemical bonds on the cell surface. Therefore, the adhesion force of cell-picker is much smaller than cell-cell due to the above two reasons. The experiment results show that the cell-probe adhesion force is much smaller than cell-cell force at each contact time, which agrees with the above analysis. Furthermore, in our experiment design, the nano picker inserted to the contact area between the two cells, which meant there was almost no contact between the nano picker and the single cell surface on substrate. Therefore, the real adhesion force between cell and picker should be even smaller than this measured result. It indicates that the influence of the nano-picker during the cell-cell adhesion force measurement can be ignored.

There is no obvious cell deformation during the adhesion measurement when using the nano picker in our experiment. Comparing to the method using the dual pipette, the physical damage to the cell can be reduced. On the other hand, comparing to the cell adhesion force measurement method under AFM, the previous cell immobilization is not required. Therefore, the influence of chemical glue to cell morphology and activate can be avoid. Furthermore, the cell on the nano picker can be changed after the experiment, which means the force measurement between arbitrary two cells can be performed as well. The nano picker was fabricated from AFM cantilever by using FIB etching technique. It is capable for different manipulation tasks by replacing the cantilever to another one with larger or smaller spring constant.

8.8 Electrical Measurement of Single Cell Using Dual Nanoprobe

In single cells analysis, it is very important to differentiate between live and dead cells. One of the main features separating dead from live cells is the loss of the physical integrity of the plasma membrane [87]. Because of this, cell viability is usually determined through examination of membrane integrity with colorimetric or fluorescent dyes [88].

However, the dye-exclusion based assays lack the capability to produce instantaneous and quantitative cell viability information, which is particularly important in the study of dynamics of cell death. For an instance, Kucsera et al. [89] proposed simple method for detecting dead and living colonies, however, this method requires several days for its completion and also requires the usage of additional several chemicals which may effect the physiology of the cells.

The birth of nanotechnology has provided numerous interesting ways which may overcome the limitation of the current conventional methods of any possible field. Single cells analyses which before seemed difficult to be realized are now showing rapid growth in research endeavors. The ability to perform characterization task at single cells level gives more better understanding of the biological cells which before realized on the data from the population averaging-based approach.

In order to successfully perform characterization of single cells, at least two main requirements are needed. First, the ability to observe a single cell at high resolution (in the range of few nanometers). Second, the ability to make a contact (directly, e.g. mechanical contact or indirectly, e.g. optical/laser contact) with a single cell. In response to these requirements, we have developed a system called E-SEM nanomanipulator system.

8.8.1 Analysis of the Single Cells Electrical Measurement Using Dual Nanoprobe

Figure 8. 45 shows a schematic of a single cell electrical measurement using dual nanoprobe inside E-SEM. The electrical measurement is performed under E-SEM mode (600 Pa and 0 °C) which provides 98 - 100 % humidity level of an environment around the cells. High humidity condition is needed in order to sustain the physiological condition of the cells.

The electrical measurement is performed by applying 2 V single pulses voltage on a single cell and measuring single pulses current that across that cell. Single pulses voltage sourcing and current measuring technique has twofold purposes. First, single pulses voltage sourcing is expected to greatly minimize the thermal noise which may influence the accuracy of the measurement data. Second, damage to the cells due to the Joule heating from the tip of the nanoprobe, which normally happen under continuous voltage sourcing technique, can be greatly minimized, if not eliminated.

In order to obtain useful single pulses current data from cells which have strong cell wall, such as *W303* cells, several considerations are required. Figure 8. 46 shows the schematics of the electrical model of a *W303* cell. It is expected that the resistance of the cell wall and membrane cell of the *W303* cells are very high as compared to the resistance of the cytoplasm [90]. Therefore, in order to obtain more reliable single pulses current data from *W303* cells, the cytoplasm is chosen as a medium of measurement.

Due to this, in order to obtain a significant electrical characterization on single cells, the electrical measurement must be done at the intracellular area of the cell, i.e. cytoplasm, by penetrating a dual nanoprobe inside the cell. This can be done with the help of the nanomanipulator unit which can provide a high resolution of linear motion for the dual nanoprobe to slightly penetrate a cell. Deeper penetration by a dual nanoprobe will burst the cells as the S area that covers region on the cell surface between first and second nanoprobe, will have high stress level during the penetration. The issue, i.e. cell bursting due to weak S area during penetration, is solved by performing short penetration. In this configuration, only tip of the dual nanoprobe is in contact with the intracellular area of the cell. By using this strategy, the electrical property of the single cells could be obtained.

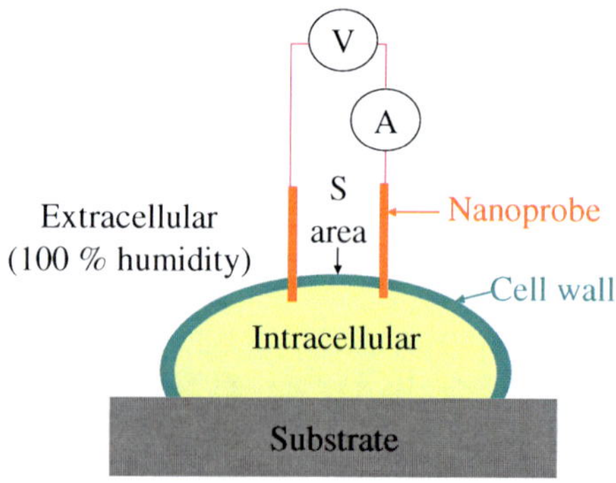

Fig. 8.45 Schematic of the single cells electrical measurement using dual nanoprobe

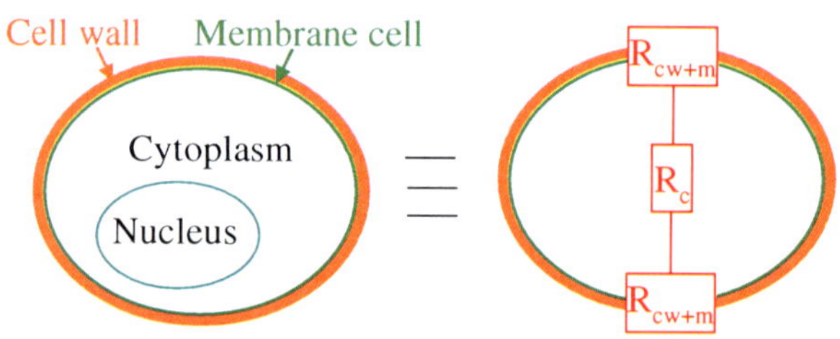

Fig. 8.46 Schematic of the electrical model of a single biological cell

Several studies reported that biological cells have a tendency to recover from small wounds. Therefore, it is expected that the measured cells are still alive after being slightly penetrated by the dual nanoprobe. Furthermore, in order to prevent excessive damages to the cells during the penetration, small diameter of the tip of the dual nanoprobe in the range of 246 nm is used.

The sample's cell sizes used in this experiment are from 4 to 6 μm. It is expected, however, that the cell size of the sample does not influence the measurement of the electrical property of the cytoplasm. The cytoplasm can be analogous to an electrolyte solution which contains ions. Ions act as corpuscles that carry electricity. The conductivity of the cytoplasm depends on the type and concentration of ions present [91], the temperature of the electrolyte solutions [92], the cross sectional area of the electrode and the distance between the electrodes [93]. In our experiments, the cross sectional area of the electrode and the distance between the electrodes are constants, i.e. 1.46 μm and 0.06 pm^2, respectively. The temperatures of the environment during the measurements are kept constants, i.e. for high vacuum mode (~ 17 °C / 0 % humidity) and E-SEM mode (8. 1 °C / 98-100 % humidity). It is suspected that when the integrity of the cell membrane is deteriorated, the ionic concentration of the cytoplasm cannot be maintained as the result of the ions or cytoplasm lost.

8.8.2 Fabrication of the Dual Nanoprobe

The nanoprobe was fabricated by using focused ion beam (FIB) etching and deposition processes on a commercial available piezoresistive cantilever. First, the FIB etching (8. 30 kV at 8221.5 × 10^{15} ion dose) was performed to remove the insulator layer (SiO_2) of the piezoresistive cantilever. Then, the FIB deposition (8. 30 kV) was executed to deposit tungsten (*W*) material on the conductive layer (doped Si) of the cantilever. Then, the FIB etching was performed to etch the cantilever plate into two separate plates where each plate contains the *W* deposition. Finally, the FIB etching was use to trim the *W* deposition into nanoneedle shape. In order to apply the nanoprobe under high humidity environment (~ 100% humidity level), the conductive layers which located on the PCB board of the nanoprobe were coated with the parylene material between steps. The thickness of the parylene coating was around 200 – 300 nm. Due to the trimming steps, the parylene coating was removed at the nanoprobe area.

The parylene coating is not only functioned as an insulator that protects the conductive layers from the interaction with water molecules under high humidity environment but also can provide a support to the base area of the nanoprobe which enhance the rigidity and strength of the nanoprobe during the contact with single cells. Figure 8. 47 shows the SEM images of the fabricated dual nanoprobe. The typical diameter and length of the nanoprobe are about 200-300 nm and 10-17 μm respectively.

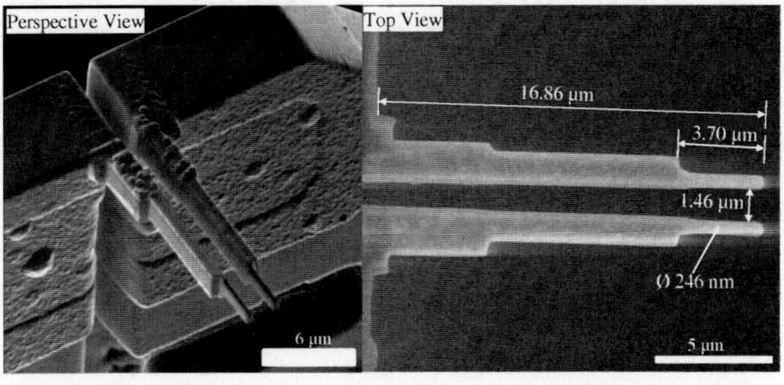

Fig. 8.47 SEM images of the dual nanoprobe at (a) perspective and (b) top views

Electrical Measurement Setup

The nanoprobe was attached to the nanomanipulator unit by inserting the nanoprobe inside a home-made connector which was firmly screwed on the nanomanipulator unit. The home-made connector provides a connection with the sub-femtoamp sourcemeter which located outside the E-SEM's chamber. The sourcemeter was further connected with the data acquisition (DAQ) card inside the computer (PC). The home-made graphical user interface (GUI) on the PC was developed using LabView source code.

The measurement protocol is based on the single pulse source voltage (8. 2 V) and single pulse measured current. The pulse width of the source voltage is 0.22 seconds. The pulse width (*pw*) of the source voltage is 0.22 seconds which was obtained from the following equation, $pw = (c / f) + s$, where c, f and s are number power line cycles, power line frequency and source delay. The values for c and s were programmed to be 1 and 0.2, respectively, while the value for f was 60 Hz.

Preparation of the Dead Cell Colonies of *W303* Cells

The preparation of the dead cell colonies was done by following the protocol proposed by [84]. The dyes Methylene Blue was added to the 25 ml YPD agar at 10 μmol/l. Plates were inoculated by streaking a yeast strain prepared from about 2-day-old agar culture and incubated for about 48 hours at 30 °C. In a model experiment, cells were killed within colonies growing on plates by applying a 80 μl drop of Amphotericin B (8. 1 $mg.ml^{-1}$) on the surface and allowing it to spread. Dye uptake from the medium was examined after 48 hours incubation at 30 °C.

Figure 8. 48 shows the images of dead cell colonies and live cell colonies on the same agar plate culture. The dead cell colonies have light blue color as the cell membrane is not functioning and therefore unable to prevent the die from coming inside the cells. The live cell colonies have a white color. y E-SEM observation (98 % humidity), as expected, most of the dead cells showed shrinkage morphology while the living cells showed round shape morphology.

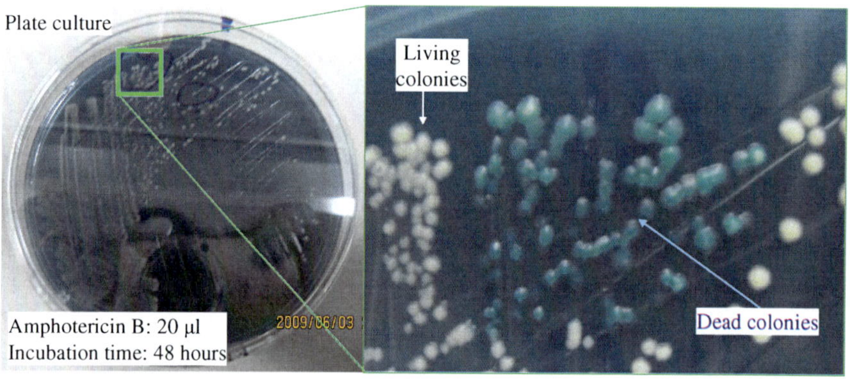

Fig. 8.48 Image of the W303 colonies at two conditions, i.e. dead (blue color) and live (white color)

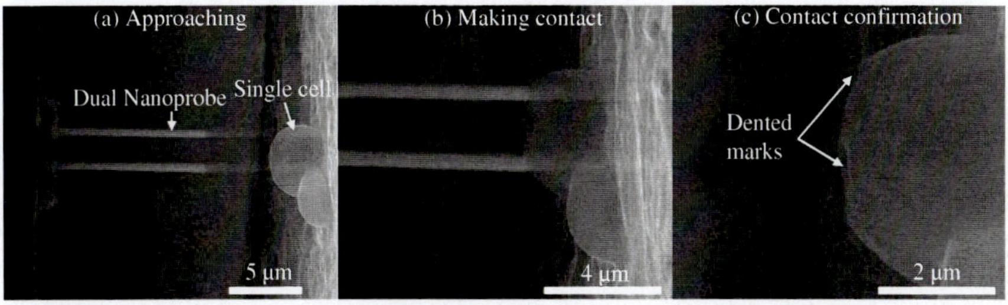

Fig. 8.49 Typical SEM images during the electrical probing on single cells using nanoprobe, (a) Before contact and (b) during contact. Successful of contact can be confirmed from the dented marks on a cell surface as shown in (c).

8.8.3 Electrical Measurement of Dead Cells under HV Mode

Figure 8. 49 shows the SEM images of a typical electrical probing on single cells by using a dual nanoprobe. The measurement was recorded when the dual nanoprobe penetrated slightly the cell wall of the single cells as shown in Figure 8. 49 (b). Once the contact was established, the electron beam was turned off and the single pulse voltage of 2 V with 0.22 s pulse width was applied. The electron beam was turned off to eliminate the influence of noise from the electron beam source. The single pulse current was then recorded. The single pulse voltage was selected than the continuous source voltage as it can reduce the effect of the noise interferences and also to minimize the effect of an excessive electrical potential to the physiology of the cells. The contact marks on the cell can be seen as shown in Figure 8. 49 (c). Figure 8.50 shows the typical result of electrical probing on single cells using nanoprobe under HV and ESEM.

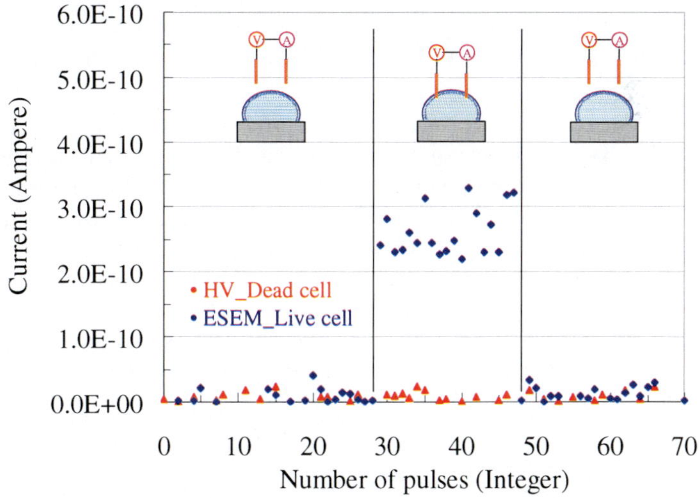

Fig. 8.50 Results of the electrical probing on single cells at two environmental conditions, i.e. HV condition for dead cell (red points), (b) E-SEM condition for live cells (blue points)

Figure 8. 51 shows the result of the electrical probing on single cells using nanoprobe under HV mode and E-SEM mode. It is interesting to notice that under HV mode, there was hardly any current flow through the cells. The recorded current was at the same level with the thermal noise level, i.e. to indicate there is no additional current flow during the measurement. Interestingly, when the cells were placed under E-SEM mode, the conducted electrical measurement found electrical current flow through the single cell, i.e. 2.63E-10 ± 1.8E-11 A. The statistical analysis was based on the Student's *t* test. The confident level of 95 % was used. The Student's *t* test is normally used when the mean population of the sample was unknown and the number of tested sample was small.

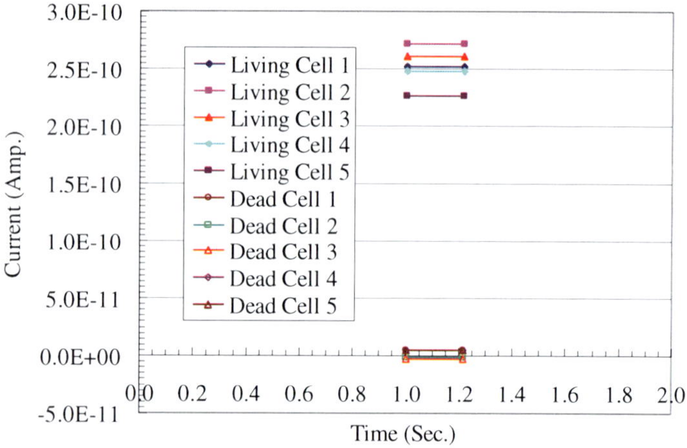

Fig. 8.51 Results of the electrical measurements on living cells (n = 5) and dead cells (n = 5) under ESEM mode

8.9 Automation System of Single Cell Analysis Using Nanotools

We developed the auto manipulation system inside E-SEM. A video capture card is used as the feedback module to build a control circle for the system. The flow chart of the automatic control is shown in Figure 8. 52. Firstly, a real-time image was got from the E-SEM system. Secondly, the position of the endeffector and the target cell was recognized based on image processing. The difference of these two position values is taken as the signal to control the movement direction of the nanomanipulator. Thirdly, the endeffector moved to the target cell automatically driven by the nanorobotic manipulator according to the position information. After the movement, the updated image was captured again to calculate the new position of the endeffector.

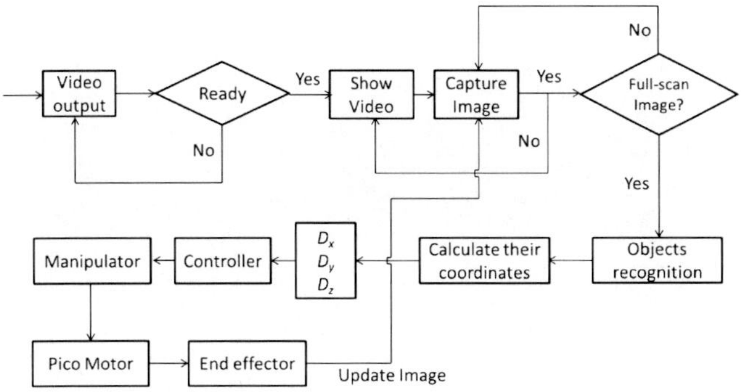

Fig. 8.52 Flow chart of the automatic manipulation system

8.9.1 Image Processing

The key part of the automatic control is based on the real time images. Therefore, the first step is to catch and save the E-SEM image. Here, we used two methods for the image capture, i.e. manually saving and auto saving with a given frequency. The manually saving is used when we set the template for the endeffector, and the automatic capture method is used during the movement of the endeffector.

The first step for image processing is to reduce the noise. There are two primary forms of noise in a EM: noise due to the statistical nature of electron collision and emission, and noise due to the SEM detector and signal processing electronics. The number of electrons hitting the specimen and then the corresponding emission of secondary electrons is inherently a statistical process. This results in shot noise extending up to high frequencies in the image. Shot noise is usually assumed to be both zero mean and independent. Noise can also originate from stray high-energy backscattered electrons colliding with either the detector or the chamber walls and emitting additional secondary electrons. Generally, the E-SEM system can eliminate these noises by itself using the hardware and software design. However, it is hard to deal with the noise from the outside, including the movement of the motor, and the external voltage potential inside E-SEM chamber. For example, the periodical noise is appeared when we using the oscillation knife to cut the single cell (Figure 8. 53), which is considered to be caused by the piezo actuator. Therefore, a median filter is used to reduce the noise in E-SEM image.

The second step for imaging processing is to find out the interesting regions. In our case, we are interest in two parts, i.e. single cell and the endeffector for single cell analysis. The endeffector is identified by the template matching technique as shown in Figure 8.51. In the nano knife indemnification, the tip of the nano knife was taken as a template due to its obvious shapThe nano knife template matching result. It demonstrates that the nano knife can be identified exactly. Next, the position of the nano knife tip can be obtained based on the image pix.

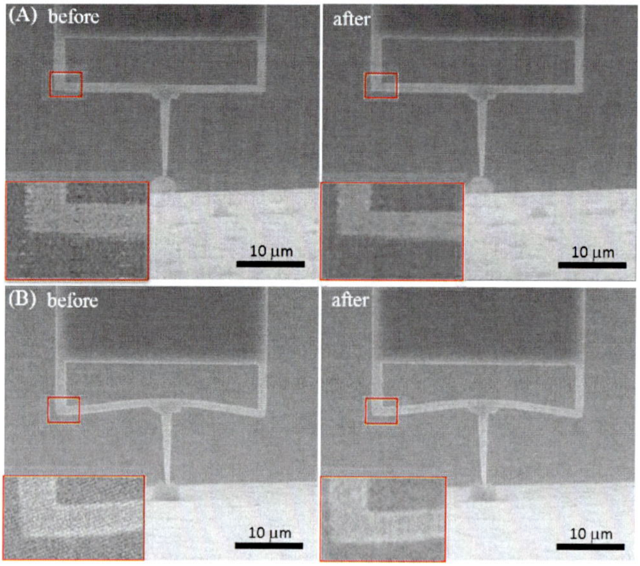

Fig. 8.53 Images comparison before and after the image filter. (A) and (B) are two examples at different experiment, the left images are the original images and the right images are the ones after filter. The left-bottom shows the enlarge image of part of the nano knife.

Yeast cell usually kept a semicircle shape inside E-SEM due to its adhesion to substrate, and the size and shape of cells are slightly different from each other. It hard for the definition what kind of cell do we want to analysis. Meanwhile, the interesting points of the cell for researcher are also different according to the analysis purposes. For example, the interesting area is the top part when we want to evaluate the cell wall's stiffness and the interesting area should change to the side part when we want to evaluate the shear adhesion force of the cell to substrate. Therefore, the single cell which will be used for analyzed is decided and chosen by the researcher. The program was set to recognize and save the position of the cell once people point out the cell and click the mouse on it. By using this human-robot corporation methodology, the analysis system can work more efficient.

One piezo motor driver can only drive one piezo motor each time. In order to increase the movement speed, the movements of the nanomanipulator to three directions are controlled by three piezo motor drivers respectively. Therefore, the endeffector driven by the three motors can move to the target directly. There are two kinds of movement mode for the piezo motor, i.e. step mode and speed mode. The piezo motor will move a certain distance with a given speed at the step mode. On the other hand, the piezo motor will move continuously at a certain speed without stop until setting the stop flag to true.

We designed two types of movement, i.e. the manually control at step mode and the automatic control at speed mode. The manually control is used for the cell manipulation purpose and the automatic control is used for the end effector positioning. In the automatic movement control, the distance between the endeffector and the cell is used as the control signal to set the stop flag value. The stop flag will be true when the distance is zero. Therefore, the piezo motor will move to the single cell automatically until the distance becomes zero.

The movement speed of the piezo motor ranges from 0 to 20 μm/s. As shown in Figure 8. 54, in order to reduce the movement time, the movement speed is set to the maximum value 20 μm/s when the distance between the end effector and the single cell is far than 30 μm. On the other hand, to prove the movement accuracy, the speed was set to 1 μm/s when the distance is less than 5 μm. The movement speed reduces linearly when the distance is between 30 μm to 5 μm.

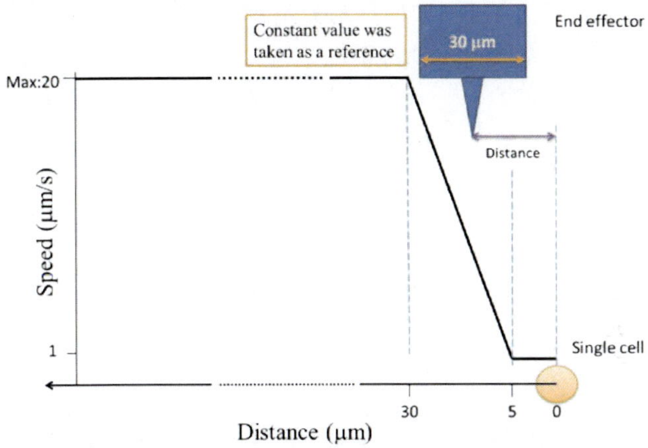

Fig. 8.54 Speed control of the piezo motor based on the distance between the end effector and the single cell

The distance between the end effector and the single cell is calculated based on the image pix. First, the width of the cantilever beam is taken as a scale bar 30 μm. the number of pixel is counting by image processing. Second, the distance value of each pixel is calculated. Finally, the distance between the end effector and the single cell can be got in real time.

8.9.2 *Experiment Result*

The automatic movement of the end effector to singe cell was demonstrated by taking the single cell cutting as an example. The nano knife was fixed to the nanorobotic manipulator, and the cell was arranged on the substrate inside E-SEM. The real time E-SEM video was captured and then the position of the objects was used as the control signal. The automatic movement of the nano knife was controlled based on real time image processing.

Figure 8. 55 shows the captured image during the automatic movement of the endeffector. The tip of the nano knife was recognized and marked with a green circle. The motors were controlled based on the position information. It can move to the target cell by itself. Form 0 s to 40 s, the nano knife moves to the first target cell automatically. Next, we choose another cell (target cell 2) for testing. It can be seen that the nano knife can move to the second cell automatically. The total time for the whole experiment is around 70 s. After that, single cell cutting experiment can be performed once the nanoknife tip touches the single cell. Therefore, we only need to focus on the single cell cutting analysis with the supporting of this automatic manipulation system.

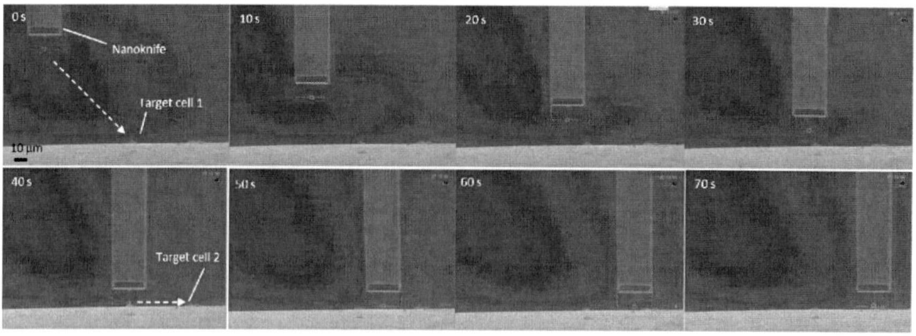

Fig. 8.55 Application of the automatic manipulation system for single cell cutting

8.10 Local Measurement and Injection to *C. elegans* Using Nanotools

We proposed to apply for the manipulation for the he *Caenorhabditis Elegans* (*C. elegans*). The *C. elegans* were widely investigated as model organism for understanding molecular and cellular mechanism of human genes, because they have many homologues genes with the human disease genes. Some studies have been done on the homologues of human disease genes, for example lrk-1 (a Parkinson's disease gene homolog) [94], and wnk-1 (a familial hypertension gene homolog) [95]. Their complex inner and outer structures are revealed by optical microscopes. For example, the lateral alae are the surface mark of seam cell body [96]. For example, the lateral alae are the surface mark of seam cell body as micro-nano scale fine structure [97]. As following section, some applications for *C. elegans* are introduced for the direct observation, local stiffness evaluation and local injection.

8.10.1 Direct Observation of C. elegans by E-SEM

Figure 8. 56 shows their optical microscopic images taken by a stereoscopic microscope. The lengths of *C. elegans* are from 100 μm to 1 mm depending on their growth phases. The wild type *C. elegans* were cultured on the BP plate (0.3% NaCl, 1.5% agarose, 0.25 % peptone) for around 3 days at room temperature. The *C. elegans* are picked up from the cultured plate, then they are dispersed in pure water. Several micro liter of the solution is dropped on the temperature controlling stage of E-SEM manipulation system by a micropipette.

Their E-SEM images of observation of *C. elegans* are shown in Figure 8. 57 at the conditions of 0 °C and 600 Pa [98]. Their shapes keep the round shape structures by water-contained condition with ESEM operation. The cell surface and inner structures are clearly observed depending on the water absorption by E-SEM (non-coating/non-dyeing/non-special treatment). As shown in the Figure 8. 57 (b), the "lateral alae" lines are also clearly observed. Their fine structures of *C. elegans* were directly observed by E-SEM, such as "Lips" and "Excretory pore".

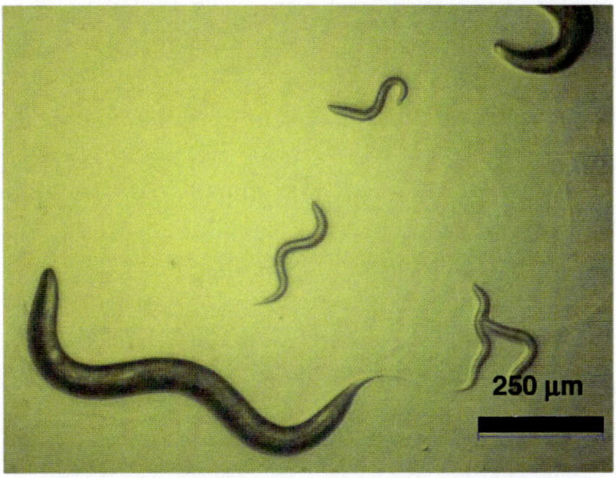

Fig. 8.56 Optical Microcsope of C. Elegances

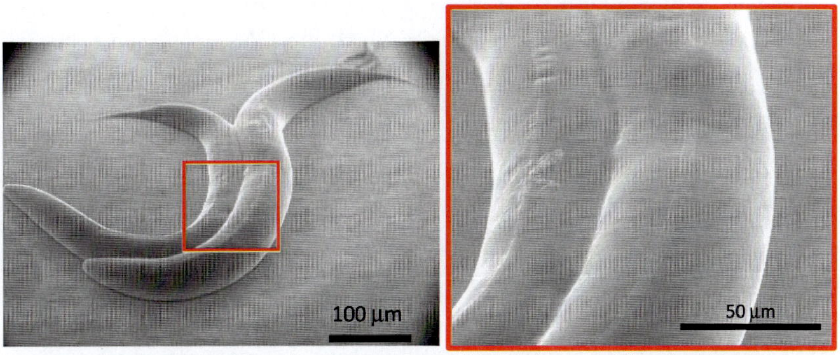

(a) Low magnification images (b) High magnification images

Fig. 8.57 Typical E-SEM image of *C. elegans* and its lateral alae

8.10.2 Viability Test of C. elegance though E-SEM

An alive condition is important to manipulate and evaluate of biological species. We evaluate the viability of *C. elegans* inside E-SEM from the motion of *C. elegans* before and after setting inside the E-SEM [98].

The *C. elegans* sample is set n the culturing medium directly. The medium is set on the temperature controlling stage with an aluminum plate. In the experiment, the pressure conditions (400, 500, and 600 Pa) are tested in 0 °C. From the experimental result, we confirm that all *C. elegans* moved before and after the E-SEM setting under all pressure conditions. The phenomenon was observed that the motion of *C. elegans* is slightly weakened after the low pressure.

Preliminary Experimental Results for Viability Test of *C. elegans*
Alive conditions are quite important to evaluate biological species by E-SEM nanorobotic manipulation system. To evaluate an alive condition of *C. elegans* inside E-SEM, the specimen chamber pressures are changed to determine the viability condition of *C. elegans* by E-SEM observation [99].

We used a *C. elegans* which was expressed eGFP was used to evaluate the viability of it. In the normal condition, the *C. elegans* under 20 °C shows the eGFP in only neuronal cells, hence it shows the concentrated and strong fluorescent intensity around the wall of *C. elegans*. On the other hand, the *C. elegans* under 40 °C shows the eGFP in body of *C. elegans* not only the neuronal cells, hence its whole body shows the fluorescent intensity un-emphatically. From this phenomenon, it is possible to determine the viability of *C. elegans* in this work.

In this paper, the pressure conditions (400, 500, 600, and 700 Pa) are tested (temperature: 0 °C). From these images, the *C. elegans* under 500, 600, and 700 Pa shows the concentrated fluorescent images of eGFP. On the other hand, *C. elegans* under 400 Pa shows the un-emphatical fluorescent images of eGFP. It is concluded that the observation of *C. elegans* under higher pressure than 500 Pa is enough to keep it in an alive condition for around 40 min.

The irradiation effect by electron beam is determined by hybrid microscope for *C. elegans* sample. In this paper, we set the chamber pressure in 700 Pa and irradiated up to 30 min by E-SEM. The observation condition of E-SEM is Acc. 15 kV and emission current 100 μA. The E-SEM and OM fluorescent images are shown by irradiating electron beams in each terms (0, 10, 20, 30 mins). From the experimental result, the fluorescent intensity of *C. elegans* is changed at 20 min un-emphatically. It is concluded that the observation of *C. elegans* under E-SEM observation in 10 min can keep it in alive condition.

8.10.3 Local Stiffness Evaluation of C. elegans Using Nanoprobe Inside E-SEM

The local stiffness of *C. elegans* was evaluated using nanoprobes fabricated by FIB etching at the tip of AFM cantilever [100]. The surface local stiffness of *C. elegans* is evaluated by applying the forces using nanoprobes form the deformation area. The deformation area of surface of *C. elegans* is depended on the stiffness of sample. When the "h" depth is applied by probes at the tip position, the deformation diameter of the high stiffness case "W_h" is smaller than that of low stiffness case "W_l". Hence, the stiffness value is evaluated from the measured diameter of deformation area by applying nanoprobes from the images of E-SEM.

Figure 8. 58 (a) show the schematic of experimental set-up of local stiffness evaluation of *C. elegans*. The nanoprobes are fabricated at the tip of AFM cantilever. It is manipulated by nanomanipulation system to evaluate various points on *C. elegans*, such as head, body, and tail. The measurement points can be arbitrarily controlled by nanomanipulation system with nano-scale revaluation from E-SEM imaging.

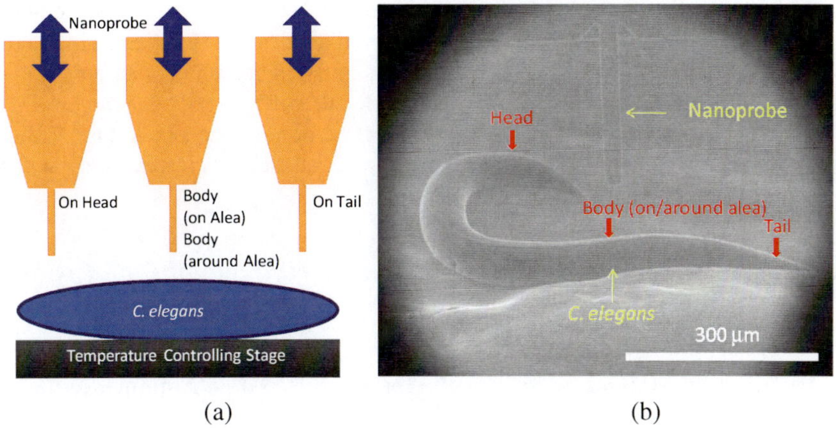

Fig. 8.58 Local stiffness measurement of *C. elegans* by nanoprobe, (a) schematic of experimental configuration, (b) Low magnification image during experiment

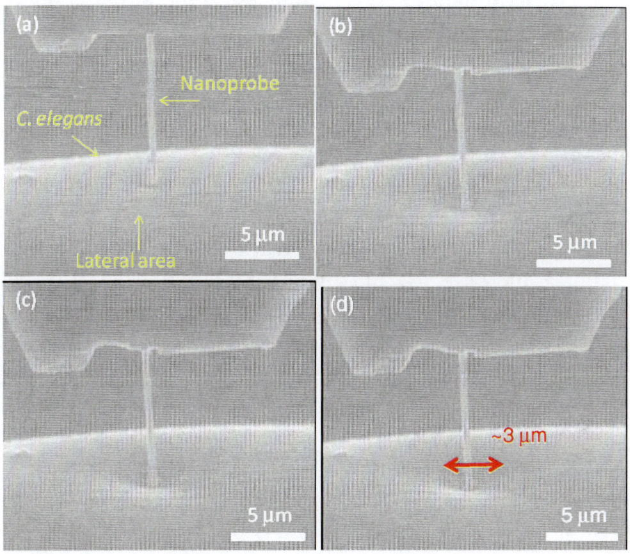

Fig. 8.59 Stiffness evaluation on the lateral area of *C. elegans* by a nanoprobe

Figure 8. 59 shows experimental set-up inside E-SEM with *C. elegans*. The fabricated nanoprobe is 449 nm in diameer and 9.95 µm in length as shown in Figure 8. 59 (b). In this experiment, the nanoprobe is inserted ~2.5 µm into the *C. elegans* surface.

The local stiffness of *C. elegans* was measured based on the E-SEM observation of their fine structures such as on or around lateral alae. The nanoprobe is set on the lateral alae of *C. elegans*. Then, the local force is applied on it by nanoprobe. On the other hand, the nanoprobe is set around the lateral alae of *C. elegans*. Then, the local force is applied on it by nanoprobe. From these results, the deformation of the cell surface around the lateral alae is approximately 3 times larger than it on the lateral alae under ~350 μN. The difference is considered to be caused by the inner structure on the lateral alae. The surface of *C. elegans* is much thicker than other area. This is the first experimental evaluation on the local stiffness of *C. elegans*. The quantitative measurement of cell stiffness will be done by cell modeling. Though the proposed evaluation method, the wider area on the *C. elegans* body will be measured.

8.10.4 Local Injection into C. elegans Using Nanoprobe Inside E-SEM

Nano-Injection Method by Nanoprobe Insertion Based on Nanomanipulation

We proposed a novel nanoinjection method by nanoprobe based on E-SEM nanomanipulation system [101]. Figure 8. 60 shows the schematic of nano-injection procedure by nanoprobe insertion. The nanoprobe has a neck to break the nanoprobe as shown in Figure 8. 60 (b) after inserting into biological specimen. At the tip of nanoprobe, the indicator or injection material can be attached and covered by sealing materials. The nano-scale small size probe is important to make the damage to biological specimen as minimum. The hybrid microscope can observe and determine the injection point and viability condition of biological sample from OM imaging.

The nanoprobe was fabricated by silicon material as shown in Figure 8. 61. The width of nanoprobe is ~600 nm, and the width of neck is ~ 200 nm. The height of nanoprobe is ~ 400 nm. The nanoprobe has a sharp tip to insert the probe into biological sample smoothly. The tip angle is ~45 degrees. The probe is fabricated by focused ion beam (FIB) process at a tip of commercially available AFM cantilever.

The nano-injection is demonstrated for the *C. elegans* using fabricated nanoprobe. In this experiment, a wild type of *C. elegans* (N2) is used. Figure 8. 62 the typical experimental results of nanoprobe insertion to *C. elegans*. We confirmed that the nanoprobe, which has ~200 nm width, is succeeded to be inserted inside the *C. elegans* body. From the bending of nanoprobe, the applied force for nanoprobe P_p to break the nanoprobe is estimated by the following equation.

$$P_p = \frac{E_p b_b h_p^{\,3} \delta_p}{4 l_p^{\,3}} \tag{8.57}$$

where, E_p is Young's modulus of nanoprobe, b_p is width of nanoprobe, h_p is height of nanoprobe, δ_p is the deflection of nanoprobe, l_p is length of nanoprobe from inserted point. The δ_p is measured as~2.5 μm from E-SEM image, and the applying force is calculated as ~700 nN by assuming E_p: ~100 GPa [102]. From this experiment, we confirmed that it is possible to insert the nanoprobe inside *C. elegans* by nanomanipulation inside E-SEM.

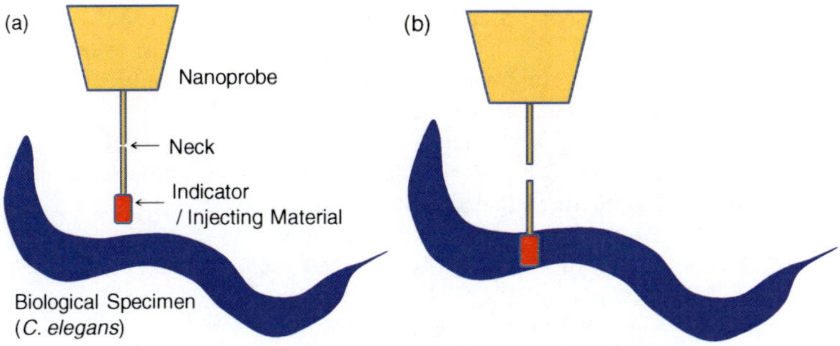

Fig. 8.60 Schematic of nanoinjection using nanoprobe based on nanomanipulation

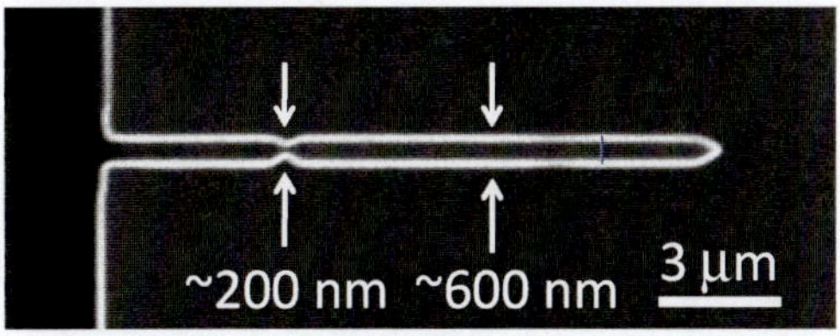

Fig. 8.61 ESEM image of nano-injector made by silicon material

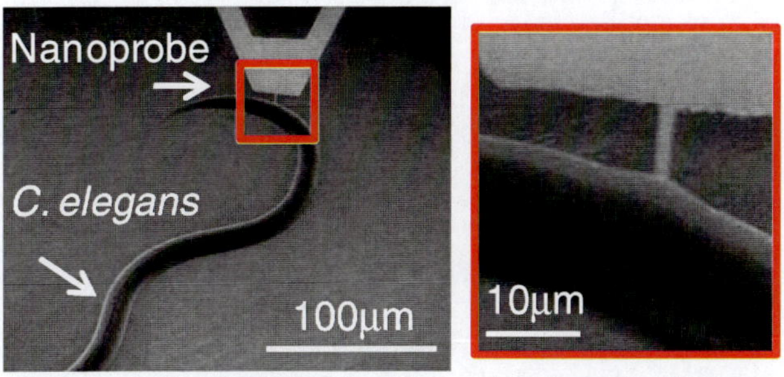

(a) Low magnification image (b) Magnified image

Fig. 8.62 Experimental results of nanoprobe insertion based on ESEM nanomanipulation system

References

[1] Fields, S., Johnston, M.: Cell biology: Whither model organism research? Science 307, 1885–1886 (2005)
[2] Fox, M.A.: The case for animal experimentation: An evolutionary and ethical perspective. University of California Press (1986)
[3] Botstein, D., Chervitz, S.A., Cherry, J.M.: Yeast as a model organism. Science 277, 1259–1260 (1997)
[4] Foury, F.: Human genetic diseases – a cross-talk between man and yeast. Gene 195, 1–10 (1997)
[5] Mager, W.H., Winderickx, J.: Yeast as a model for medical and medicinal research. TRENDS in Pharmacological Sciences 26, 265–273 (2005)

[6] Hartwell, L.H.: Yeast and cancer. Biosci. Rep. 22, 373–394 (2002)
[7] Misirli, Z., Oner, E.T., Kirdar, B.: Real imaging and size values of saccharomyces cerevisiae cells with comparable contrast tuning to two environmental scanning electron microscopy modes. Scanning 29, 11–19 (2007)
[8] Fischer-Cripps, A.C.: Nanoindentation, 1st edn. Springer (2002)
[9] Stephen, N.G.: Macaulay's method for a timoshenko beam. International Journal of Mechanical Engineering Education 35, 285–292 (2007)
[10] Sader, J.E.: Parallel beam approximation for v-shaped atomic force, microscope cantilevers. Review of Scientific Instruments 66, 4583–4587 (1995)
[11] O'Mara, W.C., Herring, R.B., Hunt, L.P.: Handbook of semiconductor silicon technology. Noyes Publications, UK (1990)
[12] Hertz, H.: On the contact of elastic solids. J. Reine und Angewandte Mathematik 92, 156–171 (1882)
[13] Pharr, G.M., Oliver, W.C., Brotzen, F.R.: On the generality of the relationship among contact stiffness, contact area, and elastic modulus during indentation. J. Mat. Res. 7, 619–627 (2002)
[14] A-Hassan, E., Heinz, W.F., Antonik, M.D., D'Costa, N.P., Nageswaran, S., Schoenenberger, C.-A., Hoh, J.H.: Relative microelastic mapping of living cells by atomic force microscopy. Biophysical Journal 74, 1564–1578 (1998)
[15] Klis, F.M., Boorsma, A., Groot, P.W.J.D.: Cell wall construction in saccharomyces cerevisiae. Yeast 23, 185–202 (2006)
[16] Pelling, A.E., Sehati, S., Gralla, E.B., Valentine, J.S., Gimzewski, J.K.: Local nanomechanical motion of the cell wall of saccharomyces cerevisiae. Science 305, 1147–1150 (2004)
[17] Sader, J.E.: Parallel beam approximation for v-shaped atomic force, microscope cantilevers. Review of Scientific Instruments 66, 4583–4587 (1995)
[18] Mendels, D.-A., Lowe, M., Cuenat, A., Cain, M.G., Vallejo, E., Ellis, D., Mendels, F.: Dynamic properties of afm cantilevers and the calibration of their spring constants. Journal of Micromechanics and Microengineering 16, 1720–1733 (2006)
[19] Aksu, S.B., Turner, J.A.: Calibration of atomic force microscope cantilevers using piezolevers. Review of Scientific Instruments 78, 1–8 (2007)
[20] Brehm-Stecher, B.F., Johson, E.A.: Single-cell microbiology: Tools, technologies, and applications. Microbiology and Molecular Biology Reviews 68, 538–559 (2004)
[21] Harold, F.M.: Molecules into cells: Specifying spatial architecture. Microbiology and Molecular Biology 69, 544–564 (2005)
[22] Elowitz, M.B., Levine, A.J., Siggia, E.D., Swain, P.S.: Stochastic gene expression in a single cell. Science 297, 1183–1186 (2002)
[23] Shapiro, H.M.: Microbial analysis at the single-cell level: Tasks and techniques. Journal of Microbiological Methods 42, 3–16 (2000)
[24] Fung, Y.C.: Biomechanics: Mechanical properties of living tissue, 2nd edn. Springer (2005)
[25] Smith, A.E., Zhang, Z., Thomas, C.R., Moxham, K.E., Middelberg, A.P.J.: The mechanical properties of saccharomyces cerevisiae. Proceedings of the National Academy of Sciences 97, 9871–9874 (2000)
[26] Obataya, I.: Nanoscale operation of a living cell using an atomic force microscope with a nanoneedle. Nano Letters 5, 27–30 (2005)
[27] Lanero, T.S.: Mechanical properties of single living cells encapsulated in polyelectrolyte matrixes. Journal of Biotechnology 124, 723–731 (2006)
[28] Donald, A.M.: The use of environmental scanning electron microscopy for imaging wet and insulating materials. Nature Materials 2, 511–516 (2003)

[29] Obataya, I., Nakamura, C., Han, S.W., Nakamura, N., Miyake, J.: Nanoscale operation of a living cell using an atomic force microscope with a nanoneedle. Nano Letters 5, 27–30 (2005)

[30] Lanero, T.S., Cavalleri, O., Krol, S., Rolandi, R., Gliozzi, A.: Mechanical properties of single living cells encapsulated in polyelectrolyte matrixes. Journal of Biotechnology 124, 723–731 (2006)

[31] Touhami, A., Nysten, B., Dufrene, Y.F.: Nanoscale mapping of the elasticity of microbial cells by atomic force microscopy. Langmuir 19, 4539–4543 (2003)

[32] Sneddon, I.N.: The relation between load and penetration in the axisymmetric boussinesq problem for a punch of arbitrary profile. Int. J. Engng. Sci. 3, 47–57 (1965)

[33] King, R.B.: Elastic analysis of some punch problems for a layered medium. International Journal of Solids and Structures 23, 1657–1664 (1987)

[34] Dao, M., Chollacoop, N., Van Vliet, K.J., Venkatesh, T.A., Suresh, S.: Computational modelling of the forward and reverse problems in instrumented sharp indentation. Acta Materialia 49, 3899–3918 (2001)

[35] Tortora, G.J., Funke, B.R., Case, C.L.: Microbiology: An introduction, 8th edn. Benjamin Cummings (2003)

[36] Bausch, A.R., Moller, W., Sachmann, E.: Measurement of local viscoelasticity and forces in living cells by magnetic tweezers. Biophysical Journal 76, 573–579 (1999)

[37] Gueta, R., Barlam, D., Shneck, R.Z., Rousso, I.: Measurement of the mechanical properties of isolated tectorial membrane using atomic force microscopy. Proceedings of the National Academy of Sciences 103, 14790–14795 (2006)

[38] Guck, J., Ananthakrishnan, R., Mahmood, H., Moon, T.J., Cunningham, C.C., Kas, J.: The optical stretcher: A novel laser tool to micromanipulate cells. Biophysical Journal 81, 767–784 (2001)

[39] Thoumine, O., Ott, A., Cardoso, O., Meister, J.J.: Microplates: A new tool for manipulation and mechanical pertubation of individual cells. J. Biochem. Biophys. Meth. 39, 47–62 (1999)

[40] Leary, S.P., Liu, C.Y., Apuzzo, M.L.I.: Toward the emergence of nanoneurosurgery: Part iii - nanomedicine: Targeted nanotherapy, nanosurgery, and progress toward the realization of nanoneurosurgery. Neurosurgery 58, 1009–1025 (2006)

[41] Bianco, A., Kostarelos, K., Prato, M.: Applications of carbon nanotubes in drug delivery. Current Opinion in Chemical Biology 9, 674–679 (2005)

[42] McKnight, T., Melechko, A., Griffin, G., Guillorn, M., Merkulov, V., Serna, F., Hensley, D., Doktycz, M., Lowndes, D., Simpson, M.: Intracellular integration of synthetic nanostructures with viable cells for controlled biochemical manipulation. Nanotechnology 14, 551–556 (2003)

[43] Brigger, I., Dubernet, C., Couvreur, P.: Nanoparticles in cancer theraphy and diagnosis. Advance Drug Delivery Reviews 54, 631–651 (2002)

[44] Labhasetwar, V.: Nanotechnology for drug and gene theraphy: The importance of understanding molecular mechanisms of delivery. Current Opinion in Biotechnology 16, 674–680 (2005)

[45] Stylios, G.K., Giannoudis, P.V., Wan, T.: Applications of nanotechnologies in medical practice. Injury. Int. J. Care Injured 36, 6–13 (2005)

[46] Goodman, M.B., Lumpkin, E.A., Ricci, A., Tracey, W.D., Kernan, M., Nicolson, T.: Molecules and mechanisms of mechanotransduction. The Journal of Neuroscience 24, 9220–9222 (2004)

[47] Brigger, I., Dubernet, C., Couvreur, P.: Nanoparticles in cancer theraphy and diagnosis. Advance Drug Delivery Reviews 54, 631–651 (2002)

[48] Han, S.W., Nakamura, C., Obataya, I.: A molecular delivery system by using afm and nanoneedle. Journal of Biosensors & Bioelectronics 20, 2120–2125 (2005)

[49] Raimann, P.R., Hochgatterer, N.S., Korepp, C., Moeller, K.C., Winter, M., Schroettner, H., Hofer, F., Besenhard, J.O.: Monitoring dynamics of electrode reactions in li-ion batteries by in situ esem. IONICS 12, 253–255 (2006)

[50] Smith, A.J., Atkinson, H.V., Hainsworth, S.V., Dong, H.B., Haghayeghi, R.: In situ environmental scanning electron microscopy (esem) of semi-solid samples. Semi-Solid Processing of Alloys and Composites Solid State Phenomena, 116–117 (2006)

[51] Johnston, G.C., Pringle, J.R., Hartwell, L.H.: Coordination of growth with the cell division in the yeast saccharomyces cerevisiae. Experimental Cell Research 105, 79–98 (1977)

[52] Thomas, K.C., Dawson, P.S.S., Gamborg, B.L.: Differential growth rates of candida utilis mother and daughter cells under phase cultivation. Journal of Bacteriology 141, 1–9 (1980)

[53] Tan, Y., Sun, D., Huang, W., Che, S.H.: Mechanical modeling of biological cells in microinjection. IEEE Transactions on Nanobioscience 7, 257–266 (2008)

[54] Sokabe, M., Sachs, F., Jing, Z.: Quantitative video microscopy of patch clamped membranes stress, strain, capacitance, and stretch channel activation. Biophys. J. 59, 722–728 (1991)

[55] Radmacher, M., Fritz, M., Hansma, P.K.: Imaging soft samples with the atomic force microscope: Gelatin in water and propanol. Biophysical Journal 69, 264–270 (1995)

[56] Ahimou, F., Touhami, A., Dufrene, Y.F.: Real-time imaging of the surface topography of living yeast cells by atomic force microscopy. Yeast 20, 25–30 (2003)

[57] De-Deus, G., Paciornik, S., Mauricio, M.H.P., Prioli, R.: Real-time atomic force microscopy of root dentine during demineralization when subjected to chelating agents. International Endodontic Journal 39, 683–692 (2006)

[58] Engel, A., Müller, D.J.: Observing single biomolecules at work with the atomic force microscope. Nature Structural Biology 7 (2000)

[59] Timoshenko, S.P., Gere, J.M.: Theory of elastic stability, 2nd edn. McGraw Hill, New York (1961)

[60] Ahmad, M.R., Nakajima, M., Kojima, S., Homma, M., Fukuda, T.: In situ single cell mechanics characterization of yeast cells using nanoneedles inside environmental sem. IEEE Transactions on Nanotechnology 7, 607–616 (2008)

[61] Wu, H.W., Kuhn, T., Moy, V.T.: Mechanical properties of l929 cells measured by atomic force microscopy: Effects of anticytoskeletal drugs and membrane crosslinking. Scanning 20, 389–397 (1998)

[62] Li, X., Bhushan, B.: A review of nanoindentation continuous stiffness measurement technique and its applications. Materials Characterization 48, 11–36 (2002)

[63] Radok, J.R.M.: Viscoelastic stress analysis. Quart. Appl. Math. 15, 198–202 (1957)

[64] Lee, E.H., Radok, J.R.M.: The contact problems for viscoelastic bodies. J. Appl. Math. 27, 438–444 (1960)

[65] Lee, E.H.: Stress analysis in visco-elastic bodies. Quart. Appl. Math. 13, 183–190 (1955)

[66] Cheng, Y.-T., Cheng, C.-M.: Relationships between initial unloading slope, contact depth, and mechanical properties for spherical indentation in linear viscoelastic solids. Materials Science and Engineering A 409, 93–99 (2005)

[67] Sato, M., Ohshima, N., Nerem, R.M.: Viscoelastic properties of cultured porcine aortic endothelial cells exposed to shear stress. J. Biomech. 29, 461–467 (1996)

[68] Thoumine, O., Ott, A.: Time scale dependent viscoelastic and contractile regimes in fibroblasts probed by microplate manipulation. J. Cell Sci. 110, 2109–2116 (1997)

[69] Stenson, J.D., Thomas, C.R., Hartley, P.: Modelling the mechanical properties of yeast cells. Chemical Engineering Science 64, 1892–1903 (2009)

[70] Lee, E.H., Radok, J.R.M.: The contact problem for viscoelastic bodies. Journal of Applied Mechanics 27, 438–444 (1960)
[71] Cheng, L., Xia, X., Yu, W., Scriven, L.E., Gerberich, W.W.: Flat-punch indentation of viscoelastic material. Journal of Polymer Science: Part B: Polymer Physics 38, 10–22 (2000)
[72] Lu, H., Wang, B., Ma, J., Huang, G., Viswanathan, H.: Measurement of creep compliance of solid polymers by nanoindentation. Mech. Time-Depend. Mater. 7, 189–207 (2003)
[73] Cheng, Y.-T., Cheng, C.-M.: Relationships between initial unloading slope, contact depth, and mechanical properties for spherical indentation in linear viscoelastic solids. Materials Science and Engineering A 409, 93–99 (2005)
[74] Gibson, C.T., Watson, G.S., Myhra, S.: Determination of the spring constants of probes for force microscopy/spectroscopy. Nanotechnology 7, 259–262 (1996)
[75] Mancini, M., Moresi, M.: Rheological behaviour of bakers yeast suspension. Journal of Food Engineering 44, 225–231 (2000)
[76] Lo Presti, S., Moresi, M.: Development of a novel integrated membrane process for citric acid and production by yarrowia lipolytica. In: Fourth Italian Conference on Chemical and Process Engineering (IChea P4), Florence, Italy, pp. 411–414 (1999)
[77] Speer, R.A., Durance, T.D., Tung, M.A., Thou, J.: Colloidal properties of flocculent and nonflocculent brewing yeast suspensions. Biotechnology Progress 9, 267–272 (1993)
[78] Tolic-Norrelykke, I.M., Munteanu, E.-L., Thon, G., Oddershede, L., Berg-Sorensen, K.: Anomalous diffusion in living yeast cells. Physical Review Letters 93, 1–4 (2004)
[79] Cheng, Y.-T., Cheng, C.-M.: Relationship between contact stiffness, contact depth, and mechanical properties for indentation in linear viscoelastic solids using axisymmetric indenters. Struct. Control Health Monit. 13, 561–569 (2006)
[80] Hwang, D.S., Sim, S.B., Cha, H.J.: Cell adhesion biomaterial based on mussel adhesive protein fused with rgd peptide. Biomaterials 28, 4039–4046 (2007)
[81] Klotz, S.A., Drutz, D.J., Zajic, J.E.: Factors governing adherence of candida species to plastic surfaces. Infection and Immunity 50, 97–101 (1985)
[82] Kim, M., Ahn, J.W., Jin, U.H., Choi, D., Paek, K.H., Pai, H.S.: Activation of the programmed cell death pathway by inhibition of proteasome function in plants. Journal of Biological Chemistry 278, 19406–19415 (2003)
[83] Chu, Y.S., Thomas, W.A., Eder, O., Pincet, F., Perez, E., Thiery, J.P., Dufour, S.: Force measurements in E-cadherin?mediated cell doublets reveal rapid adhesion strengthened by actin cytoskeleton remodeling through Rac and Cdc42. The Journal of Cell Biology 167, 1183 (2004)
[84] Puech, P.H., Poole, K., Knebel, D., Muller, D.J.: A new technical approach to quantify cell-cell adhesion forces by AFM. Ultramicroscopy 106, 637–644 (2006)
[85] Abbott, J.J., Nagy, Z., Beyeler, F., Nelson, B.J.: Robotics in the small, part I: Microbotics. IEEE Robotics & Automation Magazine 14, 92–103 (2007)
[86] Dembo, M., Torney, D., Saxman, K., Hammer, D.: The reaction-limited kinetics of membrane-to-surface adhesion and detachment. Proceedings of the Royal Society of London. Series B. Biological Sciences 234, 55–83 (1988)
[87] King, M.A.: Detection of deal cells and measurement of cell killing by flow cytometry. J. Immunol. Methods 243, 3–12 (2000)
[88] Bonorat, A., Mares, D.: A simple colorimetric method for detecting cell viability in cultures of eukaryotic microorganisms. Current Microbiology 7, 217–222 (1982)
[89] Kucsera, J., Yarita, K., Takeo, K.: Simple detection method for distinguishing dead and living yeast colonies. Journal of Microbiological Methods 41, 19–21 (2000)

[90] Raicu, V., Raicu, G., Turcu, G.: Dielectric properties of yeast cells as simulated by the two-shell model. Biochimica et Biophysica Acta 1274, 143–148 (1996)
[91] Bianchi, H., Fernandez-Prini, R.: The conductivity of dilute electrolyte solutions: Expanded lee and wheaton equation for symmetrical, unsymmetrical and mixed electrolytes. Journal of Solution Chemistry 22, 557–570 (1993)
[92] Kuyucak, S., Chung, S.-H.: Temperature dependence of conductivity in electrolyte solutions and ionic channels of biological membranes. Biophysical Chemistry 52, 15–24 (1994)
[93] Bester-Rogac, M., Habe, D.: Modern advances in electrical conductivity measurements of solutions. Acta Chim. Slov. 53, 391–395 (2006)
[94] Moriguchi, T., Urushiyama, S., Hisamoto, N., Iemura, S., Uchida, S., Natsume, T., Matsumoto, K., Shibuya, H.: WNK1 Regulates Phosphorylation of Cation-Chloride-Coupled Kinases, SPAK and OSR1. J. Biol. Chem. 280, 42685–42693 (2005)
[95] O'Kane, C.J.: Modelling human diseases in Drosophila and Caenorhabditis. Seminars in Cell & Developmental Biology 14(1), 3–10 (2003)
[96] Sakaguchi-Nakashima, A., Meir, J., Jin, Y., Matsumoto, K., Hisamoto, N.: LRK-1, a C. elegans PARK8-related Kinase, Regulates Axonal-dendritic Polarity of SV Proteins. Curr. Biol. 17, 592–598 (2007)
[97] Wood, W.B.: The Nematode Caenorhabditis Elegans, p. 38. Cold Spring Harbor Laboratory Pr. (1988)
[98] Nakajima, M., Ahmad, M.R., Kojima, S., Homma, M., Fukuda, T.: Local Stiffness Evaluation for Alive C. elegans by Environmental-SEM Nanorobotic Manipulation System. In: Proc. of the 2009 Int. Symposium on Micromechatronics and Human Science (MHS 2009), pp. 638–643 (2009)
[99] Nakajima, M., Akimoto, H., Hirano, T., Kojima, M., Hisamoto, N., Homma, M., Fukuda, T.: Biological Specimen Viability Analysis by Hybrid Microscope Combined Optical Microscope and Environmental-SEM. In: Proceedings of 2010 IEEE Nanotech. Materials and Devices Conference (IEEE-NMDC 2010), pp. 315–320 (2010)
[100] Nakajima, M., Ahmad, M.R., Kojima, S., Homma, M., Fukuda, T.: Local Stiffness Measurements of C. elegans by Buckling Nanoprobes inside and Environmental SEM. In: Proc. of the 2009 IEEE/RSJ Int. Conf. on Intelligent Robots and Systems (IROS2009), pp. 1849–1854 (2009)
[101] Nakajima, M., Hirano, T., Kojima, M., Hisamoto, N., Homma, M., Fukuda, T.: Direct Nano-injection Method by Nanoprobe Insertion based on E-SEM Nanorobotic Manipulation under Hybrid Microscope. In: Proc. of the 2011 IEEE Int. Conf. on Robotics and Automation (ICRA 2011), pp. 4139–4144 (2011)
[102] O'Mara, W.C., Herring, R.B., Hunt, L.P.: Handbook of Semiconductor Silicon Technology. Noyes Publications (1990)

Chapter 9
Conclusion

9.1 Summary

In this book, we introduced the advanced technology from the basics and applications aspects of micro-nanorobotic manipulation system with their applications and related technologies. This book is organized as follows:

In Chapter 1, we introduced the overview and introduction of micro-nanorobotics and manipulation technology with some historical background and current achievements with related works.

In Chapter 2, the physics in micro-nano fields was explained in the scaling effects, mechanics, electronics, fluidics, surface interaction, and optical physics.

In Chapter 3, the related technologies in micro-nano scale were explained about the material and science, microscopes, fabrication technology, sensing and actuation, control techniques, and assembly techniques.

In Chapter 4, biomanipulation system under optical microscope was explained under optical microscope by on-chip micro-fluidics devices, micro-tools, and laser micro-nano manipulations.

In Chapter 5, a rotational speed control of single bacterial flagellar motor is introduces as one of the applications of micromanipulation system under optical microscope.

In Chapter 6, nanomanipulation system was introduced under electron microscopes, with configurations of nano-robotics and manipulation system, hybrid nanorobotic manipulation system inside FE-SEM/TEM, nanorobotic manipulation system inside E-SEM, nano-tool exchanger system under hybrid microscope, automation of nanorobotic manipulation system inside E-SEM.

In Chapter 7, measurement, manipulation, and assembly of carbon nanotubes were explained under FE-SEM and TEM based on the nanomanipulation system. The techniques of picking up, cutting, peeling, bending, fixing of carbon nanotubes are presented as one of the applications of nanomanipulation system under optical microscope.

In Chapter 8, measurement and analysis of biological cells were explained under E-SEM for single cell nano-surgery system, especially mechanical property characterization, adhesion force measurement, electrical measurement, and cutting of single cell using various nano-tools such as nanoprobe/ nanofork/ nanoputter/ nanopicker/ dual nanoprobe with automation system.

In Chapter 9, we give a description of summary and applications of micro-nanorobotic manipulation and their applications finally.

9.2 Future Aspects

As introduced in this book, the micro-nanorobotic manipulation technologies have borderless potential for various applications. Currently, the research direction is going to synthesis from analysis in micro-nano scale.

In dry environment, the atomic scale assembly technique will be the cutting-edge way to realize "assembler" as proposed by Dr. Kim Eric Drexler for example. The automation technique is important to realize large scale structure, then the self-assembly technique needs to be integrated to micro-nano manipulation system.

In wet environment, the tissue or medical applications will be important to realize tailor-made regeneration medicine, in-vitro medical simulator, health monitoring and so on (Figure 9.1). The biological cell manipulation technique is provided from engineering technology to science appellations to achieve highly integrated sensing and actuating technologies. The multi-scale technique is key technique to realize complex tissue structures.

In the future, the micro-nanorobots would be programmed for the specific therapeutical tasks and injected into the blood to work at the cellular level to repair the tissue, clean arteries, attack the cancer and the viruses, and reverse the aging process etc. DDS with the micro-nanorobots have advantages of an improved concentration of the therapeutic agents at the targeted area and a decrease in the systemic side effects compared with the traditional chemotherapy-based interventions for such as tumor targeting including target chemotherapy.

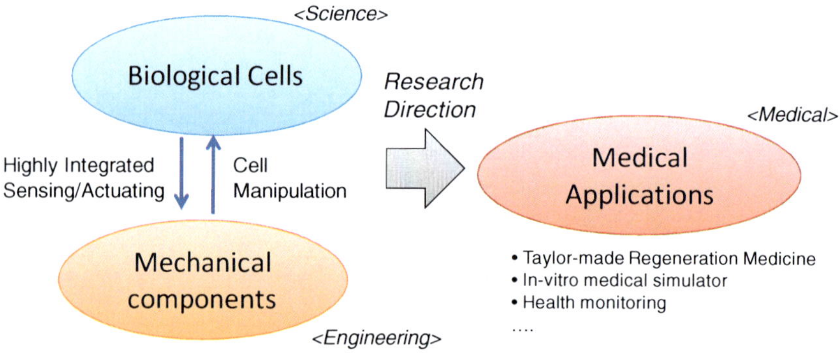

Fig. 9.1 Importance of medical application in wet environment

As social demands, as next-generation engineering and technology, green and life innovations are engaged for research and development. As shown in Figure 9.2, the micro-nano technology is important to contribute based on the combination of different fields. As summarized in Table 9.1 for green innovation, technological challenges are required to solve some difficulties on natural resources, environmental pollution, energy development and also food/agriculture fields. On the other hand, to achieve the life innovation, various technical challenges are expected in the medicine for life biology-analysis/synthesis as summarized in Table 9.2.

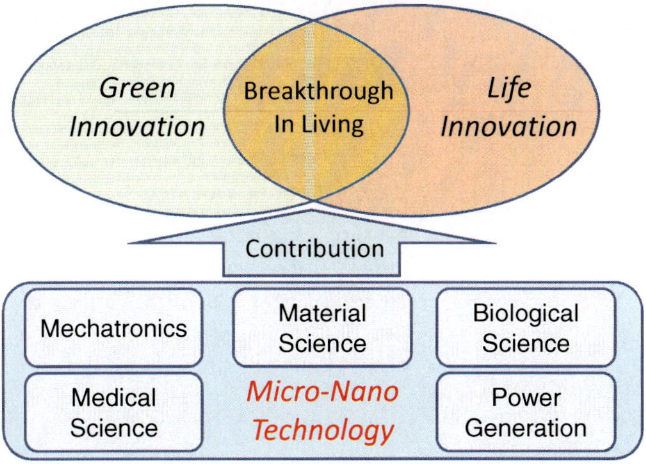

Fig. 9.2 Global and social issues for Green and Life innovations based on micro-nano technologies

Table 9.1 Examples of Green Innovation by Micro-Nano mechatronics and robotics

Green innovation	Technological Challenges
Natural Resources	Micro/nano devices for Discovery of oil resource, Management of water resources, Prevention of forest destruction, etc.
Environmental Pollution	Monitor, control and management of Environment, Distributed sensing & control, Pollution control, Green vehicle, etc.
Energy Development	Energy saving, harvesting and alternatives , Energy grid and management, Power control and green electronics, etc.
Food and Agriculture	Safety, testing and tracing, Efficient harvesting, nutritious products, genetically-modified products, etc.

Table 9.2 Examples of Life Innovation by Micro-Nano mechatronics and robotics

Life innovation	Technological Challenges
Medicine for life	Inspection and diagnosis , Re-generative medicine, Gene therapy and life science, monitoring diseases, Neuro Science, In-situ diagnostics, Cell diagnosis and surgery, New drug and medicine, DDS, Minimally invasive surgery, Rehabilitation, Techno-care, Wearable robots, Cyborg, QoL, etc.
Biology – Analysis and Synthesis	Sensing , manipulation and automation, New species, DNA diagnosis & manipulation, Cell screening, transport, cultivation, and function and differentiation control, Artificial cell, Life in chip, Cloning of stem cells, etc.

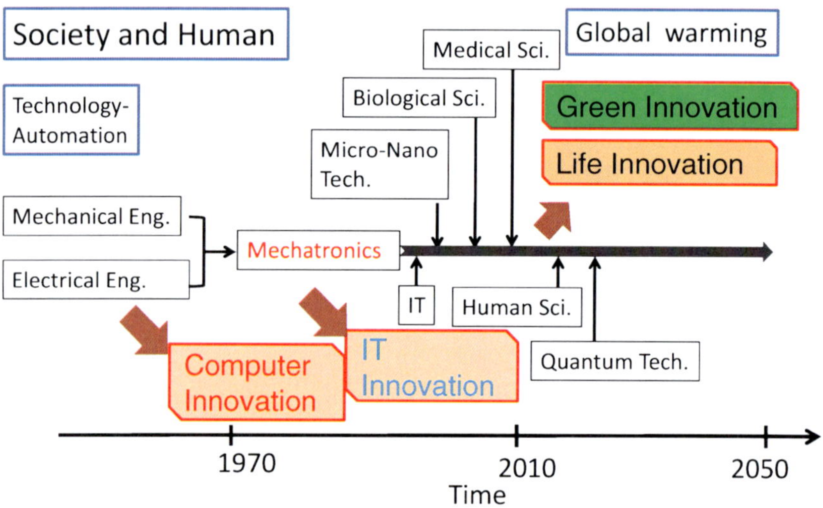

Fig. 9.3 Micro-Nano mechatronics is important technology for Green and Life Innovations

As shown in Figure 9.3, these innovations will be in early edge of twenty first century to follow the computer and IT innovations. Great opportunities are waiting ahead for us based on micro-nanorobotics and manipulations.

Subject Index

MIX
Papier aus verantwortungsvollen Quellen
Paper from responsible sources
FSC® C105338

Printed by Books on Demand, Germany